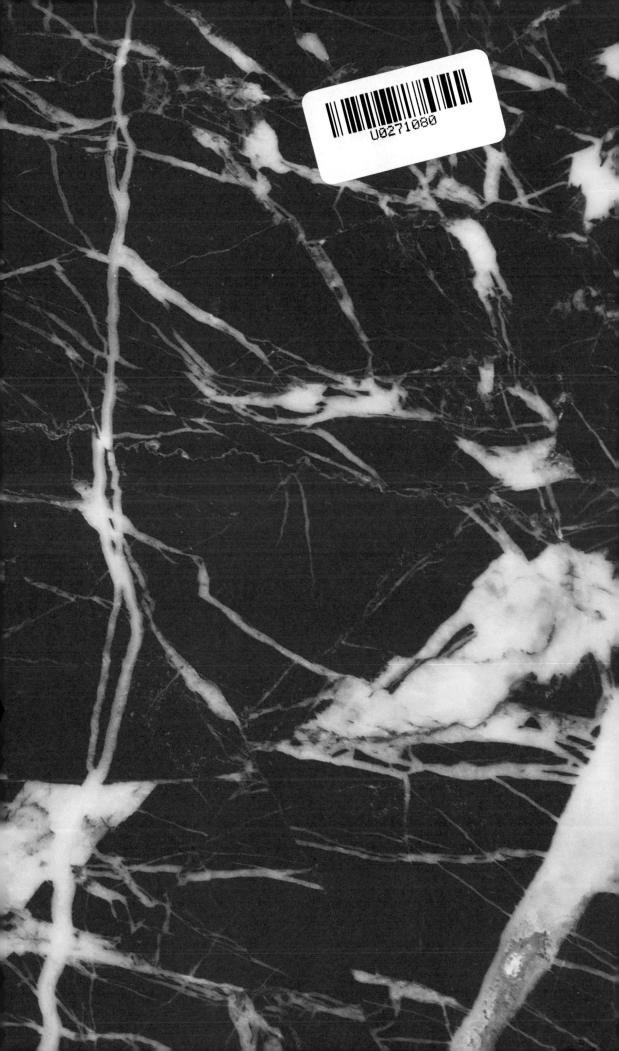

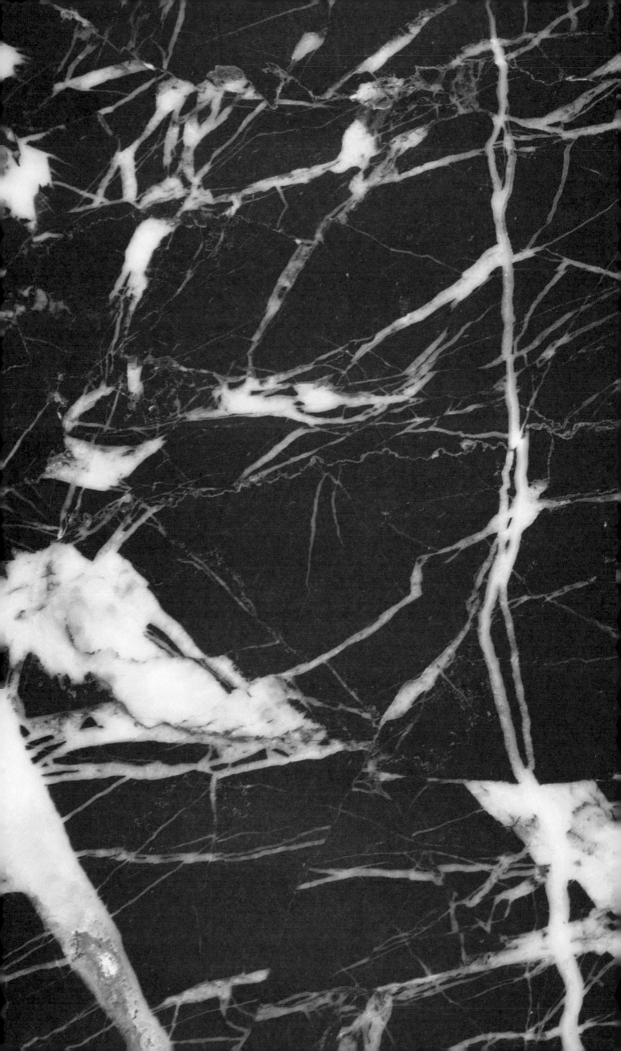

ARCHITECTURE STUDIES
建筑研究

ARCHITECTURE STUDIES
建筑研究

02

Chief Editors: Mark Cousins, Chen Wei Deputy Chief Editors: Li Hua, Ge Ming
School of Architecture of Southeast University
Architectural Association School of Architecture
Shanghai Xian Dai Architectural Design Group

主编：[英] 马克·卡森斯 陈薇 执行主编：李华 葛明
东南大学建筑学院 [英] 建筑联盟学院 上海现代建筑设计集团

Topography & Mental Space
地形学和心理空间

CHINA ARCHITECTURE & BUILDING PRESS
中国建筑工业出版社

PROLOGUE

序幕

A NOTE ON SYMBOLS
符号注释

○

Explanatory Notes & Commentary

●

References

✳

Figures

○

注释和说明

●

参考文献

✳

插图

Oscar Wilde once cabled his publisher, asking about sales of a new novel.

'?'

The publisher exuberantly replied.

'!'

Sam Roberts, 'Dot-Dot-Dot, Dash-Dash-Dash, No More', *Week in Review*, The new York Times, 12 February, 2006.

卷首语

《建筑研究》为年刊,是"当代建筑理论论坛"的研究成果。"当代建筑理论论坛"是一项长期的研究计划,旨在搭建建筑理论研究跨文化交流的平台。它包含三个部分的活动:理论著作的翻译,讨论著作中相关议题的国际研讨会,以及《建筑研究》的出版。

论坛关注的核心有两个:作为现代知识形式的建筑学,和作为探索、质疑与丰富这一知识构建条件的中国。在建筑研究的边界不断扩展,建筑解读与讨论越来越多地进入跨学科质询的同时,建筑学自身的构建依然是一个问题:如何返回建筑,如何将更广泛的议题转化为建筑问题,并由此重构建筑知识,并与建筑实践相关联。这一问题,或许在实践领域里正经历着快速扩张和变化,而理论建构异常薄弱的今日之中国,比其他地方更为紧迫。

整个论坛,以"建筑"作为研究诸般现象的起点。它是思考当代世界的方式,也是漫长旅程的归宿。

整个论坛,以主题"词"为索引。它们是建筑自身在现代性条件下面临的问题,同时它们与人类的建造活动和更广泛的文化议题密切相关。我们期望,借助它们,引发不同建筑文化间的交流和争议,并为激发新的思想打开通道。

《建筑研究》首先是对论坛年会及翻译的记录。但不仅于此。作为一种媒介,我们期望它不仅是特定时刻与地点的延伸,而且在传播的过程中,能触发和产生更多的思考与讨论，并由此获得自己的生命。虽然论坛的年会主要在中国举行，但其面对的是所有对相关建筑议题和中国建筑感兴趣的同仁。为此，《建筑研究》以中英文双语出版，在中国和海外共同发行，其出版形式本身便蕴含了以两种语言为基础的不同建筑文化间的对话。

第二期《建筑研究》的内容，主要来自2011年5月25日至5月29日，在南京和上海两地举行的第二届"当代建筑理论论坛"国际研讨会。研讨会围绕两个中心议题，"地形学"和"心理空间"，分别以戴维·莱瑟巴罗的著作《地形学的故事：景观与建筑研究》和安东尼·维德勒的《建筑异样》为基础，并包含了一系列与之相关的公开讲座和名为"山水"的展览。

与通常的情况一样，本刊中各类文献的内容与研讨会上的演说与讨论并不完全对应。在尽量保持当时气氛的同时，出版的过程也伴随着大量的重写、编辑和来自与会者的反馈。细心的读者也许会发现，修改后的文献与当时的语境之间存在着某些间隙，但我们以为，这正是读者了解和观察会议中争论及会议影响的一种方式。

整个论坛及《建筑研究》的出版有赖于三个机构的相互合作：来自伦敦的"AA"建筑联盟学院，南京的东南大学建筑学院和上海的现代建筑设计集团。这一合作本身即蕴含着我们的组织意图，建立一个理论与实践相关联而非分离的国际交流与碰撞的平台。

EDITORIAL

Architecture Studies is an annual publication. It is an outcome of the Forum of Contemporary Architectural Theories, which is envisaged as a long-term project and aims to build a bridge for the cross-cultural exchange of architectural ideas and research. It consists of three activities: translating theoretical books from English—mainly but not exculsively—into Chinese, discussing related issues brought forward by the books in symposia and conferences, and publishing Architecture Studies.

The whole project has two central concerns: architecture as a modern form of knowledge and China as a condition to explore, question and contribute to it. While the boundaries of architectural research have been extended, and interpretation and discussion on architecture has become interdisciplinary, architecture itself remains a question. How to return to architecture? How to convert broader issues to architectural questions? And consequently, how to rebuild architectural knowledge and re-establish its link with architectural practice? This matter is probably more urgent in contemporary China than anywhere else since while architectural practice has been booming in China, work on architectural theory is considerably weak.

The Forum takes architecture as the departure point for our studies, as a way of understanding the world, and as a point to return to from our journey.

The Forum is structured by thematic words, which are indexes to the questions that architecture faces in itself under the condition of modernity, and closely link to human building activities and broader cultural issues. We hope the Forum will enable exchange and dialogue between different architectural cultures and open up a channel for stimulating new ideas and thoughts.

Architecture Studies, primarily, is a record of the conferences of the Forum and the translation process of the theoretical books. But more than that, as a media, it is expected not only to be an extension of particular moments and places, but also to generate thinking and discussion. It should create its own life. Although the Forum is mainly held in China, it intends to appeal to those who are both interested in the architectural issues as well as the urban situation of China. Hence, the AS series is published in both English and Chinese, and circulates both inside and outside of China. The format itself forms a dialogue between different architectural cultures based on different languages.

The second issue of Architecture Studies stems from the Second Symposium of the Forum of Contemporary Architectural Theories held in Nanjing from 25—29 May 2011. There were two main themes: 'Topography' and 'Mental Space', in relation to the translation of David Leatherbarrow's book *Topographical Stories: Studies in Landscape and Architecture*, and Anthony Vidler's *The Architectural Uncanny: Essays in the Modern Unhomely*. There were also a series of public lectures on related issues and an exhibition entitled 'shanshui'.

The papers presented in this issue are not identical to the lectures and talks delivered at the Symposium. While maintaining the original discussion, flow and atmosphere, there has been extensive rewriting, revising, editing and reflection from participants. Readers may find certain variation between the revised papers and the context of discussion. For us, this could be a way for readers to grasp and observe the arguments in the discussion, and to reflect on the effects of the Symposium.

The organisation of the Forum and publication of AS relies on the collaboration of three institutions, the Architectural Association School of Architecture in London, the School of Architecture at Southeast University in Nanjing, and Shanghai Xian Dai Architectural Design Group in Shanghai. The cooperation itself indicates our intention to build an exchange where theoretical thinking can be linked to building practice.

ACADEMIC STRUCTURE
学术架构

Chair of Academic Committee
学术委员会主席

MARK COUSINS
Architectural Association
School of Architecture
马克·卡森斯
"AA"建筑联盟学院

CHEN WEI
Southeast University
陈薇
东南大学

Academic Committee
学术委员会

STANFORD ANDERSON
Massachusetts
Institute of Technology
斯坦福·安德森
麻省理工学院

ADRIAN FORTY
Bartlett School of
University College London
阿德里安·福蒂
伦敦大学学院

MICHAEL HAYS
Harvard University
迈克尔·海斯
哈佛大学

DAVID LEATHERBARROW
University of Pennsylvania
戴维·莱瑟巴罗
宾夕法尼亚大学

BRETT STEELE
Architectural Association
School of Architecture
布雷特·斯蒂尔
"AA"建筑联盟学院

ANTHONY VIDLER
The Cooper Union
安东尼·维德勒
库柏联盟

LIU XIANJUE
Southeast University
刘先觉
东南大学

WANG JUNYANG
Tongji University
王骏阳
同济大学

LI SHIQIAO
University of Virginia
李士桥
弗吉尼亚大学

WANG JIANGUO
Southeast University
王建国
东南大学

DONG WEI
Southeast University
董卫
东南大学

ZHANG HUA
Shanghai Xian Dai Architectural
Design Group
张桦
上海现代建筑设计集团

SHEN DI
Shanghai Xian Dai Architectural
Design Group
沈迪
上海现代建筑设计集团

Translation Advisors
翻译顾问

STEPHAN FEUCHTWANG
London School of Economics &
Political Science
王斯福
伦敦政治经济学院

JIANFEI ZHU
University of Melbourne
朱剑飞
墨尔本大学

XING RUAN
University of
New South Wales
阮昕
新南威尔士大学

DELIN LAI
University of Louisville
赖德霖
路易威尔大学

Executives
执行

LI HUA
Southeast University
李华
东南大学

GE MING
Southeast University
葛明
东南大学

CONTENTS

目录

CONTENTS

PROLOGUE

Editorial 14
Academic Structure 16

TOPOGRAPHY

IS LANDSCAPE *ARCHITECTURE*? 25
David Leatherbarrow

TOPOGRAPHY, HISTORY & 43
THE LIE OF THE LAND
John Dixon Hunt

A VIEW OF TOPOGRAPHY FROM 73
MY REAR WINDOW
Liu Dongyang

TERRAIN & XING SHENG 107
Chen Wei

MISTY RIVER, LAYERED PEAKS: 153
A CONCEPTION OF CHINESE SHANSHUI
Lin Haizhong

BUILD A WORLD TO RESEMBLE NATURE 173
Wang Shu

Discussions at the Symposium 213
28 May 2011

MENTAL SPACE

PROJECTION 237
Mark Cousins

THE FIGURE AS HOME 253
Li Shiqiao

A NARRATIVE STRUCTURE OF CROSS-CULTURAL ARCHITECTURE Zou Hui	269
BLACK Ge Ming	301
Discussions at the Symposium 29 May 2011	329

DOCUMENTS

Schedule of the 2nd Symposium	348
WHERE IS ART? Mark Cousins	350
THE MODERNITY OF THE PICTURESQUE John Dixon Hunt	358
PASSAGES IN THE LANDSCAPE OF TIME David Leatherbarrow	368
Photos of the 2nd Symposium	379

REVIEW

ON «THE ARCHITECTURAL UNCANNY» AN INTERVIEW WITH DR. ANTHONY VIDLER He Weiling	388

EPILOGUE

Biography	402
Glossary	404
Acknowledgements	406
Colophon	408

目录

序幕

卷首语 ·· 12
学术架构 ·· 16

地形学

景观是建筑吗? ·· 25
戴维·莱瑟巴罗

地形学,历史与土地的伸展／谎言 ····································· 43
约翰·狄克逊·亨特

从我家后窗看见的地形学 ·· 73
刘东洋

地形与形胜 ·· 107
陈薇

从《烟江叠嶂图》看中国山水画 ······································ 153
林海钟

建造一个与自然相似的世界 ·· 173
王澍

会议讨论 ··· 213
2011年5月28日

心理空间

投射 ·· 237
马克·卡森斯

象与家 ··· 253
李士桥

| 交叉文化建筑的情节结构 | 269 |
| 邹晖 | |

黑 ⋯⋯⋯⋯⋯⋯⋯⋯⋯⋯⋯⋯⋯⋯⋯⋯⋯⋯ 301
葛明

会议讨论 ⋯⋯⋯⋯⋯⋯⋯⋯⋯⋯⋯⋯⋯⋯⋯⋯ 329
2011年5月29日

文献

论坛日程安排 ⋯⋯⋯⋯⋯⋯⋯⋯⋯⋯⋯⋯⋯⋯ 349

艺术在何处？ ⋯⋯⋯⋯⋯⋯⋯⋯⋯⋯⋯⋯⋯⋯ 351
马克·卡森斯

风景如画的现代性 ⋯⋯⋯⋯⋯⋯⋯⋯⋯⋯⋯⋯ 359
约翰·迪克逊·亨特

时间景观中的推移／移逝 ⋯⋯⋯⋯⋯⋯⋯⋯⋯ 369
戴维·莱瑟巴罗

论坛照片 ⋯⋯⋯⋯⋯⋯⋯⋯⋯⋯⋯⋯⋯⋯⋯⋯ 379

评论

由《建筑的异样》展开安东尼·维德勒访谈 ⋯⋯ 389
贺玮玲

尾声

人物简介 ⋯⋯⋯⋯⋯⋯⋯⋯⋯⋯⋯⋯⋯⋯⋯⋯ 403
词汇表 ⋯⋯⋯⋯⋯⋯⋯⋯⋯⋯⋯⋯⋯⋯⋯⋯⋯ 404
致谢 ⋯⋯⋯⋯⋯⋯⋯⋯⋯⋯⋯⋯⋯⋯⋯⋯⋯⋯ 407
尾署 ⋯⋯⋯⋯⋯⋯⋯⋯⋯⋯⋯⋯⋯⋯⋯⋯⋯⋯ 408

TOPOGRAPHY

地形学

IS LANDSCAPE *ARCHITECTURE?*

景观是建筑吗？

戴维·莱瑟巴罗
David Leatherbarrow

TRANSLATOR
Shi Yonggao

翻译
史永高

I was not the first to pose this question; it was first asked in print by Garrett Eckbo, one of the most important landscape architects in America in the 20th century.• One could equally reverse the question and ask: 'Is Architecture Landscape?'. In either formulation the question is about the relationship between two arts that are normally understood as separate professions these days. In point of fact Eckbo was not the first to puzzle over this issue, even if his exact formulation had no antecedents. The question had already been posed in the 19th century, when landscape architecture emerged as a distinct discipline. The early theorists of the field, Humphrey Repton and John Claudius Loudon in England, Quatremère de Quincy in France, and Andrew Jackson Downing in the USA, all wondered about the relationships between these two practices—if they were two.• The professional accrediting and licensing bodies that were formed subsequently tried to settle the matter and institutionalise the distinction. But the question may be older, for it is possible to say that the distinction between these disciplines, at least the suggestion of fundamental differences, was debated even earlier in the 18th century. The cases I have in mind include the Abbé Laugier and William Chambers; the first compared the routes through a forest to the streets of a town, while the second used landscape aesthetics to evaluate the merits of a building's façade. Despite this tradition and indeed maybe because of it, the questions these theorists asked have not disappeared in our time.

• 1

• 2

我不是第一个提出这个问题的人；20 世纪美国最重要的景观建筑师之一，加雷特·埃克博首先以书面方式提出了这一点。°这一问题显然也还可以倒过来问，即"建筑是景观吗？"而任何一种表达方式，焦点都是在于景观和建筑之间的关系问题。纵观历史，埃克博其实也并非第一个对这一问题感到困惑的人，即便此前从未有人作出类似的明确表述。19 世纪，景观建筑学开始成为一个区别于建筑学的新学科，二者之间的关系问题便也应运而生。早期的景观建筑理论家°都曾思考过这二者之间的关系 —— 如果说那时它们还是两者的话。°其后，职业行会以及注册机构相继成立，并试图把二者之间的差异体制化。进一步看，这一问题其实更为古老，因为甚至早在 18 世纪，人们已经在争论这两个学科之间的差异，起码也是对于这种差异的暗示。其中包括了洛吉耶神父以及威廉·钱伯斯，前者把穿越森林的小道比作城镇的街道，后者则借助园林美学对建筑立面的审美价值来进行评价。尽管这一问题历史悠久，°建筑与景观的关系问题在当代也并未消失。

○ 1

○ 英国的汉弗莱·雷普顿和约翰·迪厄斯·劳登，法国的夸特梅尔·德·昆西，美国的安德鲁·杰克逊·唐宁

○ 2

○ 或许也正因这一传统

My aim, in the study that follows, is not to give a final answer but to introduce a term of comparison, topography, that will help us see and describe what landscape and architecture have in common. Questions concerning technique and working materials will enter into the issue, as will matters of standing or rank. But a more important topic for me will be the historical and cultural content of land and building form; I believe and shall try to show that the conception of land and materials 'in themselves' is an unhelpful abstraction that abbreviates the real subject matter of design in landscape, architecture, and cities, thereby reducing its relevance and risking its future.

THE DISCIPLINES OF LANDSCAPE & ARCHITECTURE

Subordinate standing has often been conferred on landscape architecture because its professional history is much shorter than architecture's and its theory correspondingly slighter—three centuries of theorising landscape in the Western tradition, three to four times that for architecture. Recent writings and projects, however, show that the dependency suggested by this lineage is false. Today there is wide agreement that the traditional subordination of landscape architecture to architecture is no longer acceptable.• If in the past landscape and urban design turned to architecture for ideas and methods, in our time, concepts and techniques that were thought to be proper to large-scale landscape design have been appropriated by architecture: phenomena of process or temporal unfolding, 'registration' prompting articulation, 'mapping' as a survey technique, and so on. The strong impact of these ideas and methods on architecture cannot be denied, nor can it be explained as a result of ecological consciousness, to which it nevertheless relates. Recent arguments for urban ecology, landscape urbanism, and ecological urbanism suggest that thinking about landscape° will enable architects and planners to re-conceive the nature and task of designing buildings and cities, as if the whole of which buildings are part is territorial, environmental, or topographical, that is, not only architectural.

• 3

○ in its several senses

This realignment of the disciplines has given rise to much enthusiasm, and exaggeration, among contemporary landscape theorists. Frequently we are told that the coupling of landscape design and city making—landscape urbanism—offers contemporary culture a coordination of previously unrelated disciplines. That this observation is patently false is apparent to anyone with a modicum of historical knowledge, as I have already suggested and will consider further below. Current practice presents not a coupling of previously unrelated arts but a new understanding of what binds them together. For centuries landscape and architecture have been two interrelated ways of articulating culture, shaping and showing typical situations. In their expressivity gardens and buildings resemble other forms of cultural production: poetry, philosophy, or politics,

但是，在接下来的讨论中，我的目的并非要对这一问题给出一个最终答案，而是希望通过引入一个可资比较的术语 —— 地形学 —— 来更好地发现和描述景观和建筑之间的共性。这自然会涉及技术与材料，以及地位或等级的问题。但是于我来说，更为重要的则是土地和建筑形式所具有的历史和文化内涵。我相信，并且我也将展示，那种把土地和材料只看作其自身的观念，实在是一种毫无助益的抽象概念，因为它缩略了景观、建筑、城市设计中的真正主题，并因此约减了它的关联，°使其前景堪忧。

○ 它与生活世界的关联

景观与建筑的学科性

景观建筑学在学科上常被认为从属于建筑学，因为其专业历史相较而言要短很多，相应地，其专业理论也相对薄弱，——在西方，景观学的理论历史只有三个世纪，而建筑学则是它的三至四倍。然而近期的著作和实践却表明，这种谱系所暗示的从属关系并不真实。如今，人们已经取得更多的共识，认为那种把景观学从属于建筑学的观念是不可接受的。•如果说，以往景观学和城市设计从建筑学那里获取思想和方法的话，如今，那些曾被认为是适用于大尺度景观设计的一些概念和技术，经过转化而反过来应用于建筑设计当中：过程性以及随时间而变化的现象、由°"记录"而来的表达、作为一种测量方法的"地图术"等等。不能否认景观思想与实践对建筑学的强烈冲击，但也不能认为这源自我们对于生态思想的自觉，尽管生态思想确实与建筑不无关联。从近年对于城市生态学、景观都市主义以及生态都市主义的倡导可以看出，对于景观°的考虑会使建筑师们重新认识建筑和城市设计的本质和任务，似乎把建筑置身其中的那个整体是领域性的、环境性的，或者说是地形学的，亦即，不仅仅是建筑性的。

• 3

○ 对于外界影响的

○ 在其多重含义上

这种层级上的重新排序，往往使当代景观学理论家们热情澎湃，甚至难免夸张。我们常常听到所谓景观设计与城市设计的结合 —— 即景观都市主义 —— 为当代文化赋予了一种纽带，联系起两个此前并无多少关系的领域。然而正如我之前已经指出的，对于任何了解一点景观学与建筑学历史的人来说，这种观点的错误都是显而易见的。事实情况是，今天的实践所展示出来的，并非什么连接了两个此前分离的学科，而是对于把它们结合起来的那些因素如今有了一种新的理解。在许多个世纪里，景观学和建筑学都一直是两种相互

for example. The practise of these spatial arts, like that of the others, assumes criticism: creative thought proposes alternatives to existing arrangements once the latter have been judged to be inadequate.°

° inadequately useful or expressive

THE ROLE OF TOPOGRAPHY

I propose using the word topography to name the topic, theme, or framework that architecture, urban design, and landscape hold in common. It not only establishes their similarity but also provides them with the grounds for their contribution to contemporary culture. The task of landscape architecture, urban design, and architecture as topographical arts is to provide the prosaic patterns of our lives with durable dimension and beautiful expression. Obviously, this sense of the word is wider than conventional usage, which equates topography with the land. Topography incorporates terrain, built and un-built, but more than that, for it also includes traces of practical affairs ranging from the typical to the extraordinary. Our English term derives from ancient Greek: Τοπογραφεώ, which meant description or determination of a place. Nowadays we might say 'writing the site'. Strabo used the word in this sense in his Geography. So conceived, topography certainly refers to something that is physical, but accents its 'written' or legible aspects, by virtue of which the footprints on its shores invite, sustain, and represent the performances of everyday life. Marks of this kind are not exactly vestiges, for that would suggest they only indicate something that occurred in the past. Were this all that topography made legible, both the present and the future would be foreclosed. Topographical inscriptions do, indeed, give evidence of previous enactments but they also indicate those that are still occurring and that may unfold in future. A topographical trace is an outline or proposal that is taken up in an act of making or inhabiting that has no obligation to its past other than the preservation of a tension between its forms and those projected out of the present. Topography gives itself to perception, experience, and knowledge as both a representation and an accommodation of prosaic and practical purposes, historically formed and re-formed.

联系的方式，来表达我们的文化，并塑造和反映典型的人类境遇。从表现性而言，它们与其他文化生产方式，例如诗歌、哲学、政治都有相似之处。这些空间艺术的实践，也与其他实践方式一样，有一种批判性：一旦现有的表现方式被认为不再充分，°新的创造性思想便会起而代之。

○ 可能是用途上也可能是表现上

地形学的角色与地位

这里，我提出以"地形学"一词来命名这一建筑学、城市设计、和景观学所共享的话题、主题，或者说框架。这一概念不仅建立了它们之间的相似性，而且为它们提供了对当代文化作出贡献的必要基础。作为地形学艺术的景观建筑学、城市设计、建筑学，其任务在于为我们的庸常生活赋予一种持久的向度，并提供一种美好的表现。显然，这一意义上的地形学，已然超出其传统意义——传统上，"地形"这个概念等价于土地。固然，地形学包含了地貌——无论是人工创造的还是自然形成的，但是它又远不止这些，因为地形同样也容纳了各种各样的人类生活实践的痕迹。°英文的这一词汇地形来源于希腊文的"Τοπογραφεώ"，意为对于一个场所的描述或是定义。在今天，我们则可能会用"场地书写"这样的话语来表达这一含义。斯特拉波在其著述《地理学》中便是在这一意义上来使用的。如此理解的地形学指向的当然是某种物质性的东西，但是它强调的方面则在于这种物质性的可"书写"性或可辨识性，就如那些沙滩上的脚印一般，邀请、保持、再现了日常生活的情景。并且，这些印记并不完全是一种过去的遗存，否则，就会暗示它们所表达的只是一些已然逝去的东西。假如这些便是地形学所彰显的全部，那么，"现在"和"将来"便都会被排除在外。地形学的铭刻确实呈现了过往的活动，但是它们也反映了当下仍在发生的以及在未来将要展现的。地形学印迹是一种轮廓或者提议，涌现于制作或者住居这样的行为，而这些行为对于过去并不负有任何义务，除了在那些过往的以及那些发端于现在的形式之间，保持一种张力。地形学是知觉性的、体验性的，也是知识性的，它是对于庸常的和实用性目的的再现和适应，一直被塑造和重新塑造着。

○ 既有那些非常典型的，也有那些异同寻常的

When discussing a building and its site or surrounding terrain we typically oscillate between geometry and physical fact, shape and substance. No one believes, however, that a choice between these two makes sense. Doesn't that mean we should question their distinctness? Concern for topography cannot mean focus only on the profile, compass, or configuration of a given plot or stretch of land, for the project's realisation and expression depend equally on the materiality, colour, thickness, temperature, luminosity, and texture of physical things. Further, when considered in its temporal aspects, it is plain that land is not only soil but all that is hidden beneath it and emerges from it, as well as the several agencies that sustain that emergence. Yet, on the other hand, the turn toward terrain would lead nowhere if it meant interest solely in the physical aspects of land. Attention to the qualities of built landscape naturally leads to reflection on their associations and thus to their expressive power, even their potential for representation. Interest in the physical aspects of land can also lead to an awareness of its functional potentials, what the materials of a site can do, how they can act or perform in service of some purpose other than expression or representation. Attention to the performative aspects of topography—in addition to its pictorial qualities—also invites recognition of its expected and unexpected events, the latter revealing the limits of both foresight and design intelligence. For this reason, it is hardly surprising that the theme of 'programming' is so strong in current work. While land is obviously physical, it is also clearly spatial. There is no reason to deny or to neglect this, even if the perspectivity of the typical approach tends to favour frontality and pictorialism. Attention to the spatial aspects of a place—its enclosures, continuities, and extent—can thus lead to interpretations of its potentials for occupation and use, which are not only, or not essentially pictorial but practical.

In addition to the material, spatial, and practical aspects of the topography, there is a third characteristic to which I alluded earlier that is just as important as these first three: its temporal quality. Seen over time, the materials of landscape continually renew themselves. A site's metabolism is key to its capacity for continued relevance. Time is also the medium of one's experience of landscape, for terrain is known most fully in the duration of spatial passage or movement, its delayed or accelerated sequences, as well as its repetitions and inaugurations. Temporality opens an essential dimension of spatial sense.* In sum: topography is important to landscape, • 4

在对于建筑与其基地或是周边地貌的讨论中，我们通常都会徘徊于其几何性与物质性，其形状与实质之间。但是，没有人会认为在这样的二者之间硬要作出一个抉择会有什么意义。这难道不是在暗示，我们应当质疑它们之间的分别吗？对于地形的关注，不应仅仅意味着聚焦于给定地块的外形、边界以及其结构性关系，因为任何项目的最终实现和表现出什么样的效果，都同样还取决于那些实实在在的要素的材料、色彩、厚度、温度、光度、肌理。并且，若进一步考虑它们在时间向度中的变化，则显而易见的是，土地不仅仅是土壤，还包括了那些隐藏其中的以及将会从中不断涌现的东西，同样还包括那些能够促使这种涌现得以发生的诸多力量。不过，从另一方面来说，如果向着地貌的转向仅仅意味着对于土地物质性层面的兴趣，则亦难有什么前途。关注建成景观的质量自然会使我们思考那些与之相关的东西，并由此思考它的表现力，甚至是它们再现性潜力。对于土地物质性层面的兴趣，也会使我们意识到其功用上的潜力，场地中各种各样的材料都能发挥什么作用，以及在表现和再现以外，它们如何为着其他目的来行动和产生效能。对于这些地形的图像性层面以外的效能层面的关注，也还意味着对于那些预期中的以及预期外的自发性事件的承认，其中，后者恰恰暴露了设计者的预见力以及智慧所具有的局限性。这样一来，在当今的那些作品中，"计划"这一主题是如此强烈也就不足为奇了。虽然土地显然是物质性的，但同时它也还是空间性的。没有任何理由可以否定或者忽视这一点，即便通常的取向更为喜欢正面性以及它的画面性。因此，对于场地的空间性层面 —— 它的围护性、连续性以及延展性 —— 的关注，可以把它的潜力转化为居住和使用，而这些目的都不仅仅是 —— 或者说主要不是 —— 图像性的，而是实践性的。

在地形的材料、空间和实践层面以外，我之前还提到过另一个特征，它与这三者有着同等的重要性：即地形的时间性。在时间的长河中，材料不断改变着自身的面貌。场地自身的新陈代谢，对于它与周围环境建立起持续的关联性也至关重要。不仅如此，时间还是我们体验景观的一种媒介，因为只有在空间和运动中，在其序列的延缓或是加速中，在它的重复和开始中，才能获得对于地貌最为丰满的感知。事实上，时间性开启了空间的一个非常根本性向度。总结而言：地形对于景观、建筑和城市设计都是重要的，因为对于地貌的材料性、空间性、实践性以及时间性的关注，展示了那种

• 4

architecture, and urban design because attention to the materiality, spatiality, practicality, and temporality of terrain shows how alternatives to the pictorial or aesthetic approach can increase the cultural content of the building, landscape, and city, as it° is assumed and renewed in the practice of everyday affairs; which is to say the ethical sense of praxis, as it has been examined in recent political philosophy.*° For my part° I would like to show that contemporary design work points to the decisive role played by the situations and institutions of practical life in the restructuring of urban, architectural, and landscape topography.

- ° cultural sense
- • 5
- ° in the writings of Hans-Georg Gadamer, Jürgen Habermas, Richard Rorty, and Giorgio Agamben
- ° with these writers in the background

PRECEDENT

The expanded sense of topography I have elaborated may be new to current usage but it is not without precedent in the histories of architectural, landscape, and urban design, as I suggested at the outset. Considering a couple of the exemplary projects of the past several years, such as the Olympic Sculpture Park in Seattle or the Freshkills Lifescape in New York,° it is plain that proposals that use landscape as the basis for urban transformation are not without precedent. The most prominent example from the 1980s is Bernard Tschumi's Parc de la Villette, with its combination of programmes, involvements with the pre-existing urban infrastructural environment, phased development, and so on. But that project had its own antecedents. If one looks back another twenty years there are other well-known designs that combine landscape and architecture in new urban order. An obvious case is Le Corbusier's plan for the Venice Hospital.° That project's interwoven seascape and cityscape anticipated the 'mat buildings' of Peter and Alison Smithson, which were envisaged as horizontal 'ordering systems' that synchronised architectural structure with the flows of climate.° The practise of folding architecture into the environment was not an innovation of 1965, however. The same style of working can be seen in Le Corbusier's designs from earlier decades.°

- ° Field Operations
- ° 1965
- ° as in the Kuwait project of 1969
- ° before Chandigarh, the Radiant Farm, the Co-operative Village, and so on

Le Corbusier was not the only architect in the 1930s to integrate superstructure and substructure. Consider, for example, the Parkway project in Philadelphia by Jacques Gréber and Paul Cret. Although planned at exactly the same time as Le Corbusier's Co-operative Village, it demonstrates a very different understanding of urban terrain. Gréber's way of working was closer to the 'landscape urbanism' of Tony Garnier, whose Cité Industrielle° elaborated a spread-out, horizontal landscape of buildings, public space, and infrastructure, precisely what Le Corbusier rejected in some but not all of his urban designs. Before Garnier and throughout the 19th century there were many projects that coordinated landscape with city design, none more fascinating than Ledoux's design for the saltworks and 'urban' settlement at Chaux. In that project architecture, law, and morals were proposed as outgrowths of the natural world,

- ° 1917

不同于图像化和美学化取向的方法如何增益了建筑、景观、城市的文化内涵，因为，正是在日常生活当中，我们才能不断地获取和更新这种文化内涵；也就是说，实践的伦理意义，很多晚近的政治学者都对此作过论述。至于我，我希望能够表明，当代的一些设计作品说明，在城市、建筑和景观地形学的重组中，生活实践的境遇和机制都发挥了决定性的作用。

- 5
- 诸如伽达默尔、哈贝马斯、罗蒂、阿甘
- 以所有这些学者作为学术背景

先例

相对于现在通用的"地形学"概念，我在这里提出的扩展意义可能是一种新的概念，但是就如我在开始时已经暗示的，它在建筑、景观、城市设计的历史上并非没有先例可循。想一想过去几年中落成的几个代表性项目，例如位于西雅图的奥林匹克雕塑公园，或是由"场地手术"事务所完成的位于纽约的弗莱士河公园项目，我们就会发现，以景观设计的思想来进行城市更新的提案早已有之。而 20 世纪 80 年代伯纳德·屈米的拉维莱特公园则是最为重要的案例，它承载了多种多样的需求与计划，并与现有的城市环境基础设施以及分期开发等要求相结合。不过，这一项目其实也有自己的参照先例。如果我们再往回 20 年，则发现还有别的在新的城市条件下把景观与建筑结合起来的著名设计。勒·柯布西耶 1965 年的威尼斯医院便是很明显的一例。这一项目中海景和城市景观的交织，预示了后来史密森夫妇的"毯式建筑"，这种"毯式建筑"被构想为水平性的"秩序系统"，使建筑与气候流相同步。然而，这种把建筑叠进环境的做法也并不是 1965 年才有的发明，同一类型的设计在柯布西耶更早年的设计中也已出现。

- 例如 1969 年的科威特项目
- 在昌迪加尔项目以前，例如光明农场，协作村等等

在 1930 年代，勒·柯布西耶并非唯一一位把地上部分和地下部分相结合的建筑师。这里，我们想一想雅克·格雷贝尔和保罗·克瑞在费城完成的帕克大道项目。虽然与柯布西耶的协作村设计恰在同时，却显示了一种与之非常不同的对于城市的理解。格雷贝尔的方式更接近托尼·加尼尔的"景观都市主义"，加尼尔的工业城发展了一种由建筑、公共空间、基础设施共同构成的水平延伸的景观，而这恰恰是柯布西耶在他的一些城市设计中加以拒绝的方式。在加尼尔之前，贯穿整个 19 世纪，有许多项目都结合了景观设计和城市设计，其中当数勒杜的皇家盐场和绍村的"都市"聚居地最为令人着迷。在这一项目中，虽然勒杜的身影可见

- 1917 年
- 但也并非全部

although Ledoux's hand is evident at all scales and in all parts of the buildings, the programmes, and the representations of nature. Architects in the preceding century were equally sensitive to the potentials of landscape when envisaging urban form: a representative case could be Pierre Patte's plan for siting monuments in Paris,° for ° 1765 each public space was designed as if it occupied a clearing in a park or forest, as Laugier had recommended. So, too, for architects in the baroque, mannerist, and renaissance periods. Well-known cases include Versailles, the Rome of Pope Sixtus V, Raphael's Villa Madama, and Rosetti's designs for Ferarra—not only the piazzi he planned, but the large public campo. That last term, meaning both square and field, shows that the interweaving of landscape and cityscape is long standing in projects, techniques, concepts, and vocabulary. The current agenda for 'landscape urbanism' or 'ecological urbanism' cannot be understood apart from this history, nor should it be seen as radically new. In fact, the uniqueness of current work only emerges in contrast with what preceded it.

One way to identify the singularity of today's projects for the urban landscape is to focus on their rejection of what can be called the two-world hypothesis, the notion that artificial and natural terrain are categorically distinct, that landscapes that have been made are radically dissimilar to those that generate themselves. Another way to view this opposition is to separate the two parts of the English word landscape; for the first half indicates material that is naturally given and the second indicates the look, appearance, or image of what has been made—although we also see unaltered terrain as picture-like. In contemporary theory this apparently obvious distinction is rendered problematic by questioning performance or operation of topography, putting issues of agency and meaning out of play, at least for a while. With respect to the issue of terrain as image—the 'scape' of landscape—no interpretation has come under more forceful criticism than the supposed pictorialism of past practices. Instead of images or pictures, contemporary landscapes are intended to offer *effects*, which are not matters of form, but the visible aspects of operations. With respect to the other contested term, land, the matter is more complicated; for giving it the same treatment as image, rejection would deprive project making of its proper subject matter. Denied instead are several common conceptions of land: first, the notion that land is inert; second, that it is property. No longer still or static, soil, water, sky, and vegetation are seen as vital processes, the decisive characteristic of which is their developmental nature, their metabolism. The common conception of land as property has also been denied in current theory, at least the idea that ownership requires the articulation of visible boundaries. In the minds of contemporary designers, landscape extends beyond the plot the client happens to own, approaching the horizon. Here, too, we have a new rendition of an old conceit, specifically the 'borrowed landscape,' so often discussed and seen in Asian

于建筑、功能、以及自然之再现的不同尺度和不同部位，但是，建筑、法律、道德等却是被认作自然世界的自然结果。在此前的 18 世纪，建筑师们在构想城市形态时同样对于景观非常敏感：代表性的例子有皮埃尔·帕特位于巴黎的纪念坐像，○在这里，就如陆吉埃所推荐的那样，每一个公共空间的设计都好像它占据了公园或森林中一块空地。更早的，像是对于巴洛克、手法主义、文艺复兴时期的建筑师，也是同样。著名的作品则有，巴黎的凡尔赛，教宗西斯都五世治下的罗马，拉斐尔的马达马别墅，以及罗塞蒂为费拉拉所做的设计 —— 不仅仅是他设计的小广场，而是那个更大范围的公共广场——"public campo"。这后一个词"campo"有着广场和土地的双重含义，表明在构想、技巧、概念和语汇中，景观和城市其实互相交织在一起。如今所谓的"景观都市主义"和"生态都市主义"的理解绝不能脱离历史，也不能被视作什么平地而起的崭新思想。事实上，它们的独特性仅仅体现在与其之前一小阶段的设计的对照之中。

　　要发现如今这些关于城市景观的项目的独特性，一个途径便是聚焦于它们对于那种两分世界假设的拒绝。这种假设认为，人工地貌和自然地貌截然不同，人为制造的景观和那些自然长成的绝无相似。另一种观察这种二元对立的方法，是把"landscape"这个词分开，其前一部分"land"指的是一种自然给定之物，后者"scape"则表示已然成型之物的外观、外貌或形象 ——虽然我们有时也会用"图画一般"来形容未经改变的自然地貌。在当代理论中，这种人工与自然之间看似显而易见的区别，却由于对地形所具有的效能以及运行的质问，而显得不无问题，并把力量和意义等问题放到了一边。○在把地貌视作图画 —— 地景的"景色"—— 这一方面，没有什么会比过去那些实践中的图像主义受到更为严厉的批判了。当代的景观学希望创造的，不再是图画和图像，而是效果，并且这种效果并非形式问题，而是运行问题的可视化方面。至于另一个词汇"土地"，问题则更为复杂。因为，如果以对于图像问题同样的方法来对待，来加以拒绝，则可能剥夺了设计中真正恰当的主题。因此，被否定的倒是有关土地的几个通常认识：首先，认为土地是惰性的；其次，认为它是一种财产。土地、水、天空、植物，都不再被视作静止的和不变的，而是生气勃勃的过程，其最重要的特征便在于发展的本性，它一直处在新陈代谢的状态之中。在当代理论中，那种把土地当做一种财产的观点同样也被否定，起

○ 1765 年

○ 至少可以暂时如此

landscapes,*° as well as the so-called 'informal' gardens of 18th-century Europe.* Nevertheless, landscape as image and property are rejected in current theory and practice because both envisage terrain as static, as if it were like a building, neglecting the fact that it is always and inescapably developmental, dynamic, or metabolic in character.

• 6
○ *shakkei* in Japanese, *jiejing* in Chinese
• 7

Despite the centrality of the concept of change in current discourse, the realisation that landscape is subject to both growth and deterioration is hardly a recent discovery. It is likely impossible to find a single landscape theorist from any period that did not recognise this rather conspicuous fact. Nor is it new to observe that the norms and structures of society vary through time; as long as there have been histories, there has been awareness that the institutions, laws, morals, and habits of cultures develop and disintegrate over time. The key question concerns the relationships between changes in each of these realms — nature and culture — or what is taken to be the order of both.

To document their design for the Olympic Sculpture Park, Weiss and Manfredi developed a number of renderings that visualised the spaces and schedules by which the project could meet its several programmatic requirements.* The slowest configuration appears to be the one that gave the project its name: the ensemble of artworks. Although the collection will grow over time, one can suppose that the works placed within the landscape will preserve their standing for decades, at least. Similarly slow will be the pace of change in the buildings and 'infrastructural networks,' especially when contrasted with the movement organised by the several paths that cross the site. Faster still are the transportation routes that cut or edge the terrain.° The schema for environmental remediation introduces still another chronology into the landscape, a very long one, ranging from the unrecorded prehistory of the land's misuse, to the time of 'capping,' and then into the unending time of 'monitoring.' In addition to the site-times that are local to project other temporalities serve as wider frames of reference: the schedules of business and leisure in the town that are structured by the workweek and the hours of the day, the seasons of the year that regulate the growth and deterioration of the plantings, and the slow and steady rhythm of the waves against the shore. Despite the fact that I have listed these places, events, and schedules separately, their 'internal' workings require reciprocity. The same dependencies° can be seen in the

• 8

○ a trucking route, rail line, waterfront trolley circuit, bicycle path, and ferry line

○ eccentricities

码是那种认为所有权需要有一个明确边界的观点。在当代设计师的概念中，景观超越了地块的边界，°而是一直伸向远方的天际。这里，我们又重新演绎了那个关于"借来的景观"的古老概念，无论是在亚洲的景观学中，•°还是在欧洲18世纪被称作"非正式"的园林中，•这都是一个常被讨论的话题。无论如何，在当代理论和实践中，那种把景观当做一种图像或是财产的观念都被摒弃，因为二者都把地貌视作一种静止的东西，似乎它就像一栋建筑那般，从而忽略了这样一个事实：景观的本性无可避免地总是发展的、运动的或者说新陈代谢的。

    尽管像"变化"这样的概念已然居于当下话语的中心，但是意识到景观不仅在不断生长，同时也在不断衰败，却并非近来才有的发现。在任何时代的景观学理论家中，我们都很难发现有哪位没有意识到这一显而易见的事实。同样，社会的结构与规范也在时间的长河中不断变化，这也不是什么新的发现。自打有历史以来，人们便认识到，各种不同文化中的典章、法律、道德、习惯等也在发展和解体之中。关键问题在于，这些不同领域——自然与文化——内的变化之间到底是一种什么样的关系，或者什么才被认为是二者共有的秩序。

    为了说明奥林匹克公园雕塑的设计，魏斯和曼弗里迪绘制了一系列渲染图，来表现为满足任务要求而来的空间效果和进度安排。•其中最慢的是这一项目之所以被称为"雕塑公园"的工作：那些艺术作品。虽然这些收藏会随时间而增加，我们仍然可以假定，被置入景观当中的这些作品至少会在那儿呆上好几十年。同样缓慢的还有发生在建筑和"基础设施网络"上的变化，尤其是当与场地上组织出的几条穿过性道路相比较时更是如此。更快的则是切过场地或是在场地边缘的交通线。°这个环境修整方案还为景观体系引入了另一个更为长期的时间考虑，从未作记录的史前时期对于土地的滥用，到"哄抬"时期，再到没有休止的"管制"阶段。

    在这些与项目密切相关的场地时间以外，其他的时间性则作为更为广泛的参照：城市中由工作周以及每天工作时间来决定的工作和休闲安排，控制植物的生长与衰败的季节变化，以及缓慢而坚定地涌向海边的浪花。尽管我把这些场所、事件、日程分别列出，但是它们的"内在"运行机制却要求彼此之间的互动。在我提到的其他项目中也有着同样的依赖性，°像是弗莱士河公园项目，以及更近一些的，同样由詹姆

○ 它的主人只是碰巧拥有了这一区域而已
• 6
○ 日文称作"shak-kei"，中文称作"借景"
• 7

• 8

○ 卡车运输线、轨道交通线、滨水轨道线、自行车专用线以及轮渡线

○ 离心性

other project I mentioned, the Freshkills Lifescape, as well as the more recent and impressively successful High Line project in New York City, also the work of James Corner and Field Operations. In these two and many others of our time the hackneyed distinction between natural and cultural history has been refused.

CONCLUSION

By virtue of their multiple and reciprocal temporalities, these projects allow us to see beyond the idea that the city is something 'for us' and the natural world is something 'in itself.' Presented instead is an ensemble of situations or institutions that give shape, schedule, and orientation to the patterns of everyday life, constituting a world that is so successfully urban that it is taken to be natural, despite the contradiction this usage implies. I would like to describe this condition as cultural ecology, a topography in which human affairs are synchronised with natural processes: business over a breakfast table brightened by morning light, musical performance in the quiet of a cool evening, waterfront leisure on warm weekends, retaining walls resisting erosion while supporting a viewing terrace, and so on. Because the situations of urban life are enmeshed in the both the histories of a culture and the processes of the natural world, cities are not the result of several architectural projects alone but also the legible face of the social practices they accommodate, when scheduled according to ecological processes, in acknowledgement of their several claims, anticipated and unforeseen.

Plato's last dialogue, Laws, begins with three old thinkers climbing Mount Ida toward Zeus' birthplace. Both their pace and that of the dialogue follow the slow course of the sun on the day before the summer solstice. Four well-worn paths cross° as they move toward the cave of beginnings: the philosopher's path, the dialogue's, the sun's, and the year's. In this myth, as in the stories proposed by the better projects of our time, a new urbanity is to be synchronised with the cycles that both structure its rhythms and transcend them. Whether we call this architecture or landscape is a minor issue as long as the topography we have in mind sustains the kind of culture we unthinkingly assume is natural.

° and cross out their separateness

斯·康纳及其"场地手术"事务所完成的纽约城的高线公园项目。在这两个以及许多其他当代项目中，那种老套的在自然与文化之间所作的区分已被决绝地抛弃。

结论

因为它们多重而又互相影响的特质，这些作品让我们得以超越那种观念，即认为城市是"为我们"而生的东西，而自然世界则是另外某种"自善其身"的东西。相反，这里我展现的则是一种境况或制度的总合，正是它们为日常生活的模式赋予了形式与方向，并建立起一个在城市性层面如此成功的世界，以至于可以把它看作是"自然的"——尽管我这样的使用事实上带来了某种矛盾性。对于这种状况，我想可以把它描述为一种文化生态学，这是一种人类活动与自然进程相同步的地形学状况：清晨和煦的阳光下美好的早餐时光，宁静而凉爽的夜晚的音乐表演，在一个温暖的周末度过水边的闲暇，挡土墙抵御了侵蚀，同时也支撑起一块观景的平台，等等。由于都市生活既是在文化的历史之中，同时也在自然世界的进程当中，城市就不可能仅仅是建筑项目的集合，而同时也是由这些项目所容纳的社会生活与实践，这些实践根据生态进程来计划，并同时认知到那些可以预见的以及不可预见的。

　　柏拉图的最后一篇对话，《律政》篇，从三位智者向着宙斯的出生地登山讲起。无论是他们行进的步伐还是他们对话的节奏，都应和着夏至之前太阳缓慢地移动。在他们向着那个代表着起始的山洞进发时，四条路径交织在一起○：哲学家的路径，对话的路径，太阳的路径，以及岁月的路径。在这一神话中，就如我们当代那些优秀的作品一样，一种新的风雅，将与决定并超越节奏的那些循环相同步。至于我们是管它叫建筑还是景观，则并不重要，只要我们脑中的地形学依然能够保持那种我们未加考虑便认作是自然的文化。

○ 因此也破坏了它们的分离。

NOTES 注解

1 Garrett Eckbo, 'Is Landscape Architecture?', *Landscape Architecture*, volume 73, issue 3, 1983: 64–65.
2 Humphry Repton, *Observations on the Theory and Practice of Landscape Gardening*, London: T. Bensley and Son, 1803; John Claudius Loudon, *Observations on the Formation and Management of Useful and Ornamental Plantations*, Edinburgh: Archibald Constable & Co., 1804; Quatremère de Quincey, *An Essay on the Nature, the End, and the Means of Imitation in the Fine Arts*, J. C. Kent tr., London: Smith, Elder & Co., 1837; Andrew Jackson Downing, *A Treatise on the Theory and Practice of Landscape Gardening adapted to North America*, New York: Wiley & Putnam, 1841; Judith Major, *To Live in the New World*, Cambridge, US-MA: The MIT Press, 1997.
3 Mohsen Mostafavi and Gareth Doherty ed., *Ecological Urbanism*, Baden: Lars Müller Publishers, 2010; Charles Waldheim ed., *The Landscape Urbanism Reader*, New York: Princeton Architectural Press, 2006.
4 Kevin Lynch, *What Time is This Place?*, Cambridge, US-MA: The MIT Press, 1973; John Dixon Hunt, '"Lordship of the Feet": Toward a Poetics of Movement in the Garden", *Landscape Design and Experience of Motion*, Michel Conan ed., Washington, D.C.: Dumbarton Oaks, 2003: 187–213; David Leatherbarrow, 'The Image and Its Setting', *Topographical Stories: Studies in Landscape and Architecture*, Philadelphia: University of Pennsylvania Press, 2004: 200–234.
5 Hans-Georg Gadamer, *Hermeneutics, Religion, and Ethics*, Joel Weinsheimer tr., New Haven, US-CT: Yale University Press, 1999; Jürgen Habermas, *The Structural Transformation of the Public Sphere*, Thomas Burger tr., Cambridge, US-MA: The MIT Press, 1989; Richard Rorty, *Contingency, Irony, and Solidarity*, Cambridge University Press, 1989; Giorgio Agamben, *The Coming Community*, Micheal Hardt tr., Minneapolis, US-MN: University of Minnesota Press, 1993.
6 Chen Congzhou, *On Chinese Gardens*, Shanghai: Tongji University Press, 1984: 56. In Suzhou, for example, there are 'excellent examples of borrowing from out-of-city landscapes and distant temples and Buddhist pagodas.'
7 Here the famous line from Alexander Pope is once again entirely apposite: 'He gains all points, who pleasingly confounds, Surprises, varies, and conceals the bounds.' Alexander Pope, 'Epistles to Several Persons: Epistle IV', *Collected Poems*, London: Dent, 1975.
8 Marion Weiss and Michael Manfredi, *Surface/Subsurface*, New York: Princeton Architectural Press, 2007.

6 陈从周《说园》，上海：同济出版社，1984年，"五"。如在苏州："至若能招城外山色，远寺浮屠，亦多佳例。"

TOPOGRAPHY, HISTORY & THE LIE OF THE LAND

地形学，历史与土地的伸展／谎言

约翰·狄克逊·亨特
John Dixon Hunt

TRANSLATOR
Shen Wen
Kong Dezhong

翻译
沈 雯
孔德钟

Topography is the science or practice of describing a particular place. An older English usage—*topograph*—simply referred to the same attempt to represent or describe a locality. The original Greek signifies that we write or draw a place.°

○ *draw*: graphein
　place: topos

In regard to landscape architecture, we need to understand the topography of a given place—or we should do!—before any intervention in made there. And it is unlikely that any place is without some history—whether as a result of its geology, its planting, or its early inhabitants and their activity. Presumably, in any good landscape design, that intervention takes the topography and history into account. Sometimes that may mean accepting what is there without any emendation of the site. Sometimes it will involve intervening in such a way that the site's topography is changed, made clearer to visitors, and generally enhanced or even radically altered.

Here I need to draw out a pun that is embedded in the English word 'lie'. 'Lie of the land' means primarily how the land forms itself, lies across the surface of the earth, visible to our scanning eye. But 'lie' also means a falsehood, a fibe, or a pretence, so that in this sense the 'lie' of the land means that it deceived us in some way: what we see is not necessarily the original land form or even geology. I shall play upon both these meanings in what follows.

地形学是描述一个特定地点的科学或实践。一种更古老的英语用法——地形°——指的仅是关于表现或描绘一个地点的类似企图。最初的希腊语——地方°表达的是我们去书写或描画°一个地方。

○ topography
○ topo
○ graphein

关于景观建筑，在任何介入于其中发生之前，我们需要理解一个给定场地的地形——或者说我们应当要。一块场地不太可能会没有一点历史——无论是作为其地质、植被，还是那里的早期居民及其活动的结果。大概在任何好的景观设计中，介入时都会考虑地形和历史。有时候那也许意味着不对场地作出任何修改而接受已有的一切。有时候它需要某种介入，以某种方式改变场地的地形，使之对于参观者更为清晰，并在总体上获得提升甚至根本性更改。

这里我需要特别指出英语单词"lie"的双关含义。"lie of land"首先意味着可以被我们双眼看到的土地自身的形成，及其在地表展开的形态。但是"lie"同时也意味着谎言，"fibe"或者伪装，因此在这个意义上土地的"谎言"意味着它以某种方式欺骗了我们：我们所看见的并不是原初的土地形式与地质。在后文中，我会同时在这两种意思上周旋。

I am interested now in the way that the history of a site is made clear, is realised and communicated to others: this history has two main narratives,° these are sometimes entwined, sometimes separated. It can be the history of a geology and an ecology.* And/or it can be the history of something that has happened there, whether or not it has changed the site itself. That kind of history impacts our understanding or reception of a site, and in its turn may lead to significant reworking of the topography to foreground that historical event or circumstance.

° for all history is necessarily narrative

* FIGURE 1

I am here specifically interested in how history or historical circumstance is inscribed upon a given topography. Again, this has two modes: one in which the historical event is local, indigenous and can be made visible, or one in which some suitable history is imported into the site and subsequently made to appear suitable and even 'true' there. One history is—shall we—'real' history, the other is a 'fabricated' or 'invented' history.

I will give one specific instance of an English topography in which the lie of the land, how the land is shaped and spreads itself across the ground, is both accepted and also changed with a view to enhancing our understanding and appreciation of that site. The place is Blenheim in Oxfordshire.*• It was originally a royal place, called Woodstock Manor, and it was set in a parkland devoted to hunting, where little was done to modify the place other than, presumably, to wall it around to keep the deer in. A river ran through the valley, and the land sloped gently towards its route. At some point in the 12th century one of the Kings, Henry II, kept a mistress there and built a bower on the land for her accommodation, some trace of which still survives.* In the 18th century the site was given, by a grateful nation, to the Duke of Marlborough, the general who had defeated the French at the battle of Blindheim in Bavaria. The architect and landscape gardener, Sir John Vanbrugh, was called upon to design both an imposed palace, Blenheim, and to make an impressive route through the park, over a colossal and also imposing bridge to the palace itself. Vanbrugh's wish to keep the old Woodstock Manor in the new landscape as a gesture to the older history was denied by the Duchess of Marlborough and destroyed. The site was further 'improved' later in the century by the famous 'Capability' Brown who made one significant chance: he flooded the river valley, raising the water to immerse half of Vanbrugh's colossal bridge, and this is what we see today in the topography of Blenheim.°

* FIGURE 2
• 1

* FIGURE 3

° on the bridge itself you can look down into the water and see the lower half of the structure.

The lie of the land is clearer, because the water has been enlarged, and the valley more expansive and much more expressive of the parkland around it. There were once ancient oaks trees, though none survives these days, but wonderful beech trees are there* and contribute largely to how we appreciate the historical topography.

* FIGURE 4

约翰·狄克逊·亨特
JOHN DIXON HUNT

对于一块场地的历史是如何被清晰化，被意识到以及传递给他人的方式，我很感兴趣：这种历史有两种主要的叙述，°它们有时候会交织，有时候会分离。它可以是一种地质学或者生态学的历史。* 它也可以是曾在那里发生过的事情的历史——无论它是否改变了场地本身。那种历史影响了我们对于一个场地的理解或者接受方式，并反过来会引发地形的显著改写，以此凸显那些历史性事件或境况。

  ° 因为所有的历史都是叙述性的
  * 图1

在这里，我对于历史或历史性境况是如何被记录在一种特定地形学之上的方式尤其感兴趣。同样，这有两种模式：第一种模式中，历史性事件是当地的，并且能变为可见的；而另一种模式中，一些合适的历史被引入场地，继而显得在那里是恰当的甚至是"真实"的。第一种历史是——我们是否能称之为——"真实的"历史，而另一种是"编织"的或"创造"的历史。

  我将举一个英国地形的具体例子。在这个例子里，土地的谎言，与土地是如何成形并在地表蔓延这两件事情同时被接受了，并且提升了我们对于那块场地理解与欣赏的方式。这个地方是牛津郡的布伦海姆。*• 它最初是一个皇家之地，被称为伍德斯托克庄园，位于一片用于打猎的草木区上，大概除了用墙将这片地围起来以圈闭鹿之外，几乎对场地没有做任何修改。一条河流从山谷之中流淌过，土地向其路线逐渐倾斜。在 12 世纪的某个时期，国王亨利二世在那里供养了一个情人，并为了其起居在此地建造了一座凉亭，它的一些痕迹仍然保存着。* 在 18 世纪，为了感激在巴伐利亚的布伦海姆战役中击败德国的莫尔伯勒公爵，英国将这块地赠予了他。建筑师与景观园艺师，约翰·范布勒爵士，受邀在场地上设计一座宫殿——布伦海姆，以及一条穿越公园的令人难忘的路径，它越过一座巨大壮丽的桥而到达宫殿。范布勒的希望是能在新的景观中，将老的伍德斯托克庄园作为一种对旧有历史的姿态而保留下来。莫尔伯勒公爵夫人否决了这个想法，并且拆除了庄园。在 18 世纪晚期，著名的"能人"布朗对这片地作了进一步"改善"，他创造了一个意义重大的机会：他淹没了河谷，通过提升水位而浸没了范布勒的巨大桥梁的一半，这便是我们今天在布伦海姆的地形中所见到的。°

  * 图2
  • 1

  * 图3

  ° 在桥梁上你可以往下在水中看到这个结构较低的一半。

地形学，历史与土地的伸展／谎言
TOPOGRAPHY, HISTORY & THE LIE OF THE LAND

The history of the place is both indigenous and invented: it is indigenous because the topography has been enhanced without being destroyed and the ecology largely flourishes there too. It is invented, because the faraway Battle of Blindheim in Germany is commemorated here in England, and the lines of the original battle were apparently laid out on the land in the form of clumps that signalled the battle lines under the Duke of Marlborough. The huge bridge, now accepted by most people as more acceptable now it has been submerged, provides a suitable triumphal causeway to the Palace. The topography therefore is a combination of the place itself, manipulations by different landscape architects, and the insertion into it of an event that had nothing originally to do with this place.°

○ and the destruction of the old Woodstock Manor effectively eliminated any earlier history that would rival the Marlborough's recent ambitions.

Now I need to turn to the role of history in topography. In the 17th century a writer named Robert Burton maintained: 'how lifeless all history is without topography'. I want to turn this around and argue that all topography is lifeless without history. And to argue this I want to look at a selection of sites that reveal *both* the articulation of a topography because some landscape architect has brought it into full view, augmenting the given history of that place, *and* how landscape architects have tried—sometimes successfully, sometimes not—to insert history into a topography.

But first a few preliminaries on the role or function of history in landscape architecture. History is never just the past, but an interpretation of the past. This requires a historian who narrates that past. That can happen in a book or novel, where a narrator guides the story; it can also happen in a film or a painting where the filmmaker or painter has also 'formed' his or her formal presentation. But in landscape architecture the author or narrator is never there; he may have designed the whole thing, but he must leave it to others to seize the shape or drift of his story, whatever that is, and words are usually rarely used, as they are in books or even pictures with a title and a story attached to them. Sometimes, modern landscape architects use words like Larry Halprin's Memorial for Franklin Delano Roosevelt in Washington,* but many architects these days dislike them.° So the historian is *not* a direct participant in the landscape. But nonetheless his presence can be palpable and in many cases his authorial voice can be as clear as an authorial voice would be in a book or film.

* FIGURE 5
○ Chinese architects however are happy to continue to rely upon verbal inserts into gardens.

Let us take a splendid landscape project: Paolo Bürgi's Cardada site in the Swiss canton of Ticino.* At the summit of the mountain is his Geological Observatory*—a name that makes it evident that we find ourselves there in a place in order to observe the history of the adjacent topography and geology. And sure enough on the far hillside is a scar, a slippage in the land forms, an event that occurred millions of year ago when the European and African tectonic plates slipped past each other and ruptured the geology. And furthermore we had earlier noticed lower down the mountain, on the floor of the aerial promenade, that the remains of living organisms were

• 2
* FIGURE 6

地面的伸展变得更为清晰，因为水域扩大了，山谷变得更为广阔而能更好地呈现围绕它的草木地。那里曾经有古老的橡树，虽然今日已无一存活，但是美丽的山毛榉树依然如旧，*这大为增进了我们对于这种历史地形的欣赏之情。

这片场地的历史既是当地的，又是被创造的：这种历史是当地的，因为它的地形免遭破坏而得以增强，并且那里的生态大为繁盛起来；这种历史也是被创造的，因为遥远的德国布伦海姆战役是在这里开始的，最初的战线被以土地的形式明晰地展现在土地上，标记了莫尔伯勒公爵所指挥的战线。那座巨大的桥梁，现在被大多数人认为是更可以接受的，已经浸没而成为一条通往宫殿的凯旋之路。这种地形因而是一种结合体，包括了场地自身，不同景观建筑师的操作，以及一件最初与这块场地无关的事件的介入。○

现在我谈的是地形学的历史角色上。在17世纪一个名为罗伯特·伯顿的作家声称："没有地形学，所有的历史将是多么毫无生命力。"我想反过来说，去论证如果没有历史的话，所有的地形学都是没有生命力的。为此，我想观察一下这些场地的选择：这些场地既揭示了一个地形的表达——因为一些景观建筑师已将之完全呈现于视野中，证明了那块场地的已有历史；也揭示了景观建筑师是如何尝试的——或成功或失败——将历史插入到一种地形中。

首先我想说一下历史在景观建筑中的角色或作用。历史从来都不只是过去，而是一种对于过去的阐释。这就要求有一个可以叙述那种过去的历史学家。这可以在一本书或者一部小说中发生，叙述者在其中引导着故事；这也可以在一部电影或一幅绘画中发生，电影制作人或画家同样也"形成"了他或她自己的形式表达。但是在景观建筑中，作者或叙述者从来都不会在那里；他可能设计了整个建筑，但必须让别人自己来把握他故事的轮廓或趋向——无论是什么样的故事，并且通常很少使用词语，不像书籍或图片中那样会附加以一个标题或故事。有时候，现代的景观建筑师会使用词语，就像为富兰克林·德拉诺·罗斯福所做的华盛顿赖瑞哈普林纪念馆那样，*但是当今的很多建筑师不喜欢那种做法。○因此，历史学家不是景观的直接参与者。尽管如此，但他的存在还是可以被感知到，在很多例子中，其叙述的声音可以像书籍或电影中的叙述声音那般清晰。

让我们看一个极好的景观项目案例：保罗布基在瑞士提契诺州的卡尔达达基地。●在山脉顶峰是他设计的地质观测

* 图4

○ 老伍德斯托克庄园的毁坏，有效地消除了任何能对抗莫尔伯勒的新生雄心的先前历史。

* 图5
○ 不过中国人还是很乐意继续依赖于园林中语言的介入。

● 2

地形学，历史与土地的伸展／谎言
TOPOGRAPHY, HISTORY & THE LIE OF THE LAND

etched*—the subtle recognitions of DNA and the mysterious workings of the physical world, the recognition of distant cataclysms. Now in all of these—and indeed other elements of the project—we have clear evidence of a historian narrating not only distant, historical events, but also ones that confront us directly in this particular landscape. We can miss the etched organisms, we can even miss the scar on the distant hillside; except that on the railings of the Observatory we do have descriptions that narrate explicitly that historical event. * FIGURE 7

But many modern as well as historical projects have no such way of alerting us to their historical stories, so the chance is that we would completely miss the stories of these topographies. While one can often research and so elucidate the origins of a project and learn how the history informed the site in the first place, on many occasions even this scholarly information is no longer available. And more often than not the archival information, while valuable, makes little sense of how one sees the site nowadays. Sometimes a historical event and its narrative can be inferred from how the topography is envisaged: at the Villa Lante, that the grotto of the Deluge° sits at the top of the garden, followed by the fountain of the dolphins, may explain the whole treatment of water down the hillside in the remainder* of the garden: this would then alert visitors to how, after the biblical Deluge, human beings had developed a multitude of ways to adapt and use the elements of water. But I also imagine that many visitors, no less responsive to the Villa Lante, simply do not read that historical narrative or indeed any other, in part because there are no verbal clues to help them, and in fact there is much to see on the site that does not specify any historical narrative.• We tend to enjoy most the way the hillside site has been amplified and its local water used.   ° referring to the biblical flood.   * FIGURE 8   • 3

So, if there is no notice or other verbal clue and we do not understand what these specific sites mean, do we have any way of recalling whatever history a given site would declare? I would suppose that on many occasions, visitors keen to apprehend some story may make the narrative up for themselves. For example, in the film, *L'Année derniere à Marienbad*,* on the terrace of the garden a man explains to a woman how he would understand the meaning of a statue,* telling us what the sculptural figures are about: the male figure is holding the woman back or somehow frustrating her. It is the character's personal fiction, of course, but it supposedly informs his understanding of the landscape. But a little later in the movie a rather boring figure starts to explain the 'true historical' meaning of the pair by saying that the statue was that of Charles III, and he then proceeds to explain its 'true' history.•°   ° 1961   * FIGURE 9   • 4   ° interestingly, though, as he begins to tell that story, the film, typically, trails off and leaves him without finishing that 'history'.

So it may be that the story envisaged by the designer, or something approximating to that original design, may be a wholly different tale from the one invented by a visitor according to how he or she responds to it. So, it is often possible that the *reception* of sites and landscapes has simply nothing to do with an 'original' topogra-

台*——这个名称很明确地告诉我们,这是一个为了观察邻近的地形和地质学历史的地方。可以清楚地看到,在远处的山坡有一条痕迹,那是土地形式发生的滑移,在数百万年前,欧洲与非洲的地质板块从彼此旁边滑过并且破坏了地质所留下的痕迹。此外我们刚才就注意到了在山脉的低处,空中步道那层,有机生物的遗迹被蚀刻于上*——"DNA"的精细识别与物质世界的神秘运作,遥远的灾难清晰可辨。在所有这些痕迹之中——以及实际上这个项目的其他要素之中——我们显然看到了一个历史学家的存在,他不仅叙述着遥远的历史事件,同时也在这种特殊的景观中与我们直面相对。如果没有在观测台栏杆上对那次历史事件的清楚描述,我们可能会错过那些被蚀刻的有机物,甚至也会错过在远处山坡上的痕迹。

* 图6

* 图7

但是很多现代项目与历史项目一样,并不具备这种向我们提醒其历史故事的方式,所以很可能我们会完全错过这些地形学故事。尽管有人可以经常做研究,并且由此阐明一个项目的源头,研究历史是如何最先地影响场地,但是很多时候甚至连这样的学术讯息都无法得到。这些档案性信息尽管很有价值,但几乎对人们现在是如何看待场地没有什么帮助。有时候从地形学的想象中可以推测出一个历史事件及其叙述:在兰特别墅园,洪水形成的洞穴°位于花园的顶部,海豚喷泉在其旁边,在花园的遗迹中这或许能解释水冲泻下山坡的整个过程,*这将提醒观者,在圣经的洪水之后,人类已经发展出了众多适应及利用水的方式。但是我同样可以想象,很多对兰特庄园同样有感触的参观者,并没有去阅读那种历史叙述,一部分是由于没有任何语言线索来辅助他们,而实际上在场地上可以看到很多并不包含任何历史性叙述的东西。•我们最愿意去做的是欣赏山坡地被扩展,以及当地水源被使用的方式。

○ 这里指的是圣经中的洪水。

* 图8

• 3

因此,如果没有任何提字或其他言语的线索,并且我们不理解这些特殊场地的意义是什么,我们会不会有一种方式去回忆一个特定场地所宣称的历史?我猜想在很多时候,热衷于去理解这些故事的参观者可能会自己编造一种叙述。例如,在电影《去年在马里昂巴德》°中,在花园露台上一个男人向一个女人解释他对一座雕像意义的理解,*告诉我们雕塑像在做什么:那个男性雕像正在向后拉那个女性雕像,或者说在某种程度上阻挠着她。这当然是电影角色自己的个人编撰,但也表达了他对于景观的理解。但是在电影的后面,

○ 1961年

* 图9

phy; visitors now respond to whatever they find to say about it. This is highly problematical and indeed complex, and we have not really begun to grasp how to confront this very complex matter of what I have called the 'afterlife of gardens'.• This is *not*, however, to be entirely rejected. How we respond to an invented topography and its history is ultimately and importantly part of the site.

• 5

If nothing on the site immediately recommends itself historically, it may be that some relevant knowledge of that site which has been 'borrowed' for that particular place. In that context, would we not better think of 'memory' rather than 'history' as the more useful term, a memory that may or may not be strictly 'historical' and related to the specific site, but nonetheless suits the topography. But a memory, even one fabricated from the resources of one's imagination, could have the *effect* of a historical story or narrative that may be as compelling as any other. It is this—what I've just called a 'fabricated memory'—that we need to consider further.

Therefore, I suggest then that there are *two* basic roles that history might play in bringing life to a topography. The most obvious role for a landscape architect is when projects both respond to a site and reveal its historical elements; where topographical or local conditions determine its historical value and meaning, where the landscape architect has only to draw out and make truly visible the historical narrative. Let me give some examples here.

In the Renaissance, this could be the knowledge of ancient Roman villas, where a modern villa acknowledges its predecessors and also provides a contemporary example. In today's world, we are likely to reference our own building types to acknowledge the past: Gas Works Park in Seattle would be a case of a designer, Richard Haag, incorporating and displaying the disused materials of the site, and relying upon both the redundant chimneys and our knowledge of its toxic earth,° to recall an earlier industrial episode. Today, with toxic materials continuing to leech from the site, the historical reference is even more apparent and disturbing; the actual Gas Work structure itself is now fenced off and even more of a *momento mori* of a historical event; obvious as this is as an object, I suspect many virtually ignore the historical reference of what they see.

° though invisible to the naked eye

A more recent and famous instance is the Duisburg-Nord parkland in Germany,•* with the added contribution, not only of leaving the industrial remains as they were, but of drawing attention to them. The steel works remain to feature as conspicuous elements of the site, with all their powerful and sublime and technologically obsolete equipment; but now new resources have been brought into the site—athletic activities, gardens, theatrical events that take place on a stage made from the iron plates of the former foundry. These draw very specific and deliberate attention to the past of this Ruhr site; one cannot ignore them—climbing on the structures, scubadiving in them or even finding oneself in bunkers that have now been reformulated as gardens but still left with a clear sign of their

• 6
* FIGURE 10

一个相当乏味的人物开始去解释这对雕像人物的"真实的历史"意义，指出这是查理三世的雕像，并接着解释它的"真实"历史。

因此，设计者预想的故事，或者接近原初设计的故事，很可能会与参观者根据他／她自身反应而创造出的故事完全不同。经常很有可能的是，场地与景观被理解的方式与"原初的"地形学毫无关系；参观者一旦发现任何能用以去理解场地的东西，都会对其作出回应。这是极有问题的并且很复杂，对于这种我称之为"花园的后世"的复杂情况，我们还不真正知道应该如何处理。然而，这并不是完全的抵制。我们如何去回应一个被创造出的地形学及其历史，是场地最根本和重要的部分。

如果场地上没有任何东西可以直接表明自己的历史，有可能一些关于这块场地的相关知识会被"借用"过来。在这种情况下，我们最好是使用"记忆"而非"历史"作为更恰当的术语，不管这种记忆是否在严格意义上是"历史的"并与特定场地相关的，但它都与地形相契合。一种记忆，哪怕是从一个人的想象中编织出的记忆，都可以具备一种历史故事的效果，或者一种足够吸引人的叙述效果。这种我称之为"编织性记忆"的东西，是我们需要进一步思考的。

因此，我认为在为地形学带来活力这个角度中，历史可以扮演两种基本角色。对于一个景观建筑师，最显而易见的方式是项目能对一个场地作出回应，并揭显其历史元素；地形学状况和当地的状况决定其历史价值和意义，景观建筑师只需将其画出来，让历史性叙述真正可见。在此，让我们看一些例子。

在文艺复兴时期，出于对古罗马别墅的认知，一座现代的别墅既向其先辈致敬，同时也提供了一个当代的范例。在当今世界，我们可能会用我们自己的建筑类型向过去致敬：设计师瑞克·哈格设计的西雅图煤气厂公园，是运用与展示原有场地上废弃材料的案例，他依靠废弃的烟囱，运用了我们对有毒土壤的知识，来唤起对早先工业时期的回忆。现在，随着有毒材料继续侵蚀土地，这些历史参照物显得更为醒目与令人不快；原煤气厂结构本身如今已被隔开，甚至更多地是作为一个历史性事件的"死亡象征"；尽管它作为一个物体显得非常突出，我怀疑很多人都会忽略他们所见到的物体性的历史参照。

一个更为新近与著名的例子是德国的北杜伊斯堡公

- 4
  然而，很有趣的是，当他开始讲故事时，电影渐弱，留下了他未完成的"历史"。

- 5

○ 尽管肉眼不可见

origins. However, we probably have to accept that the Duisburg-Nord parkland may gradually lose some of its historical significance, as other activities dominate and as the memory of those who now use the site lose contact with its past, however strongly the physical remains survive. And then down the road, the blast furnaces are going to need care and conservation, which would bring another layer of historical meaning onto this topography.

My second version of history applied to a site, a contrary mode, is when a site has nothing at all in it that appears to connect with any historical narrative that might be conceivably imported into it. But whatever does get projected there may actually invent some wholly fictive historical narrative, a fiction that takes root and is thereafter identified with that topography, either because it uses and declares that topography, or grafts something else on to it. Such a project relies as much upon the role of memory and imagination as upon the presence of seen and understood historical facts. Yet the entanglement of memory and imagination with the treatment of a site is always crucial, and cannot be dismissed.

The first condition of historical narrative does, however, require sites that have visible and usually conspicuous, palpable, elements for the designer to use. But the second condition is often more likely — a site where no obvious or even relevant historical narrative survives. So what happens is that such a project may be wholly fabricated. Sometimes this is wonderfully and inventively accomplished, and the site becomes associated with its creation and in retrospect presents us with an acknowledgement of some invented history that, though it could logically be placed anywhere, nevertheless finds its site here. The great success of Duisburg-Nord is perhaps that it actually combines both historical ground and an invented narrative.

Three modern parks in Paris suggest the range of what may be done in those circumstances, from virtually minimal historical reference to slight vestiges and then to an invented milieu. There are the Parcs André Citroën, Bercy and the Jardin Atlantique. The site of Parc André Citroën was cleared of the former car factory; only its name and a rather forlorn and not very visible bust of Monsieur Citroën make any gesture towards the original site and its history.* The slightly sloping level ground towards the River Seine has been largely untouched, except for the insertion of gardens and fountains at the top and along the sides of the site.

* FIGURE 11

Somewhat more indicative of a historical past are the few telltale remains in the Parc Bercy, at the other end of Paris, where elements of the old site still remain and the topography is again left largely as it was before the park was established. An embankment against the River Seine protects the site on one side. Formerly this was the *dépôt des vins*,° where casks were stored, of which some are still on show, and where railway lines that helped to move these huge bulks are left embedded in the ground,* at an angle to the contemporary pathways; there is also a row of the vaulting cellarage. Maybe these

° wine storage

* FIGURE 12

园,●* 这个设计对原有场地作出了新增的贡献,不只是将工业遗迹按原样保留,而且更增加了它们的吸引力。这些钢铁作品作为场地中具有显著特征的元素而保留了下来,包括所有那些壮观有力但技术陈旧的设备。现在场地中引入了很多新资源——体育运动,花园,和在一座用先前铸造的铁板搭建成的舞台上举办的戏剧表演。这些事物有意地将人们的注意力引向这片鲁尔区场地的过去——在结构上攀爬,在其中潜游,甚至发现自己身处在昔日的燃料舱中,这些燃料舱现在已被改造为花园的形式,但仍留着其原初的清楚印记。然而,我们也许不得不接受,北杜伊斯堡公园可能会逐渐遗失掉一些自己的历史意义,因为其他的活动开始占支配地位,而现在使用者的记忆与场地的过去不再有关联,然而物质的遗物却幸存着。接下来,爆炸炉将会需要照料与维护,这将给这里的地形带来新一层面的历史意义。

● 6
* 图10

　　作为一个相反的模式,我对历史作用于场地的第二个视角发生在场地自身不存在任何东西,却与可以被令人信服地置入其间的历史性叙事相关。但无论什么被投射在那里也许实际上都创造出了某种完全虚构的历史性叙事,一种植根于并随之与那里的地形一致的虚构,要么是这种虚构因其使用并明示了地形,要么是因其在上面嫁接了其他叙事。这种项目对记忆与想象所扮演角色的依赖,与其对可见与可被理解历史事实的存在的依赖等量其观。然而记忆与想象和场地的处理之间的互相纠缠往往是决定性的,不可忽略。

　　然而,历史性叙事的第一种情形的确要求场地具有可见且通常明显、易察觉的元素,以供设计者使用。而第二种情形则通常更可能这样——一块场地上不存在任何明晰的或甚至只是相关的历史性叙事。因此经常出现的情况是,一个项目也许完全是被虚构出来的。有些时候这样的虚构被美妙而有创造力地实现了,场地也变得与其创造物有关,并在回想中向我们呈现对某种虚构历史的认同,尽管在逻辑上,它可以置于任何地方,但却在这里找到了处所。北杜伊斯堡的巨大成功也许归因于它真正将历史背景与一个虚构的叙事结合了起来。

　　巴黎的三个现代公园显示出了在不同情境下所能做的事情的范围,从几乎最少的历史参照到细小的遗迹再到一个虚构的环境。这三个公园分别是安德烈雪铁龙公园,贝西公园以及大西洋花园。安德烈雪铁龙公园的场地是清理后的以前的汽车工厂;只有它的名字和更为凄凉且不易观察到的雪铁

fragments will testify to the former history of the site, but I suspect many will miss their significance.

But one site in Paris is particularly successful, and an excellent example of a site that is accorded a historical significance because of what has been invented for it. The topography in this urban setting has been completely erased° and reinvented above the train station at the Gare Montparnasse. Over the station has been laid out the Jardin Atlantique,* created over the departure platforms of the TGV° trains that serve the northern Atlantic coast from the Gare Montparnasse. Of course, there are waiting rooms on the station below, but a better idea is to prepare the voyager with some hint of where he will be going when he reaches the Atlantic coast. So something of a new, invented topography has been constructed: there is a boardwalk or promenade, slightly tilted towards the 'water', where 'waves' advance upon the shore; there are two sets of cliff caves° to explore, seaside planting to see, and children's seaside games. There is also a weather station, which records the rainfall, the wind and temperature—all very apt information for those journeying to the seaside. The site is apt, though nothing in it derives from any historical events or items—except perhaps the vents that send the noise of the TGVs from the station below; yet the garden will certainly continue to be associated with the Gare Montparnasse and its ultimate destination. What the Jardin Atlantique does is essentially to create a historical memory—of trains, seashore and nostalgic destinations,° and then locate such a site in an anonymous location that nonetheless thereby establishes its own historical narrative.

To see the long traditions of this invented history or what might be called a memorial narrative, I'd like to go back to one of the most mysterious and creative instances of landscape architecture—the Désert de Retz. Created by a certain Monsieur de Monville in the 1780s, a period of great scientific and philosophical Enlightenment in Europe, he proposed assembling the complete sum 'de la connaissance et des curiosités de l'homme du XVIIIe siècle'. So he gathered a collection of buildings from many civilisations—the remains of a gothic church, a Pyramid, a Theatre, a Turkish pavilion, a Chinese House,* a ruined classical temple dedicated to Pan and— most notably—a huge and supposedly 'destroyed' column,* that would have been a vast and imposing structure 120 metres high!° This was and is simply a wooded valley in the forests around Paris, and what Monsieur de Monville inserted there had no local reason for their being there. But what he did has enhanced this place and made it one of the most eloquent locations that I know.

There are various explanations, responses, to this creation, but here I'll take hints from one of more acute appraisals of the site by the poet and critic, Yves Bonnefoy, published in 1993.* Bonnefoy proposed that de Monville gathered his recollections of many historical civilisations—including a horticultural garden—that would be the epitome of all cultures and religions, or maybe be a medita-

○ indeed covered over

∗ FIGURE 13

○ bullet trains

○ like theatrical scenery

○ intimations, perhaps, of *Monsieur Hulot's* holidays in the film by Jacques Tati.

∗ FIGURE 14
∗ FIGURE 15
○ it was of course constructed 'new' as a fragment.

• 7

龙的萧条景象示意着原本的场地及其历史。* 通向塞纳河的缓缓倾斜的地面已基本弃置，除了在场地顶端及场地边缘的花园和喷泉。

  在巴黎的另一端，贝西公园中，不多的几处遗迹似乎还包含了更多的历史过去的暗示，在那里原来场地的要素仍然存在，而地形基本上保留着公园建造之前的样子。塞纳河的堤岸在一侧保护着场地。以前这里是酒窖，° 那里储存着木桶，有一些酒窖依然可见，并且地面上还嵌有运载这些大家伙用的火车轨道，* 其与现在的轨道成一定角度；那里还有一排拱顶的地窖。也许这些碎片可以证明场地曾经的历史，但我怀疑其中很多将无法显现它们的意义。

  但巴黎的一处地方特别成功，是场地因虚构而被赋予了历史意义的绝佳案例。这个城市环境中的地形已被完全清除，° 并在蒙帕纳斯火车站之上被重新建构。火车站之上是大西洋花园，* 在起自蒙帕纳斯的服务于北大西洋海岸线的"TGV"° 出站台的上方。当然，车站的下部有候车室，但更好的点子是在其中为到达大西洋海岸的航行者们准备一些提示，提示他们将前往何处。因此某种新的虚构的地形被构筑起来：一条向"水域"微微倾斜的甲板或人行道，在那里"海浪"冲向海岸；有待探索的两组悬崖中的洞穴，° 可观赏的海生植物，还有儿童的海边游戏。还有一个气象站，记录着雨水、风和温度——这些对海边之旅都非常有用的信息。尽管其中没有任何东西源自历史事件或物件——也许除了"TGV"噪声的排声口，但这个场地是适宜的；而人们将继续把这个公园与蒙帕纳斯站及其最终目的地确切地联系起来。大西洋花园所做的，在本质上是创造一个历史记忆——关于列车、海岸及充满乡愁的目的地的记忆，° 然后将这样一个场地放置于某个无名的地方，却因而构建起了自身的历史性叙事。

  为了了解这种虚构历史或纪念性叙事的漫长传统，我想回顾一下最神秘且最有创造力的景观建筑实例之一——雷兹沙漠。雷兹沙漠于 18 世纪 80 年代，由欧洲伟大的科学哲学启蒙时期的蒙威利先生创造。他计划收集全部的"18 世纪者的知识与见闻"。因此他收集了来自诸多文明的各种建筑——来自哥特教堂、金字塔、大剧场、土耳其宫殿、中国住宅、* 荒废了的献给潘神的古典神庙的遗迹，以及——最著名的——一个巨大且被认为已遭"毁灭"的柱子，* 它曾经是高达 120 米的巨大且令人震撼的结构物！° 这里曾经

* 图 11

° depot des vins

* 图 12

° 事实上是被覆盖了
* 图 13
° 子弹列车

° 如剧院的布景

° 也许暗示了雅克·塔蒂影片中哈里特先生的假期。

* 图 14
* 图 15
° 它当然是被构筑为一个"新"的碎片。

地形学，历史与土地的伸展／谎言
TOPOGRAPHY, HISTORY & THE LIE OF THE LAND

tion on the decadence and immanent end of all human achievements—*désert* here signals a place of meditation, where the different topographical insertions were made. On the eve of the French Revolution, says Bonnefoy, the Désert de Retz became a sketch of history in all its grandeur and its decadence, for in this place our attention is focused on what is known from all the regions of the earth. In each case we are confronted with a true place° or what he calls a 'champ de gravitation', a real topography where ideas and life have gravitated, a significant grounds of being, and not merely an abstract idea. But the valley of de Retz is, however, a wholly fictive, a wholly invented or fabricated location, and this is what makes it an act of what Bonnefoy calls an 'authentic modernity'. ° a *lieu*

What I deduce from this inquiry is that anywhere, any location, could be selected for a project if the creative intelligence and imagination finds in it a true place or *lieu*, a centre of gravity, and not a mere abstraction. This perception, then, opens a rich but also very demanding mode for the landscape architect—to make a place out of nothing historically but where a historical narrative is invented for a particular topography—or even for a topography that is the result of massive urban transformations. The Jardin Atlantique does that, as would a selection of other recent projects.

This is of course a very modern idea. It was Charles Baudelaire who, writing about the dull work of theatrical scene painters, said they were all bad liars° because they did not lie sufficiently or convincingly. Stage designers and painters of course make a place, a topography, on whatever stage they are asked to design for. If they design good stage settings, then the 'place'° is a success and convincing. This Baudelairean model sustained one contemporary creation that Baudelaire himself might well have applauded, the park of Buttes-Chaumont in Paris*: out of a waste land, a garbage dump and the execution place of criminals was invented this fabulous place° of cliff sand precipices, a suspension bridge, caverns and grottoes, lakes and temples. And it still survives as an enchanting park where by now, 150 years later, is belies its merely imaginative projection and seems 'real' historical ground. And this transformation of place into 'true fact' is of course a measure of all great art.•

° menteurs

° i.e. the play

* FIGURE 16

° lieu

• 8

There is of course a thin or at least an uncertain line between really good lying° and the sort that convinces no one.° I offer now in the final part of this talk, a few examples of historical narratives that spring convincingly and with the aura of certainty from the bare minimum of a given topographical site. I propose here one French design, a trio of America ones, and one in Portugal.

° as with the Buttes Chaumont
° I'd instance La Villette, myself, or more obviously any Disney landscape.

Bernard Lassus was invited to make a rest area on both sides of the autoroute outside the former city of Nîmes in Provence, France.• There was little if anything to mark the autoroute that runs essentially right across France from Turin in northern Italy to Barcelona in Spain. In recreating such a rest area the appeal of a specific place was needed: why stop here if there was nothing to see, apart from

• 9

和现在都只是巴黎周边的森林中一个树木繁茂的山谷,并且蒙威利先生置入此处的东西缺乏存在于此的依据。但他的所作所为强化了这个地方而将使之成为就我所知的最为动人的地方之一。

针对这个创意存在多种解释、回应,但这里,我将引用诗人及评论家伊夫·博纳富瓦1993年对这块场地的评价,这是对这块场地更为尖锐的评价之一。博纳富瓦提出,蒙威利先生收集的是他对于诸多历史文明的回忆——包括一个园艺花园,而它们是所有文化和宗教的缩影,或者也许是对一切人类成就的衰落和固有终点的冥思,这里,沙漠暗示了一个冥思的场所,在此嵌入了不同的地形。博纳富瓦称,在法国大革命之夜,雷兹沙漠变为历史所有辉煌和衰落的一幅速写,因为正是在这里,我们的注意力集中在被地球上所有地区所熟知的东西身上。每个个例中我们面对的是一个真实的地方,一个场所,或他所称之的"引力场",一个理念和生活容入其中的真实地形,一个存在的重要基础,而不仅仅是个抽象概念。然而,雷兹山谷是一个完全虚构、完全人造或制造出来的地方,且正是这点使之成为一次博纳富瓦所说的"真正现代性"的行动。

• 7

我从这项研究中推论出的是,任何地方,任何处所,只要创造性的智力和想象力在其中找到准确的地点或场所、引力的中心、而不只是一种抽象,都可以选来做项目。这种认识将为景观建筑师打开一个丰富但要求严苛的模式——不根据任何历史塑造历史,而是为特定的地形,甚或为作为大规模城市变迁结果的地形,虚构一个历史的叙事来创造场所。大西洋花园即是这样做的,一系列其他的近期项目也同样如此。

这当然是一个非常现代的观点。查尔斯·波德莱尔在写有关剧院布景画师无聊工作时声称,布景画师都是差劲的骗子,因为他们的骗局既不充分又不可信。舞台设计师和画师当然是在被要求设计的舞台上制作一个场所,一个地形。如果他们设计了好的舞台布景,那么这个"场所"便是成功而令人信服的。这个波德莱尔式的模式支持了一个当代的创作,一个波德莱尔本人也会喝彩的作品,巴黎柏特休蒙公园:在一片废弃的土地、垃圾场及法场之上,虚构出了一个有沙崖绝壁、悬索桥、天然及人工洞穴、湖泊以及神庙的令人难以置信的地方,或场所。尽管仅凭想象的投射物和看起来"真实"的历史背景都是虚假的,在150年之后的今天,它仍然是一个充满魅力的公园。这种场所成为"真实事实"

○ menteurs

○ 即剧目

* 图16

地形学,历史与土地的伸展／谎言
TOPOGRAPHY, HISTORY & THE LIE OF THE LAND

going to the toilet?° Lassus first levelled the ground where the autoroute had been excavated and stretched a huge *tapis vert°* across the roadway, diminishing it and even posing the question as to which historically came first, the grass plateau or the autoroute. On this large carpet are installed three particular structures—a 19th century, pillared façade unwanted by Norman Foster when he re-designed Nîmes' opera house, and two steel belvederes* in the shape of the famous Tour Magne, a Roman monument in Nîmes, each with a stone maquette of the same tower.° This accumulation of different references—the tapis vert is a significant French landscaping item, notably seen at Versailles—reveals a distinct sense of Frenchness on the otherwise anonymous road from Turin to Barcelona; so much so, in fact, that citizens of Nîmes have started to come here to observe and appreciate their 'Nîmes-ness'. The place is wholly contrived—the unwanted opera house is given a new location here, and the distant Roman city and its famous structure are here miniaturised on the two belvederes from which we can glimpse the real and historical city to the west.

○ of no historical significance!
○ A green carpet
* FIGURE 17
○ the rest area can only be visited by travelling in both directions, so this doubling of the experience was necessary.

Similar proposals mark the historical interventions of two different American designers. Both Ken Smith and Martha Schwartz are inventive and often witty with their responses to topography.• Smith took an unprepossessing roof top—admittedly on the Museum of Modern Art in New York—and invented a garden for it; it can bear no weight on the roof, but the garden invokes a flat camouflage pattern not unlike some modern exhibits in the Museum below. Similarly, Schwartz, for the Civic Arena in Fort Lauderdale, Florida, substituted sixteen sculpture canopies of steel and coloured vinyl that recall the lost arboreal environment of the Everglades; the project was originally designed to contain living royal palm trees that would have commemorated the displaced topography, but when cost precluded most of the trees, she invented this new topography.

• 10

In a better, less ironic fashion Schwartz, and one that revealed some real imagination° she designed a new Exchange Square in Manchester, England.* It marks a sloping edge where the ancient city rises to meet the new shopping centre, erected where an IRA explosion took place in 1996. The meeting edge of old and new is marked by a trench of bubbling fountains and a line of river birch trees that line the boundary of the mediaeval 'hanging Ditch'; its pipes seem to pull the water from the underground water course. A new landscape, a new urbanism and a new topography.

○ in contrast to the fancy which the 19th century writer Coleridge opposed to the imagination.
* FIGURE 18

A last example comes from Portugal, where João Gomes da Silva designed his five unseen 'gardens' or Garcia de Orta Gardens* along the banks of the river in Lisbon during an international exhibition in 1998.• A series of different gardens draws upon different parts of the original historical Portuguese empire,° and bring home to modern Portuguese visitors historical memories of those distant locations by recalling their planting, their paths and vernacular constructions.

* FIGURE 19.1 & 19.2
• 11
○ though now 'unseen'

的变化当然是所有伟大艺术的一种度量。

在很好的谎言°和不令人信服的谎言°之间当然存在一道微妙或至少是不确定的界线。在这次演讲最后的部分，我想举几个历史叙事的例子，它们起自给定场地最少的信息，却带着确切的味道令人信服地呈现出来。我在这里将举一个法国的设计，一个美国的三部曲和一个葡萄牙的设计。

伯纳德·拉叙斯受邀在高速公路两侧设计一片休息区域，地点是法国普罗旺斯尼姆的城市外部。那里几乎没有任何东西告诉你这条从意大利北部的都灵到西班牙巴塞罗那的高速公路穿过了法国。在重新创造这样一片休息区的过程中，需要有一个场所的特定性：除了洗手间以外，°如果没有任何事物可被观看，为什么要停驻此处？拉叙斯首先在高速公路已被开凿的地方抬升起地面，并横穿路面做了一个巨大的绿色地毯，°减弱了路的存在，甚至提出这样的问题，即绿色的高原和高速公路哪个更早出现。这块大地毯上装置了三个特殊的结构物——一个被诺曼·福斯特在重新设计尼姆剧院时舍弃的19世纪时的柱廊立面，以及两个钢制的以著名尼姆市罗马纪念物马涅塔为形状的观景楼,* 每个结构物都相伴着相同塔楼的石制模型。°这种不同参照物的堆积——绿毯是法国重要的景观元素，在凡尔赛宫尤为常见——揭示了都灵到巴塞罗那之间独特的法国味道，否则这段道路将变得毫无特征；事实上，其效果十分出众以致尼姆的市民开始前来此地观看和欣赏他们的"尼姆特征"。这个场所是完全被发明出来的——被舍弃的剧场在这里被赋予了新的落脚处，遥远的罗马城及其著名的结构物在这里被微缩在两个观景楼里，身处其上，我们能够看见西面真实的和历史的城市。

相似的方案记录了两位不同的美国设计师对历史的介入。肯·史密斯和玛莎·施瓦茨都别出心裁而常常富于机智地回应着地形。史密斯采用一个不讨人喜欢的屋顶——明晰地放置在纽约现代艺术美术馆里——并为其虚构了一个花园；花园并不能承受屋顶的重荷，但采用了单调的伪装性图案，而不同于下部美术馆中的一些现代展品。相似地，在佛罗里达劳德代堡的市政舞台设计中，施瓦茨采用十六个钢制乙烯上色的雕塑化覆盖物，它们唤起了对已失去的湿地里栖息于树上的环境的回忆；项目最初的设计包含成活的皇家棕榈树，其也许将成为被取代了的地形的纪念，但当资金不允许大部分的树时，她虚构出这个新的地形。

- 8
  ° 如柏特休蒙公园
  ° 从自身的角度看，我会以拉维莱特公园或更明显的迪士尼乐园的景观为例。

- 9
  ° 不具有任何历史性意义
  ° *tapis vert*
  * 图17
  ° 休息区只有在这两个方向旅行时才会被游历，因此这种双重体验是必需的。

- 10

I do not need to tell you that there are some particularly awful attempts to improve a given topography with some addition that is presumably intended to be an enhancement of the site. It would be useful to explore failures, but this is not the place to do so. And I put aside the whole agenda of land art, interesting and very suggestive as it can be on many of its sites, for land art does not seek and often does nothing to provide social places,° which is the primary focus of landscape architecture. Land art may intervene interestingly in the topography of places,° like the western deserts of North America or the badlands of disused and toxic waste sites. But landscape architects have a different task, still responding to a given topography, but investing it with meaning and usefulness for a greater range of people.

° public or private

° often distant

在更胜一筹的而不那么反讽时尚的施瓦茨身上，以及一个尊崇某种真实想象°的施瓦茨身上，她设计了英格兰曼彻斯特的新交易广场。* 其标识了一条带缓坡的边界，这里，古老的城市逐步上升而遭遇新的购物中心，后者是建在1996年发生"IRA"爆炸的地方。新旧遭遇的边界是一长条不断喷涌的喷泉和一排河桦木，这些树界定出中世纪的"绞刑沟渠"的边界；而其管道看起来似乎要将水从地下水道牵引出来。一个新的景观，新的城市主义及新的地形。

    ○ 以此对抗19世纪作家柯勒律治用来对抗想象力的幻想。
    * 图18

最后一个例子来自葡萄牙，在那里若昂·戈麦斯·达席尔瓦于1998年的国际展览期间在里斯本延河岸设计了他的五个"不可见的花园"或者说加西亚·奥尔塔花园。*• 一系列不同的花园根据最初的历史上的葡萄牙王国°的不同部分绘制，并通过让人回想他们的种植物、路径以及当地的建筑物，将那些遥远地方的历史回忆带还给现代的葡萄牙游客们。

    * 图19.1和19.2
    • 11
    ○ 尽管如今"不可见"

我不需要说，确有一些特别糟糕的尝试，试图通过某些也许意在强化场地的添加物而改进某个给定的地形。探索失败也许有用，但现在不是这么做的场合。并且我撇开了整个大地艺术的议题，尽管在它很多场地的处理上有趣而非常有启发性，因为大地艺术并没有探索社会性场所°并常常对其的形成毫无贡献，而这点在景观建筑是最首要的关注点。大地艺术也许有趣地介入了场所的地形，°例如北美的西部沙漠或废弃而有毒的废物荒地。但景观建筑师有不同的使命，尽管同样是要回应给定的地形，但要为更大范围的人们赋予场地的意义和益处。

    ○ 公共的或私密的

    ○ 通常是偏远的

FIGURES

1

2

3

4

1 位于英格兰德比郡霍克斯通的悬崖。
  摄：约翰·狄克逊·亨特
  Cliffs at Hawkstone, England.
  Photo: John Dixon Hunt
2 位于英国布伦海姆的湖泊与部分淹没的桥梁。
  摄：艾米莉·塔库伯曼
  The lake and partly submerged bridge at Blenheim, England.
  Photo: Emily T. Cooperman
3 布伦海姆"费尔罗莎蒙德之井／凉亭"遗迹。
  摄：艾米莉·塔库伯曼
  The remains of 'Fair Rosamund's Well/Bower' at Blenheim, England.
  Photo: Emily T. Cooperman
4 费尔罗莎蒙德之亭侧畔的山毛榉树。
  摄：艾米莉·塔库伯曼
  Beech trees near Rosamund's Bower.
  Photo: Emily T. Cooperman

插图

5

6

7

8

5　哥伦比亚特区华盛顿的罗斯福纪念馆。
　　摄：约翰·狄克逊·亨特
　　The FDR Memorial, Washington, D.C..
　　Photo: John Dixon Hunt
6　保罗·布基设计的地址观测台，
　　位于瑞士卡尔达达山脉。
　　摄：约翰·狄克逊·亨特
　　Geological Observatory designed by Paolo Burgi,
　　Cardada Mountain, Switzerland.
　　Photo: John Dixon Hunt
7　卡尔达达的观测平台上的 DNA 蚀刻印记。
　　摄：约翰·狄克逊·亨特
　　Engraved impressions of DNA
　　on the floor of the viewing platform, Cardada.
　　Photo: John Dixon Hunt
8　位于意大利巴尼亚亚的兰特庄园的水链。
　　摄：约翰·狄克逊·亨特
　　The *catena d'acqua* at the Villa Lante,
　　Bagnaia, Italy.
　　Photo: John Dixon Hunt

9

10

11

9 《去年在马里昂巴德》的电影截图。
　Stills from film *L'Année dernier à Marienbad*.
10 德国北杜伊斯堡公园。
　摄：约翰·狄克逊·亨特
　The Park of Duisburg-Nord, Germany.
　Photo: John Dixon Hunt
11 巴黎安德烈·雪铁龙公园。
　摄：艾米莉·塔库伯曼
　Parc André Citroën, Paris.
　Photo: Emily T. Cooperman

12

13

14

15

12　巴黎贝西公园。
　　摄：艾米莉·塔库伯曼
　　Parc de Bercy, Paris.
　　Photo: Emily T. Cooperman
13　巴黎大西洋公园。
　　摄：艾米莉·塔库伯曼
　　Jardin Atlantique, Paris.
　　Photo: Emily T. Cooperman
14　法国巴黎沙漠的中国住宅（现已损毁）。
　　Chinese House (now destroyed) in Désert de Retz, France.
15　巴黎沙漠"损毁"的塔。
　　摄：约翰·狄克逊·亨特
　　La colonne brisée,
　　the 'Ruined' Tower in Désert de Retz.
　　Photo: John Dixon Hunt

16

17

18

19.1

约翰·狄克逊·亨特
JOHN DIXON HUNT

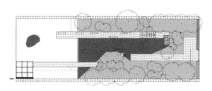
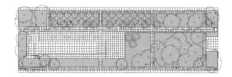
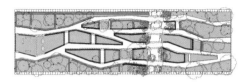
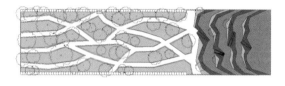

19.2
Africa Garden
Macaronésia Garden
Brasil, S. Tomé Garden
Goa Garden
Macau Garden

16 巴黎柏特休蒙公园。
    摄：艾米莉·塔库伯曼
    Parc des Buttes Chaumont, Paris.
    Photo: Emily T. Cooperman
17 伯纳德·拉叙斯设计的观景楼，位于法国尼姆市区外。
    摄：约翰·狄克逊·亨特
    Design for one of the Belvedere towers by Bernard Lassus at the rest area outside Nimes, France.
    Photo: John Dixon Hunt
18 玛莎·施瓦茨设计的交易广场，位于英格兰曼彻斯特。
    图片提供：玛莎·施瓦茨
    Exchange Square in Manchester designed by Martha Schwartz, England.
    Photo supply: Martha Schwartz
19.1 俯瞰由若昂·戈麦斯·达席尔瓦设计的五个"不可见的花园"，位于葡萄牙里斯本的奥尔塔花园。
    Aerial view of the five 'Unseen Gardens' of Garcia de Orta, designed by João Gomes da Silva, Lisbon, Portugal.
19.2 其中花园的表现图。
    若昂·达席尔瓦，加西亚·奥尔塔花园，五个"不可见的花园"，位于葡萄牙里斯本。
    Representations of the gardens.
    Joao da Silva, Garcia de Orta, five 'Unseen Gardens', Lisbon, Portugal.

NOTES

1. I rely here, in part, upon James Bond and Kate Tiller ed., *Blenheim: Landscape for a Palace*, Gloucester: Sutton Publishing Ltd, 1987.
2. Raffaella Fabiani Giannetto ed., *Paolo Bürgi Landscape Architect: Discovering the Horizon: Mountain, Lake, and Forest*, New York: Princeton Architectural Press, 2009.
3. Actually Claudia Lazzaro offers two quite different versions or narratives of the Villa Lante. First in her article, 'The Villa Lante at Bagnaia: An Allegory of Art and Nature', *The Art Bulletin*, issue 59, New York: The College Art Association, 1977: 553–60; and then in Claudia Lazzaro, *Italian Renaissance Garden*, New Haven: Yale University Press, 1990: chapter 10.
4. In fact, Alain Resnais invented the statue, partly based on a figure in a Poussin painting, for this incident: see Suzann Liandrat-Guigues and Jean-Louis Leutrat ed., *Alain Resnais: Liaisons secrètes, accords vagabonds*, Paris: Cahiers du Cinéma, 2006: 223–4.
5. See John Dixon Hunt, *The Afterlife of Gardens*, London: Reaktion Books, 2004.
6. See Lehrstuhl für Landschaftsarchitektur und industrielle Landschaft (LAI) ed., *Learning from Duisburg Nord*, Munich: Technische Universität München, 2009.
7. See Yves Bonnefoy, 'Le Désert de Retz et l'expérience du lieu', *Le Nuage rouge*, Paris: Mercure de France, 1977: 367–382.
8. I have proposed such an argument in my essay, Philippe Poullaouec-Gonidec, Sylvain Paquette and Gérald Domon ed., 'Les temps de l'histoire et l'invention du lieu', *Les Temps du Paysage*, Montréal: Les Presses de l'Université de Montréal, 2001. This claim that artists can make 'truth' out of their imaginations is what one of Shakespeare 'fools', Touchstone, says in *As You Like It*: that 'the truest poetry is the most feigning'.
9. See Bernard Lassus, *The Landscape Approach*, Philadelphia: University of Pennsylvaia Press, 1998: 164–67.
10. Both sites are illustrated and described in Tim Richardson ed., *The Vanguard Landscapes and Gardens of Martha Schwartz*, London: Thames & Hudson, 2004; and Ken Smith, *Ken Smith: Landscape Architect*, New York: The Monacelli Press, 2009.
11. See the forthcoming essay on this design and its site by João Gomes da Silva and Claudia Taborda, in *Studies in the History of Gardens and Designed Landscapes*, 2012.

注解

1. 这里我部分地参照了见英注 1。

3. 其实克劳迪娅·拉扎罗提供了两个差异很大的对兰特庄园的描述。第一个版本出现在她的论文《The Villa Lante at Bagnaia》，第二个版本出现在《Italian Renaissance Garden》。详见英注 3。

4. 事实上，阿兰·雷斯奈斯创作了这个雕像，部分以普珊一幅画作中的人物为依据。详见英注 4。

8. 我在自己的论文中称艺术家能够从他们的想象力中制造"真实"，正如莎士比亚笔下的"傻瓜"塔奇史东在《皆大欢喜》中所说："最真实的诗歌是最虚假的。"详见英注 8。

10. 《The Vanguard Landscapes and Gardens of Martha Schwartz》一书对两个场地都有配图及描述。详见英注 10。

11. 参见若昂·戈麦斯·达席尔瓦和克劳迪亚·塔波达即将发表的关于这个设计及场地的论文，将在 2012 年发表于《Studies in the History of Gardens and Designed Landscapes》。

A VIEW OF TOPOGRAPHY FROM MY REAR WINDOW

从我家后窗望见的地形学

刘东洋
Liu Dongyang

The rear window of my house presents a view of a rounded sloping ridge. It has an official name, being one of the nine hills of the East Mountain, out in the suburb of Dalian. I am used to calling it 'My Home's Back Hill', or simply the 'Back Hill',• as though it is a part of my property—having accompanied me through life's ups and downs. Calling it the 'Back Hill' conforms to my routine orientation and a subtle sense of plain ordinariness. I happen to be looking at the Back Hill through my rear window as I'm writing these words. If you were to ask me now what topography is, I would, without hesitation, point to the 'Back Hill' and say: 'That is topography!'—at least that is where I would embark on my interpretation of topography.

• 'Back Hill' is a transliteration of *hou shan*—the hill seen from the rear window of the author's house

从我家后窗望出去总会看到一道舒缓起伏的山脊。此山有大名，算此地东山九岭之一脉。我惯于叫它"我家后山"或是"后山"，好像它是我的家产，会跟我一起捱生活似的。将之称之为后山，既让它巩固着我日常起居的方位认知感，这么叫也在低调中，泛着点儿朴素和家常。此刻，当我写下这句话的时候，我正望向后窗，如果您现在问我："何谓地形？"我就会毫不犹豫地指着眼前的景色答道："它就是地形。"起码，后山是我开始讲述我对地形学认识的一个起点。

In the last ten years of my quiet and slightly dull life here, the first sign of dawn everyday has been the tireless and punctual emergence of the Back Hill from the murky night. Tinged with dark brownish green—more reddish in summer or largely cold blue in winter—the Back Hill would reveal its form quite swiftly. Is it such contours, or the differentiated but continuously changing features of the land that often come to mind when we are asked what topography is? We seem not to pay much attention to the flatness of the earth when searching for an image of 'topography'. Out of a sense of immediacy and for the ease of remembrance, there is a tendency to perceive and process landforms in a hasty and structural manner—myself included. Whenever I come across the word 'topography', I would most likely recall that gradually sharpening silhouette of the Back Hill abutting my home, and habitually believing in the power of lines in constructing the connotation of topography. Of course, there are countless lines out there, each of which presumably has its own charm. Since I could see the ridge of the Back Hill from such a close distance, I once assumed that I would pick up the details of trees and foliage quite easily. But it was not the case. The pines, elms and oaks merge into each other in darkness and what emerges from the scene are two man-made constructions—a cement retaining wall and pole. The wall runs parallel to the ridge, aligned with a path to the hilltop. It also blocks rainwater flowing toward the residential areas from the mountain in summer. As man-made constructions, the wall and the pole have been in the hill for perhaps more than twenty years. Yet at a peculiar moment each morning they exude a certain oddness of artificiality among the woods, and gradually recede into the shades of green when the sky brightens.*   * FIGURE I

Why do I, at the mention of 'topography', always think primarily of lines, a representation of some sort, instead of the Back Hill itself? I admit that my seemingly instinctive perception may be very personal, yet not at all 'intuitive'. As a frequent aerial-photographer and map user, I cannot separate the formal meaning of topographic maps from my understanding of landform. How could it be otherwise? How could I possibly share the same intimate knowledge of land possessed by farmers who have been labouring out in the corn fields for years? Similarly, how can I share the poetic imagination of ancient Chinese literati towards a tranquil scene of mountains and waters? Nevertheless, regardless of how disparate and scattered my understanding of topography, it would have to be based on my previous learning and encounters of landforms like that of the Back Hill. I am not saying that my understanding is necessarily more coherent than that of others. On the contrary, those with a similar educational background as me will probably find it hard to hold a coherent or complete understanding of 'topography'.

在我栖居后山脚下这十余年多少有些乏味的平静光阴里，每天黎明时分的第一个迹象就是这道山岭不知疲倦地准时从暗夜中浮现出来。带着深深的褐绿，在夏季有些微红，更多的是清冷的蓝，后山会在很短的片刻浮出身形来。或许一说到"地形"，我们大脑总会想起此类曲线，以及大地上细微的变化。我们似乎很少会关注平地。为了铭记和再现的方便，我们最喜欢对事物进行快速的结构性处理。我也一样。当我听到"地形"一词时，我常想到的就是后山从晦暗中渐次清晰起来的脊线，我也习惯性地用线去构建着地形感。当然，后山那里有着各种各样的线，每一种似乎都有着自己的魅力。既然我可以看到后山的脊线，既然我可以如此迫近地看到后山的脊线，我曾以为，我会在熹微中很容易辨出树和植被的身形。事实并非如此。后山成片的海松、榆树、柞树，在由暗转明的时刻仍旧彼此粘连。这时，跟着跳出来的，反而是两件人造物：一道水泥挡水墙和一根水泥柱。那堵挡水墙平行于山脊走向。它既是通往山顶的路径，也要挡住夏季从山上流向居民楼的雨水。它和水泥电线杆作为人工构筑物，已嵌入后山有二十多年了吧，却仍会在一天之中的某个时刻，在丛林中，凸现出自己骨子里的人工性，要等到天光大亮，它们才再隐退到丛林中。* 　　＊ 图1

　　怎么一提到"地形"我想的先是线呢？怎么不是后山本身而是它身上被表现出来的线呢？我得承认，我这看似直观的个人感受可能很个人，却一点都不"直观"。作为一位经常使用航拍图片和地图的人，我对地貌的认识不可能脱离地形图的意象。不是这样，又当如何？我怎么能够享有农人常年在玉米地里靠劳作获得的有关土地的切身感受呢？同样，我又真的能够享有过去文人那种寄情山水间的诗情吗？总之，不管我的地形理解怎样零乱与分散，它们总要回到我之前的学习经验中去，回到我对跟类似后山这样的地形打交道的历史中去。我可不是在说我的地形理解必然比他人的理解有着更为统一的整体性。恰恰相反，那些跟我有着相似教育背景的人反而很难拥有对于"地形"的完整理解。

Not long ago, I posted the question 'What is 'topography'?'* on a website which led to a small investigation as it elicited prompt replies from 65 online participants. Firstly, I should highlight the tricky part of this question: in Chinese, the English term 'topography' can be translated either as *dixing*—the terrain or contour of the earth—or as *dimao*—the surface or appearance of the earth. Thus, mass versus appearance. My questions were: 'What does the word *dixing* or *dimao* mean to you? Are they synonyms or not? Are they one thing or two nearly identical terms with subtle differences?' Out of the 65 participants, only one person who specialised in GIS° clearly indicated that 'in the disciplines of Land Survey, Geography or Geomorphology, *dixing* and *dimao* are mutually exchangeable terms'. The rest of the 65 participants all maintained that in Chinese, *dixing*° and *dimao*° have a few but definitive differences. To sum up: the term *dixing* speaks of the lie of a land, of altitudes or drops of the earth, close to a Cartesian notion of *res extensa*, conjuring up the image of contour lines. Moreover, *dixing* evokes a sense of vastness, a frozen moment within the processes of geological change. In contrast, *dimao* refers to flesh of the earth, something more vital, with certain sense of character or style, as in the often heard terminology 'Karsts Topography'. Therefore, *dimao* seems to be more concrete to most people, related to relief features of the earth such as mountains, forests or sand dunes. To further differentiate *dixing* from *dimao*, most participants tend to depict the former as something innate, less mutable and structural, whereas the latter as changeable and under the mercy of natural forces. One of the participants even insisted on the notion of *dixing* as a subset of *dimao*, that *dimao* could include *dixing* and its surface landscape.

Most of the online participants were probably just like me who had a general knowledge of geography, skilled in using maps, and not completely alien to the basic notions of 'contour lines' or 'Karsts Topography' but never really thought about how in geographical studies *dixing* and *dimao* could appear as synonyms. On the other hand, we the laymen somehow sense that there remain differences between these two terms. To me, it appears relatively easier to apprehend the reason why the majority of the participants made a distinction between *dixing* and *dimao*. The way I proposed my question would certainly have led people to 'find a difference' regardless. Other than that, it may largely relate back, again, to our education and experience of the world. We are living in a world of ever growing specialisation and labour division, in an age of media and abstraction, therefore, we are increasingly engaged with the 'representation' of the earth, rather than the earth itself. We have realised the occurrence of such tendency. Consequently, we often perceive the earth in its duality—half real, half virtual, or immutable in its core and mutable on its surface. This dualism is compounded by the connotations of *xing* and *mao* in Chinese. While *xing* and *mao* are generally exchangeable terms, however, *xing* can profoundly refer to

• 1

○ Geographical Information System

○ Topography as terrain or contours

○ Topography as land surface or appearance

不久前，我曾就"何谓地形？"的问题在网上发起了一次小调查。有 65 位网友自愿参与这次活动。首先，我该说说这个问题里比较"绕"的那个环节。因为在汉语中，英语单词"Topography"既可以被翻译成为汉语里的"地形"——大地的轮廓，或是"地貌"——大地的面貌。体形与面貌。我的提问因此成了"何谓地形？何谓地貌？"它们是同义词吗？还是有着细微差别的近义词？在 65 种不同的回答中，除了一位搞地理信息系统的人士明确指出"在测绘学、地理学或是地貌学中，地形和地貌基本是可以互换对等的两个词汇"之外，其他 64 位参与者都认为，汉语的"地形"和"地貌"还是存在着这样或是那样的差别的。什么差别？大家给出的解释大体如下：地形给人的印象是关于大地的走势、高差、起伏，类似几何里的广延量，一说到地形就想到等高线。还有，地形好像比较辽阔，可能跟地质构造有关，而地貌呢，则类似地表的肌肤，比之地形，地貌显得更生动，具有着某种风格或特征，例如人们常说的"喀斯特地貌"。地貌因此也显得更具体，总跟岩石、森林、沙滩有关。至于"地形"和"地貌"的关系，多数网友认为地形该是内在的、相对稳定的、结构性的东西，而地貌则是裸露给大气，总在更迭的。其中有 1 位网友坚持认为，地貌要大于地形，地貌可以包括地形加上表层。

多数参与这次讨论的朋友们可能跟我的情形差不多：都掌握着一些基本的地理常识，可以熟练使用地图，对"等高线／喀斯特地貌"这类术语并不完全陌生，但从没思考过为何在地理学里"地形"和"地貌"居然可以作为同义词出现，而我们这些非专业人士，却总能感到在两个词汇之间还是存在着一些区别的。为何这些朋友倾向于在"地形"和"地貌"之间做出区别判断呢？在我看来，这个问题相对比较好把握。首先，我提问的问题里对于"地形"和"地貌"作出的区别，肯定有暗示作用。此外，这种区别的认识可能很大程度上又是跟我们的教育和世界体验有关。我们生活在一个前所未有的高度劳动分工的世界，我们也生活在一个充满再现和抽象的时代。常常，我们对"大地的影像"比对大地本身还熟悉。我们也知道这一点。结果，我们常会感到大地也有二元特点，一半真实，一半虚拟，或者说，总是感觉它有着某种稳定的内核和常变的表面。而我们汉语里的"形"与"貌"的内涵与外延，也加重着我们这种内外有别的认识。在汉语里，"形"与"貌"多数的时候可以互换，比如"形容"与"面貌"。

yuanxing,° *xinggou,*° or *xingtai,*° such that most Chinese speakers would consider *xing* to be more complicated than *mao*, whereas *mao* is being referred to in a sense of superficiality.

I suspect that most Chinese speakers will think likewise, because the confusions and ambiguities expressed by the 65 participants are, in fact, quite closely reflected in the paradoxical definitions listed in *A New English-Chinese Dictionary*.• On page 1475, the English word 'topography' is specified as:

○ *yuanxing*: Archetype or primary form
xinggou: Configuration
xingtai: Structure & form

• 2

Dizhi	*A term that can be interpreted as either a 'local gazetteer' or 'geographical description of a place'*
Dixing	*Terrain*
Dixing xue	*Study of land terrain*
Dixing cehui xue	*Land survey;*
Jubu jiepou xue	*Sectional anatomy*
Jubu jiepou tu	*Sectional anatomical drawing*

In an earlier entry, for the item of 'topographical feature', the editors of this dictionary also adopt *dimao* as the transliteration for 'topography'. It seems that they share our view of *dimao* as 'topographical features', yet later deny any difference between *dixing* and *dimao*.

Why is this so? As the online participant with the GIS major had explicitly expressed, why is there such a complete overlap in the connotations of *dixing* and *dimao* as perceived in the discipline of Land Survey and Geography? What is the significance of such fusion or confusion of *xing* and *mao*? Could it be what a phenomenologist often expressed—'a kind of return to the world's flesh'? In any event what about the elms and pines growing on top of the Back Hill? How would a land surveyor or geographer perceive them— appearance as mass, mass as appearance?

But before we get the explanation from the field of Land Survey or Geography, I find it necessary to hear what the ancients said about 'topos', which stands as the irreducible kernel of topography. And equally, I find no better overview than the following brief commentary by Joseph Rykwert in *Topophilia and Topophobia*:•

• 3

The Greeks did use the word 'topos', generally for any place or location—in a book, in a landscape, in any discourse. Aristotle called his book on arguing 'topica'.° In medicine, it signifies the diseased points of the body as well as its secret parts, while in our generic usage, 'topos' has to do with any region or location, however big.

○ when its construction is independent of any subject.

但汉语的"形"还可以指向"原型"、"形构"、"形体"。这样一来，多数汉语使用者都会隐隐觉得"形"比"貌"要概练且复杂，而貌呢，总指向肌肤性表面。

我猜汉语使用者多会这么想，因为我们这 65 位讨论者所给出的时同时异似是而非的回答，离开专业辞书譬如《新英汉词典》里对"topography"的界定也不远。英语"topography"这个词被翻译成为：

• 2

地志；
地形；地形学；地形测量学；
局部解剖学，局部解剖图。

而在上一条"topographical feature"词条的翻译中，则直接对应着"地貌"。看来，这《新英汉词典》的编撰者们时而认同我们常人的理解——认为地貌，乃是地形上的某些物体和特征——时而，他们也会把地形学和地貌学完全等同起来。

为什么会这样？为何如那位专门从事地理信息系统的网友所言，在测绘学和地理学里，这"地形"与"地貌"几乎完全重叠呢？这种"形"与"貌"的混合或是混淆有什么意义吗？它是现象学家们常说的对于"世界肌肤的一种回归"吗？无论如何，我家后山上的那些松树和榆树们该怎么办？如何形即是貌，貌即是形，测绘学家或是地理学专家会怎样面对这片丛林？

在我们倾听测绘或是地理学里有关地形地貌的解答之前，我觉得我们很有必要先听听古人是怎样看待"topos"的。当然，我觉得，对"topography"°当中"topos"这个词根给出最简练的概述当属里科沃特在《恋地情结与怨地情结》的文字。他说：

○ 地形、地貌或是地形学

• 3

"希腊人一般会把'topos'这个词用来代指任何场所——可以是一本书中的场所，一个身体上的点，地景中的某处，或是任何一种话语中的位置。亚里士多德将自己有关论辨的书，°命名为'Topica'。在医学领域里，这个词指代着人们身体病变的部位以及身体的私处，而在我们对于'topos'一词更为宽泛的使用中'topos'常常指向一片区域，或是不管多大，指向某个地点。"

○ 是在说论辨的构成可以独立于任何具体话题的书。

Rykwert further elaborates on how the Romans translated the Greek word *topos* to *locus* — a word derived from the archaic Roman word *stolcus*. which in turn originates from the word *stare*, that is 'to stand'.*

○ 4

It is not surprising how the Greeks had used the word *topos* and extended their notion of place to describe an array of things — traversing the sky, land, human bodies and knowledge domains. I cannot help but be reminded of the military strategic typologies of *dixing* in *The Art of War*° or the exotic ethno-geographical *Dixing xun*° in *Huainan zi*. It also reminded me of *yudi yu* — territorial maps depicting lives and its multitudes — and traditional Chinese gazetteers in ancient China, in particular *Brief Notes upon Reading the Local History and Gazetteers*° by a late-Ming and early-Qing literatus Gu Zuyu.° The significance in this book's title lies in the word *shi*,° with the assumption that history, land and its multitudes are to be considered as an integrated entity. Gu once criticised his contemporary officials for 'having no true knowledge of geography and its dynamics, and yet daring to offer suggestions for state affairs'.* In Gu's opinion, if history is devoid of its specificity in space and location, the administering of national-political affairs would only lead to empty talks. As for the word *fang* in the book's title, it stands for 'land/earth', but more precisely, it is what the ancient Chinese understood as signifying the law of the land, as expressed in the sentence: 'the *tao* of the earth remains silent and everlasting, thus being named as *fang*, the square-ness of the earth. The earth is prolific to contain life, therefore, being named as *yu*, literally a chariot.'* The earth as a huge chariot carries ten thousand different things. 'Ten thousand different things' get to be quite specific when Gu Zuyu began to elaborate what the thing-ness of a place ought to include: prefectures and counties, rivers and canals, foods and goods, military farm land, military horse farms, salt and iron controls, tributes and territorial affairs based upon Sky Lodges. Gu wrote:

○ *Sunzi bingfa*
○ A Geographical Instruction

○ *Dushi fangyu jiyao*
○ 1631 – 1692
○ history

● 5

● 6

> This book will take the records of *Fangyu* from the past to present as its concern to reflect both the history and descriptions of local affairs. As history would guide the understanding of a place, a description of a place will serve history with better maps and records. Those incompatible descriptions of a place, and those unreliable historical narratives, will be eliminated from this book. I have therefore entitled this book *Brief Notes upon Reading the Local History and Gazetteers*.*

● 7

Certainly, one cannot take the fusion of history, place and 'ten thousand things' advocated by Gu Zuyu as a direct counterpart to the Greek notion of *topos* or 'topography'. Nevertheless, Gu's preface to his book and Rykwert's overview of *topos* have shed light on the inadequacy and limitation of our current understanding of

里科沃特跟着提到,"罗马人用了"locus"去翻译希腊人的"topos"。罗马人的"locus"源自古老的"stolus"——这个词又源自"stare",就是'立住'。"•

• 4

毫不吃惊,古希腊人靠着类比法则对于地形一词所含有的"场所性"曾赋予了相当广泛的使用。海阔天空,苍茫大地,知识领域,人体皱褶,他们都可以用"topos"去描述。这就不由得让我想起像《孙子兵法》里那种对于"地形"的军事策略分类,《淮南子·地形训》里对于"地形"充满奇异色彩的地理描述,当然,也会想到中国过去众多的"舆地图"、"地志"、"方志",比如顾祖禹的《读史方舆纪要》。这本书的书名就很有意味:它要把历史、大地和万物紧紧地整合成为一体。这里的"史"就是历史了。顾祖禹批判那些不懂地理与形势的大臣们"何怪今之学者,语以封疆形势,惘惘莫知。"•在他看来,历史若是落不到精确的空间位置上,治国大政就是空谈。而"方"指的就是土地,确切地说是古人感悟出来的土地之法则,所谓"地道静而有恒,故曰方。博而职载,故曰舆。"•这地上值得记载的万物诸人诸事,在顾祖禹那里也很具体,指的就是郡县、河渠、食货、屯田、马政、盐铁、职贡、分野。

• 5

• 6

> "是书以古今之方舆,衷之于史,质之以方舆。
> 史其方舆之向导乎,方舆其史之图籍乎,
> 苟无当于史,史之所载不尽合于方舆者,
> 不敢滥登也。故曰《读史方舆纪要》。"•

• 7

当然,我们不能把顾祖禹所倡导的史/地/万物一体化的整合直接对应着古希腊语境里的"场所性"或者"地形"。不过,顾祖禹的《总序》与里科沃特对"topos"微妙解析还是照见了我们对大地之"形与貌"的理解中某种贫瘠与局限。在经历了过去两三百年的科学化和专业化之后,

'mass' and 'appearance' of the earth. After two or three hundred years of scientific progress and professionalisation, would a Western and modern land surveyor or geographer today still relate a place with the human body and its mechanism? Likewise, how many Chinese scholars who have adopted the epistemology of modern Geography would insist on what Gu had—combining the study of the lands with that of people's lives and political histories? When Zhang Ziping, an early overseas-educated Chinese geologist, published his *Physical Geography* in 1923, he conveniently put all the concerns of traditional Chinese gazetteers under the panoply of 'Special Geography'. For Zhang, the field of 'Special Geography' included matters such as 'topography, climate, people, politics, industry and transportation' of a specific place—all of which he emphasised as the content of a *dizhi*.*° Through such an introduction of a modernised geographical taxonomy, a shift in the understanding of traditional Chinese knowledge of geography has started. And through Zhang's form of borrowing and reinvention, words and terms associated with 'topography' dramatically increased. Today, the traditionally-used term *dizhi* would be relegated to Special Geography, while the English term 'topography' could be translated as either the study of landforms or land appearance, or also as *dizhi*— a term that had previously been monopolised by the traditional Chinese gazetteers for thousands of years.

• 8
○ a local 'gazetteer'

Consequently, two radically different meanings of 'topography', or two different kinds of geographical knowledge, are battling for the same word—*dizhi*—in Zhang Ziping's book. Apart from the linguistic ambivalence that arose from Zhang's mis-translation, I believe, the conflicts between traditional Chinese *dizhi* and the then-newly imported modern geographical *dizhi*, to a large degree, did not originate from a linguistic problem, but a shift in the nature of the discipline of Geography. When Zhang Ziping placed the concerns of Gu Zuyu—the integration of history and geography— under Special Geography, he had carried out a double revisionist exercise on traditional Chinese knowledge of place and people, by firstly relocating and forcing traditional Chinese knowledge of place and people to relinquish the name *dizhi*, and later reinstating *dizhi* into the study of 'topography' now governed by Land Survey, Geography and Geology. He and others had then paved a way towards 'General Geography', a new discourse blessed by the rising power of modern Science. Zhang had indeed made further clarification by saying that Geography generally included Human Geography and Physical Geography as its main components.* But let us not be misled by the term 'human' in 'Human Geography', which precisely assigns humans as scientific objects, or subjects without agency, in the study of the movements and distribution of mankind on earth.

• 9

西方现代地理学或是测绘学再也不会把地形地貌跟人体病变直接联系起来了。同样，汉语世界里的学家们如果接纳了现代地理学的认识框架的话，也不会再像顾祖禹那样总把"地貌"描述跟民生与政治史划在一起。当张资平这位早期海外留学的地质学家在1923年发表《自然地理学》时，他是把我们过去"方志"所要照顾的内容全部随手划到了西式的"特殊地理学"里。"特殊地理学"这个领域包涵了某个地方的"地形、气候、住民、政治、工业及交通等之纪事，名曰'地志'"。通过这么一种现代地理学的分类，以及对传统中国地理知识的搬迁，通过借用和再造，围绕着"地志"的词汇一下子多了起来。现在，传统习惯上汉语里的"地志"被弄成了"Special Geography"；而英语里的"Topography"却被翻译成了"地志"。以前，方志们所垄断的词汇，现在成了现代地理学的地形地貌学。

　　于是，在张资平的小册子里上演了两种意义的"Topography"或者说两种不同的地理知识都在抢夺同一个词汇"地志"。除了张资平这种硬译所带来的诸多语言理解混乱之外，我以为，中国人传统意义的"地志"和当时被从西方引进的地理学的"地志"很大程度上不是语言学本身的问题，而源自地理学这个学科自身的转型。当张资平把顾祖禹的"史、地"全都放到"特殊地理学"时，张资平也对传统的中国地理知识做了二次操作：先是把我们传统的地理学知识搬走，逼着它放弃"地志"的名头，然后再在"地志"的名头下换上由土地测绘、地理和地质学所构建的地形学。这样，张资平就为那些不那么特殊的地理学腾出了现代科学名衔下的一大片空白话语空间。张资平的确说过，更为一般的地理学将包括"人生地理学"与"地文学"。但我们一定不要上当。张资平"人生地理学"中的"人生"，指的是要被"科学研究"的作为客体或者说没有能动性的主体的"人类"。"人生地理学"描述的是这样人类在大地上的流动和分布。

• 8

• 9

To be fair, 'topography' becoming a scientific field of knowledge was also what Land Survey, Geography and Geology had accomplished in Europe and America. Cartographical historians such as David L. Livingstone, in books like *Geography and Enlightenment*, have already delivered a thorough description of the modern transformation of topographical understanding.• They have demonstrated the ways and the type of crucial events that affected the development of modern Geography and Cartography, and how the nuanced meanings of the Greek *topos* gradually disappeared. I have also as in elsewhere attempted to discuss the issue of how we have moved from concrete day-to-day experience with landscape into an abstract mode of mapping the world in contemporary China. In short, since Zhang Ziping's publication of his thin brochure in 1920, we have witnessed an unmistakable end to the Chinese way of understanding the earth that had been used for thousands of years, and end of the long-held integrated mode of the traditional compiling of the gazetteer. Although gazetteers are still being published today in China, what I meant by the 'end' is, in their move toward a more 'scientific' form of geographical studies, our traditional values of history and place—namely, the roles of gazetteers as being records of the past, mirrors for the present, and knowledge of history, land and people—have been irrevocably changed and marginalised. In the face of modern Geography, Chinese gazetteers have lost their specificity to universality.

• 10

I hope the above historical account on the birth of modern topography, however brief it may be, can account for the reason of the absence of difference between *dixing* and *dimao* in today's field of Land Survey and Geography. We will read, say, in a standard Land Survey textbook, in a chapter 'The Common Terms for *dimao*' such a definition: '*dixing* stands for the general term to all earth shapes with various elevations, whereas each kind of earth shape has its more specific name'.• The author of the textbook will then explain what kind of land features can be considered as a hill, a dune, a narrow valley entry or an intersectional area of two adjacent valleys. Obviously, *dimao* and *dixing* are exchangeable in their usage. Whereas in the textbook *Landform Study*, we will read a passage like the following: 'Geomorphology is a science specialised in studying the characteristic features of the earth surface, of their formative reasons, and of their processes of changes, internal structures and distributional regularities'.• A few paragraphs later, the same author will mention the aqueduct engineering in Europe before the Renaissance and the large-scale land surveying practices of the 19th century as common examples of the origins of landform studies. Obviously, in both Land Survey and Geography, when writers do mention *dixing*, they do not intend it as a mere aesthetic and contemplative object, but something susceptible to weathering and continuous geological tectonic formation. Within such lexical context, *dixing*° and *dimao*° almost seamlessly coincide. Consequently,

• 11

• 12

○ *dixing*: topography as terrain
○ *dimao*: topography as landform

刘东洋
LIU DONGYANG

公允地讲，地形学成为一门科学化学问，也正是测绘学、地理学、地质学在欧美所取得的成就。像利维斯东这样的地图史学家以及其他地理史学家都曾在诸如《地理学与启蒙》• 10 这样一类著作里详细描述过地形学走向现代化的过程。他们给我们展示了在现代地理学和制图学的发展过程中，到底因借何种方式，经过了怎样的重要事件，希腊人的"topos"身上所具有的那些微妙含义逐渐发生剥离的。在别的文章里，我自己也曾试图讨论过我们从生动的日常世界走向偏平化地图的过程。容我此处不再赘述。简言之，我们在1920年代张资平那本小册子里已然看到历史悠久和具有整体性的方志编撰传统的终结迹象。虽说当下中国各个城市里也会出版些方志，我这里说的终结，说的是在面向"更为科学"的地理学时，我们地志里那些资治、教化、存史等传统价值观以及史方舆一体化的立场已经不可逆转地被改变和边缘化了。中国人的方志，在现代地理学面前，输给了普世性。

上述有关现代地形学知识体系在中国是如何诞生的历史可能过于潦草了些。不过，我还是希望这一概述可以帮助我们理解为何在如今的测绘学与地理学中"地形"和"地貌"不再具有区别的原因。我们会在有关地貌测绘的教材里，在标题为"地貌个部分的主要名称"章节里读到，"地形是地球表面高低起伏不规则的复杂形态的总称，其各不相同的形状，各具有一定的名称。"• 跟着，作者就开始介绍什么叫做 • 11 小山，什么叫做山丘、垭口、谷会。显然，这里的"地形"和"地貌"是可以互换的；而在地貌学教材里，我们会读到，"地貌学是研究地表形态特征及其成因、演化、内部结构和分布规律的科学。"• 然后，作者在"地貌学的发展简史"中， • 12 把地貌学在欧洲的起源跟文艺复兴之前的水工学，以及19世纪大规模的地形测量联系起来。显然，在测绘和地理这两个学科里，专家们提到"地形"时并不会仅仅把土地当成审美的风景，而是将之视为大地戏剧性的地质构造变化和气候影响下的结果。在这一语境下，"地形"和"地貌"坍塌成为一体。而地形与地貌的重叠也折射着"地貌学"身上现代性那洞穿了"场所性"深度的力量。

有趣的是，在测绘学和地理学之间，还是存在学科之间的那种劳动分工的。• 一位测绘工作者可能是极尽全力要把大 • 13 地的地形和地貌特征压缩成为一张等高线图的，这个过程我们比较了解。其中，那些被视为偶然性的事物基本上就被抹掉了，为的是要用等高线围合出来那种永久性标记。不过，

the merging of the two concepts reflects on the formation of Geomorphology as a powerful gaze of modernity that penetrates into the depth of *topos*.

Interestingly, there is still a disciplinary difference in terms of labour division between Land Survey and Geography in their visualisation of landform.* A land surveyor probably tries hard to plot the relief features and landscapes on a surface, which we are quite familiar with, and eradicate things considered to be accidental in order to enclose permanent landmarks within contour lines. But, according to the professional rhetoric of Land Survey, a land surveyor often claims his map 'to cover every kind of natural form, such as water systems, landform, soil and vegetation distribution, and all man-made forms such as settlements, roads, and buildings.'* But in reality, if there isn't a pre-requested demand, a land survey map is not likely to contain any temporary, make-shift, unimportant, or illegal building. In addition, the very nature of a survey map itself determines the portrayal of the topography as 'projective'. In other words, a land survey map would portray my home's Back Hill of its ridges and pitches in accurate measurements—with the coordinate axis, geodic references, etc.—but only label the elms and pines covering the hill as 'Woodlands'. What then is the characteristic of *dixing* according to a physical geographer? According to the prescription of Geomorphology, when *dixing* or *dimao* is being referred to any kind of landform 'from the inside to the outside, from the top to the bottom, from the nearest to the furthest of the earth', we would then understand 'topography', through the eyes of a physical geographer, as truly 'sectional'. In that sense, the studied landform belongs to a continuous geotectonic transformation process.*

• 13

• 14

* FIGURE 2.1, 2.2 & 3

The dynamic and processional view of geomorphologic topography has thus endowed 'a landform section' with a vivid sense of temporality by differentiating the stratums and layers of soils and rocks that have been changing endlessly. To get a sectional view of the earth becomes the preferable way to travel back in the geological time flow. Ten years ago, a friend of mine who studied Mining Science in Canada had once visited me in Dalian. Looking through the same rear window of my apartment, he sighed lightly, 'So barren isn't it?' 'What's so barren', I asked. My friend pointed to a facet of the Back Hill, carved out during the construction of the driving lane and said, 'You see, right at this section, the top soil doesn't seem to exist. Those sedimentary rocks are weathered severely, cracking everywhere. Even if there was plenty of rainfall, the gravel soil would not preserve water. Dalian is at a dry and windy region, and its surface water evaporates fast. How then will the top soil maintain enough water to grow vegetation?'

在测绘专业的说辞中，大地测绘是要"包括地表的各种自然形态，如水系、地貌、土壤和植被的分布，也包括人类社会活动所产生的各种人工形态，如居民地、交通线和各种建筑物的位置"。实际上，若是事先没有什么特殊的测绘要求，测绘地形图上的"建筑物"一项不会包括临时建筑、脆弱建筑、不重要建筑、不合法建筑的。此外，测绘的制图成果往往决定着测绘学对于地形的关注角度是"投影性"的。这样，面对着我家后山那道起伏的山岭，测绘工作者会记录下它的精确座标、标高、横纵，主要的沟谷和山脊，至于山上毛茸茸的松树榆树们一般被标成"林地"就算了；那么，一位自然地理学家口中的"地形"又具有怎样的特点呢？当地理学家口中的"地形"或是"地貌"指的是"大地从内到外，从上到下，从近到远"的所有形态时，我们也就明白了地理学里的"地形或是地貌"本质上就是"剖面性的"。在这个意义上，地理学所要研究的地形属于那永不停息的地质构造运动的过程。*

• 14

* 图 2.1, 2.2 和 3

因此，地貌学家对待地形的那种动态和过程化视角，真的在任何一处土地形态的剖面上都充盈进生动的时间感，那些土壤和岩石的层累，往往就是空间化了的时间。在一座小山身上切那么一刀，看到一处剖面，也就成了沿着地质时间逆流而上的最佳方式。我有一位找矿的朋友十年前曾来我家做客。也是站在这后窗前，他张口就感叹道："好贫瘠呀！"我问："什么贫瘠？"好友就指向了我家后山被凿开的一处岩层断面说："你看，这剖面上的表土几乎没有。下面的沉积岩风化严重。解理裂缝巨多。雨季即使有水流下来，很快就渗走了。没了水，又没有表土，加之这里风大，蒸发迅速，你想，这山上的树木能长高吗？"

My friend's observation reflected his own professional interests. Having been working mostly outdoors or in the wilderness, he had learned to trace the cues and clues from the land's surface to the mineral beds or coal underneath that could be exploited for practical purposes. In comparison, a physical geographer or geomorphologist may have much wider interests than his. They care to take into consideration everything from oceanography to meteorology, and more importantly, 'the structure, process and stages' of landforms.• • 15 From the eroded rocks, shrubs, and saline gravel soil on the Back Hill, they probably could go back thousands or millions of years to track down its geotectonic forces and formative process.

At this point, we may then understand my Back Hill according to the following geomorphologic description of its natural history:

> Its unified regional crystalline base was formed in the Anshan Movement and Luliang Movement. Then this area slowly entered into a relatively stable developmental stage of platform formation. Due to the earlier erosions, the land in this bay lacked sediments from the Mesoproterozoic Era, and received 5000 to 6000 metres interactive marine and terrestrial clasolite deposit and endogenetic carbonate sediments along the coastline out of the folding force in the Upper Proterozoic Era. In the Paleozoic Era, the land here was in an up-rising movement, but began to be impacted after the Mesozoic Era by the diving force of the Asian continent. Due to a strong S-SE lateral extrusion, the top stratum displayed certain ductile shearing overthrust, tectonic deforming and metamorphizing extruding effects along the crystalline base top... Then, land started to sink at the late stage of the Yanshan Movement, in accompanying eruptions of volcanoes and infill of magma. Thereafter, it had formed the Anshan rock bed of cretaceous Mesozoic group of volcanic breccias and igneous rocks. Since the Xishan Movement, there have been ups and downs for various parts of this bay, which lasted till the present moment.• • 16

But wait. Why does a report like this treat 'the present moment' in such a brief and cursory manner, as if the lives of elms and pines covering the Back Hill, planted mostly by local farmers, do not fit into the timetable of geology? Or what about when Geography converges with History to form Historico-Geomorphology — a rich and promising sub-discipline that has claimed to relay and relate the temporality of landforms with social history? Could such an interdisciplinary approach provide any room for local knowledge with regards to my Back Hill?

好友的话职业性地体现着地貌学的地形研究特点。当然，朋友是矿学出身。他们找矿的人常年外出，练就这种本事，能够从现场的环围和肌理出发，窥测到大地表层之下的秘密。他们找的是矿，很是实用。相比之下，地理学家或是地貌学家的视野要更为开阔，他们会从海洋研究到大气，当然，重要的是，描述各类地貌的"结构、过程和阶段"。从我家后山那些剥蚀的岩石、杂丛、盐甸土看将下去，地貌学专家会一路回溯百万、千万年去，去查看折叠山体的力和山体形成的过程。

• 15

在面对后山所在的这个湾时，我们常会读到地貌学家的如是描述：

> "鞍山运动、吕梁运动形成区域统一性的结晶基底之后，开始相对稳定的地台发展阶段，早期出于剥蚀状态而使中元古界缺失，自上元古界全区拗陷接受了 5000 到 6000 米厚的上元界滨海 — 浅海陆相碎屑岩和内源碳酸盐类沉积，古生代处于上升阶段，至中生代以后受到亚洲大陆的俯冲影响，盖层受到强烈的 S—SE 向侧挤压而沿结晶基地顶而发生韧性剪切推覆、构造变形及变质……燕山运动晚期本区发生烈陷，相伴有火山喷发和岩浆侵入，形成中生界白垩系安山岩、火山角砾岩、凝灰岩等堆积。喜山运动以来，发生过地域性的断块升降运动，从而形成现今的地质轮廓。"

• 16

不过，等等。为何这类地貌学报告以如此简要如此概括的方式描述这个"当下时刻"呢，就好像后山上那些农民过去种下的松树榆树的生命基本就不能计入地质学里的时间刻度中似的？或者，当地理学和历史学的确走到一起形成了一门旨在把地貌时间和社会历史融合起来、丰富和充满希望的"历史地貌学"时，这一交叉学科的内容里真会包涵有关我家后山的那些地方性知识吗？

此处，我想所做的，正是在有关地貌那很是严格的学术版图上，找到那个能够容下"地方性知识"的点。

Indeed, this is exactly what I'm attempting to do: how can we find room for the consideration of 'local knowledge' within the academically-subdivided territory of landform studies?

I have had occasional encounters that forced me to reflect upon this obvious but least talked-about professional bias on what 'topography' as a body of knowledge ought to include. There was once an occasion when I took a cab home from a shopping mall. After I told the cab driver my address, he turned his head back, looked up and down at me and asked: 'You haven't caught rheumatism, have you?' I was instantly irritated by his rudeness. But before long I was told that his village used to be where I live now. He said: 'That place is always foggy, and it gets terribly humid in the summer. We then had to live in single-storey houses, and because it was very near the sea, we were all afraid of catching rheumatism.' Having heard him, I came to realise how the experienced residents were still those who had the most intimate knowledge of the Back Hill—its climate and quirks. He then started to talk about his past experiences in the village—how the fishermen were the poorest in that area, who could only build houses out of rocks from the mountain and roofs out of sea reeds, how they could only swap two pounds of fish for two ears of corn under the state-planned-economy, and how they moved into apartment housing as compensation for the expropriation of their land for the new town development in the mid-1980s. He had been a cab driver ever since.

This warm-hearted and hospitable driver kept on briefing me on the 'topographical' experience at the foot of the Back Hill. Yet as he was sharing, I somehow felt a sense of loss. Gone is his village, his youth and all the experiences he had as a villager. So what if there is a thick fog and wet weather when air-conditioning can keep the interior at a desirable temperature these days? With such changes, it seems as though all that the driver had to deal with in terms of climate and the mountain have become things of the past. The highrise blocks built along the coastline today would hardly notice, or be affected by, these conditions brought about by the land. The microclimates and unique topographical features would no longer be factors that a design must accommodate, as they are no longer potential threats to people's health.

There is something else that has lost its charm. I remember seeing an aerial photo of this region taken in the late autumn of 1984, when the land was just about to be developed. Looking from high above, the Back Hill appeared in dark brown, and the harvested corn field in the foreground exposed a stark red surface as the gravels on the ground are rich in iron. The land here began to be 'arable' probably only after the mid-Qing Dynasty at the peak of migration. Fewer villages had a history earlier than that period. At the left side of the picture stands the Paotai Hill, named after a Qing artillery base on top of the mountain. Towards the right side, or near the shore, there was a village called Miaoxia, which obviously meant 'a village locat-

我自己会偶有奇遇，才被迫去反思在地形学里明显存在却少有人问津的专业偏见。记得有一次，我从某购物中心打车回家，司机问我地址，我就把后山的地址告诉了他。那司机故意后撤一下身子，斜着头，上上下下看了看我，问我："你没得风湿？"我即刻被他的话激怒："您这是怎么说？"司机："你住的那地方就是我们以前村子的位置。那个地方，老起雾。一到夏天就潮得要命。我们那时都住平房，靠海太近，很容易就患上风湿。"原来如此。原来在生活世界里，有经验的老户还是知道每一个山坳的小气候和小脾气的。跟着，一路上，我跟这位司机聊起了他过去的日子。司机这才跟我说，此地过去最穷的就是渔民。盖房子只能到山上拣石头，铺屋顶用的是海苇子。计划经济时代，渔民拿两斤鱼只换两穗"棒子"。°80年代中期，他们村的地被征了，整体搬迁新区，他就开始了跑出租。　　° 玉米

　　这位好心且好客的司机一路上不停地给我传授他生活在山脚下的"地形学"经验。当他讲话的时候，我多少可以感到某种失落。他的村子已经没了，他的童年以及所有他作为农民时获得的生活经验都在沸腾的开发浪潮中烟消云散了。海边岭下雾气重又如何？多开空调，室内的温度湿度不就总能让人满意？这样一来，就连司机这样的农民也都觉得，他们过去对付气候和山地的小经验，都成了过时的玩艺。如今那些沿着海岸绵延的高楼因此根本就不再顾及地貌特征。独特小地貌和小气候都不再成为设计必定承接的先决条件，当然，也不再被视为对人们健康的潜在威胁。

ed down below a temple'. And that Temple, guarded by a fresh water spring—a blessing for the fishermen there—once hosted a seat for the Sea Goddess. Not too far from the Back Hill, there was the legendary Tiger's Cave up close to the summit. No one had ever seen a tiger in this area but it seems that the earlier settlers had legends and stories about each land feature along the coast, which we latecomers have completely missed out on. *

* FIGURE 4, 5.1 & 5.2

To me, amidst the last two decades of development, not only did we neglect the geomorphologic understanding of landforms, we have severely gotten rid of the physical landmarks and memories of these lands. The local people's knowledge and familiarity of the characters of the earth could only be gained through their intense labouring on the land year after year. I personally have seen too many new town development plans that have not treated the farmers' knowledge of their villages and landscapes as foundations to the planning. I have been witness to how my home's Back Hill has been zoned and re-zoned from 'Deserted Land', to 'Woodland', then to a 'Tourist Park'. There has never been any relation to the past with each change in land use. But I have come to realise how, with each change, a profound shift has occurred to China's planning system with regards to its view of topography.

Returning again to 1984, before this piece of land was converted into an Economic Development Zone, its total planned area was about 42 km2, of which 20 km2 was of land and water; roughly 50% of the total was allocated to the new port; another 30% for an industrial park; about 10% for residential and commercial use; the rest for roads and green spaces. Back then, when real estate was a booster for the local economy, the so-called 'Plan-Ahead-of-Its-Time' planning approach often advocated more generous land use for wider roads, distance between buildings and greenery. Shenzhen with its *Qi Tong Yi Ping* measures to—'make the seven kinds of infrastructure services and levelled land available for urban development'—was the model for every new town in China. The primary attitude of politicians and planners toward terrain and landscape was to bulldoze dunes, hills, creeks to make space for industrial development, while leaving the steep mountains as parks. Due to its over 30% gradient and the need for the new town to maintain recreational space along the coastline, my home's Back Hill was spared and designated as a 'Recreational Scenic Area'.

与地貌失魅相伴的，还有另外一些东西的失魅。我记得我曾见过一张航拍照片，那是1984年深秋在这片土地即将规划成为开发区时拍摄的一张照片。从空中望下来，后山身上披着一条条有些干褐的绿色，照片的近景上，土地一片棕红。这棕红的表层土满含砾石和铁，它们可能仅仅是从清代才被开垦成为"大田"，此地几个屯子的村史只能上溯到清代。照片近景上的山头因清代设置的炮台而得名。在照片的右侧，已经出了画面，那里曾经建有一座海神娘娘庙。庙下的屯子从而得名"庙下"。它就守着一个神圣的龙眼甘泉。在濒海处，有一口淡水井，这的确够神奇。而我家后山不远处的高岭上，有过一处"老虎洞"。谁也没有真的见过老虎。但是早前的定居者们似乎给这片土地的任何一块带有特征的地貌，都赋予了一定的传说和故事。对这些传说，我们这些后来者已经完全错过。*   * 图 4, 5.1 和 5.2

对我来说，在过去二十年里，在土地开发中，我们不止忽视了有关地形的地貌学解读，还同样要命地摧毁着这片土地上来自过去乡村社会的地标和记忆。而这些当地人对于土地性格的熟稔只能通过世代劳作在土地上才能获得。我自己已经记不得到底看了多少新区的发展规划，几乎没有哪个规划会主动把农民有关乡村和土地的经验当成新城规划的基础。我自己亲眼见证了后山是怎样从荒地一次次被不断规划，改成了"林地"、"旅游区"的。每一次对于后山土地使用性质的调整都跟过去无关，却是让我从中看到中国规划体系里在面对地形时一次次深刻的变化。

时间再次退回到1984年。在这片土地变成城市前的第一张规划图上，总共42平方公里的规划区域里有20平方公里的水域和土地划给了港口，另外的30%土地划给工业，居住和商业用地不过10%，剩下的就是交通和绿地。那时的规划提倡"超前规划"。"超前"的意思就是对道路红线宽度、建筑间距和绿化用地给予慷慨的保证。作为样板，深圳和深圳"七通一平"做法成了所有开发新区效仿的对象。于是，那个时期中国新区的城市规划对待地形地貌的基本原则是，小坡、浅丘、小河，一律铲平，大山、峭壁、高岭留作公园。我家身后的这座后山，一半是因为它多数地带的坡度已经大于30%，另一半是新区要保留一处滨海休闲岸线，这才被划成了"旅游风景区"。

The period between mid-1980s and mid-1990s marked a drastic change in the city planning system in China, particularly with regards to its value system and approach to land use and development. In the 1991 edition of *City Planning Principles*,° a key planning textbook compiled by Tongji University, we will read under the section related to 'General Guidelines for Urban Land Use' some subtle but significant changes in comparison to its 1981 edition. By continually insisting on the primary criterion that land use planning must incorporate the topographical conditions of a site as its starting point, the editors responsible for this edition of *City Planning Principles* also launched a campaign for the 'social, economical and legal consideration' of land use issues in urban development.• Although this edition was by no means an exact word-for-word match to the Western land use terminology,• we still get a sense of shift in focusing on the economical and legal issues in China's state of planning. On the practical realm, the planners' treatment of topography also leaned toward a much more utilitarian direction than what was on *City Planning Principles*. In *Dalian City Planning Administrative Guidelines: References for Application*,° circulated among the bureaucrats in Dalian in the early 1990s, the requirements towards topography ended up as a set of straightforward and clean-cut procedures. They were a list of 'ought-to-dos' in a planning process regarding landform conditions:•

○ *Chengshi guihua yuanli*

• 17
• 18

○ *chengshi guihua guanli tiaoli*

• 19

4.2 An Existing Land Condition Assessment Map,° on which, the following ought to be clearly indicated;

○ *or Analysis Map*

4.2.1 The Suitable Areas for Urban Development, the Unsuitable Areas for Urban Development, and those Conditional Suitable Areas for Urban Development, i.e. Site improvements may be needed;

4.2.2 Various Land Slope Zones. Also analysis maps of drainage and swamp condition are required;

4.2.3 A Civil Engineering Assessment Map on Site's Load Bearing Capacity. Its aim is to map out the distribution of rocks, soils, and their load bearing conditions;

4.2.4 A Hydraulic Condition Analysis Map to locate the depth and flow of underground water;

4.2.5 A Map to indicate areas vulnerable to a fifty-year or a hundred-year flood and locations for dams and dikes;

4.2.6 the Construction Prohibited Area for containing underground mines or tombs

To be honest, I don't see these operative planning guidelines for land development fundamentally different in their attitudes towards topography in comparison to the planning principles taught at Tongji University in the 1990s. The latter would probably demand

从 1980 年代到 1990 年代，中国的城市规划在对待土地开发利用的问题上发生了价值观上的明显转变。在同济大学编撰的 1991 年版《城市规划原理》中，我们有关"城市用地概述"一节里就能读到某种来自规划教育工作者们微妙但深刻的变化。同济教程在一如既往地坚持城市土地利用规划必须紧密考虑土地的自然属性同时，开始提出要兼顾土地的"社会属性"、"经济属性"和"法律属性"。虽说此版《城市规划原理》中的土地经济、社会、法律属性并不等同于西方土地制度下的土地相应属性，我们还是能够从中体悟到这一时期中国规划界对土地的经济和法律问题的重点关注。而操作层面的城市规划地形学考虑要比《城市规划原理》来得直接和实用。在差不多同一时期内部发行的本市《城市规划管理条例：实施参考材料》中，有关地形问题都被明晰地罗列成为图纸要求和法规要求。譬如，《实施参考材料》会这么要求规划师如何对待地形分析：

• 17

• 18

• 19

> 4.2 城市用地评价图（或分析图），图上应标明：
> 4.2.1 适宜修建地区，不适宜修建地区和需要采取工程措施后才能修建地区的用地分析。
> 4.2.2 不同坡度地区的范围，冲沟、沼泽滑坡等地区地貌位置的分析。
> 4.2.3 岩石分布、土壤种类，分布范围及地耐力等工程地质分析。
> 4.2.4 地下水位深度及流向等水文地质分析。
> 4.2.5 50 年或是 100 年一遇的洪水淹没范围和水工构筑物的位置。
> 4.2.6 地下矿藏、地下文物等限制修建地段。

坦白地讲，我不认为这类实际规划操作中对待地形的态度真的跟《城市规划原理》中对待地形在态度上有本质的差别。只不过《原理》还会要求规划师认真调查城市用地的地质构造、地震带走向、断层、水系、林带，《实施参考材料》把这种调查改成了一种跟测绘或是地质专业之间的接力跑。规划师谨记"100 年一遇的洪水淹没范围"、"不适宜修建地区"、"耐力分析结果"就够了。概言之，在操作层面的城市规划地形学就是一种旨在划定"不许"、"切忌"、"不能"

planners to do more geotectonic research on issues such as earthquake zones, faults, regional water systems and the ecology of forestry, whereas the *Dalian City Planning Administrative Guidelines* treated land use planning as a passive receiver of the instructions provided by engineers and geologists—to mark down 'a hundred-year flooding zone' or 'unsuitable area for development due to steepness or low capability of load bearing' will just do. In other words, planning in practice treated topography as an array of what-to-dos and what-not-to-dos. Beyond the red line of the undevelopable zone, planners could take the site almost as a *tabula rasa* for anything to happen.

In that context, the sole reason that my home's Back Hill survived the waves of development owed much to this 'negative methodology' of planners towards hills and cliffs. To put it more precisely, the reason for the Back Hill to maintain its status as a scenic spot was partly due to a planning decision that left it as 'leftover green land'—after all the levelled lands have been zoned for industrial use and the Dalian New Port. However, people were quick to discover the precious value of such green land. In the 1990s, developers, planners, mayors, or real estate agents had all learned the dictum—'a panorama view is worth a million dollars'—everyone has mastered the economics of real estate development in just ten years. Under such circumstances, developers built high-rises almost halfway towards the ridge of the Back Hill. What had really stopped my apartment building to go any further towards the hill was in fact a tunnel owned by the Port Authority. A few years after my neighbourhood was completed, the summit of the Back Hill, right above the Tiger's Cave, was transformed into a humungous platform for tourist sightseeing, and the fresh water spring guarded by the Sea Goddess Temple turned into a high-end spa. At a frenzied pace, the previous land features and remains of the old villages in this area have all been levelled and erased. The legends related to fishermen's lives and landscapes, too, disappeared.

It is these changes occurring at such a momentum that often leave me perplexed. Besides the 'not-to-dos', could there be a more 'proactive planning' model in dealing with topography that takes into account and preserves the interwoven texture of landscape and humanity as its basis? Also, could the cab driver's personal life story or villagers' past experiences tied intimately to the place have any implications for planners and designers who are constructing people's home and lives? Is the topography we refer to only a landform belonging to vast geological time, which leaves no room for villages and individuals?• Or is topography only a contemplative object viewed through the lens of aesthetics? Are those haunting legends from the past and rural world of no use at all in the development of this new town?

• 20

建筑区域的消极或被动做法。言外之意是，在"禁建区"之外，在"宜于建设的土地"之内，规划师尽可以画上最新最美的图画。

我家后山之所以能作为新区的风景林地活了下来，如前所述，部分地归功于当时规划师们面向土地地貌时的"消极规划"。准确地说，它是一块剩地，是在平地都被划给了工业和海港之后，剩下来的绿地。不过，人们很快就认识到了这类"剩地"的宝贵价值。仅仅过了10年，无论是开发商、规划师、市长，还是售楼处里的销售代理都能讲出"无敌海景、寸土寸金"的名言来。全国人民在1990年代用了10年时间自学了房地产经济学。在这样的形势下，开发商终于把房子建到了后山的山坡上。而楼房停下来的位置，就是港口铁路专线地下隧道的位置。是铁路，阻止了楼房的爬行。又过了10年，"老虎洞"上方建成了一个巨大的观景平台，下庙村守着的那口神泉变成了洗浴中心。这片土地的乡村地貌被抹平或完整地修整了一遍，那些跟村民和渔民生活休戚相关的地貌传说，也都消失殆尽。

这样迅猛的变化时常令我困惑：中国的城市规划除了"消极地处理地形地貌"的"禁建模式"之外，有没有思考过一种"积极面向地貌"，因地制宜，保护土地肌理的城市规划模式呢？还有，那位原住民司机所讲述的乡村经验以及农人对于微观地貌的切身体验，真的不会教给城市规划师什么东西吗？难道城市规划的地形学充其量只是地貌学意义上的地质构造地形学吗？·或者，充其量只是视觉美学的景观设计吗？那些旧日和乡村世界的迷人传说真的就在新世界里没有了一丝一毫的价值吗？ • 20

我不知道我们的地形学该走向何方。但是我很怀疑，我们当下规划和设计的那些手段和方式会有什么好的结果。所以，如果您现在问我："何谓地形？何谓地貌？"我会犹豫一下，然后再指着后山："在它的身上，我看到我们的地形学正在失去和遗忘我们祖先留给我们的教训。"* ＊图6和7

此刻，当我从后窗望出去的刹那，一眼望到了后山上两株老榆树，以及树上的三个半喜鹊窝。在我逐渐学着从平面到剖面、从地质到社会、从宏观到微观对于后山的面目进行观察中，我也逐渐意识到还存在着其他类型的"地形知识"——我说的，不是人类的地形学，而是植物和动物们身体力行的"地形学"。喜鹊筑巢的大树基本是沿着挡水墙生长的。在雨季，挡水墙是人工力量改变地貌之后的山体汇

I am not quite sure which direction our topographical studies will go. But I deeply suspect there could be any good outcome from the current ways and schemes that characterise our planning and design practices. So now, if you were to ask me 'What is topography?' I will hesitate for a moment and point to the Back Hill and say, 'From learning about the Back Hill, I realised that our current study of topography has missed and forgotten the lessons that our ancestors have left behind.'*

* FIGURE 6 & 7

At the moment, just as I look out from the rear window, I see two huge elms on the Back Hill, and on top of them, in between the branches, three and a half magpie nests. As I myself gradually learn to view this hill from plan to section, from Geomorphology to Sociology, from macro to micro conditions, I also come to realise that there are perhaps other types of topographical knowledge, topographical experiences lived not by humans but plants and animals. The elms, on which the magpie nests are built on, grow along the retaining wall built to block the gravel and rocks from falling onto the drive lane, while also redirecting run-off water from the hill. In a place as dry as Dalian, where winds from the sea always blow along the foot of the hill, this engineered convergence of rainwater brings an advantage to the vegetation along the water course. Trees aligned with the wall tend to grow bigger and taller, and the magpies instinctively build their safer nests on these taller trees. Magpies are smart birds—they did not move when the fishermen relocated, they stayed and found the new housing compound able to provide them with more food in the garbage cans. But they also lacked preparation and anticipation. With more people moving in, they brought with them cats, some of which were later deserted by their hosts. The cruising cats threatened the lives of the magpies, small ones in particular. That may explain the abandonment of the half-built nest on top of one elm tree, under threat from the cats the magpie family has moved either to taller trees, rooftops of buildings, or somewhere higher. To those birds, when they fly over the woods, through pines and oaks, certain heights and spots in reach of the cats must mean danger and risk. On the contrary, from the point of view of the crouching cats, their territorial 'construction' somehow overlaps with that of the birds or other preys. A land that embraces all lives thus affords both dangers and opportunities for animals in its niches and heights. Only we humans have again and again forgotten the vitality that resides in each change of *topos* or *dixing* in Chinese.

Note: This article was originally published in June 2011 issue of *The Architect*

刘东洋
LIU DONGYANG

水线。沿着这条线，在海风经常抚摸的山坡上，微小淡水持有量的差别，在树的身躯上产生了差别。喜鹊自然会选那些比较高大的树上筑巢才安全。喜鹊很聪明。它们没有跟着渔民一道搬迁。在新居住区的垃圾箱里，它们找到了之前所没有的丰富食源。喜鹊也没有太多预见。随着居民越来越多搬到此地，几年前，小区里开始出现到处游走的流浪猫。对于小喜鹊来说，靠近猫会丧命的。在求生且追求享乐的本能驱使下，喜鹊一家把巢从前面的树搬到了后面的树上。今年，似乎筑到一半的时候索性搬到了隔壁楼房屋顶上的什么地方去了。对于喜鹊一家来说，从天空俯瞰下来，某些标高，某些海拔，某些地形特征充满了凶险；反过来对于猫们来说，它们的领地"建构"恰好与喜鹊一家的领地"建构"既逆转又交叠。因此，拥抱了所有生命的土地提供着各种生存的机会或是危机。倒是人类常常只记得"地形"的轮廓，遗忘了潜藏在"地形"那起伏变化的细微中的生命力量。

注：本文已在《建筑师》杂志
2011年6月刊上以同名文章发表

FIGURES

1

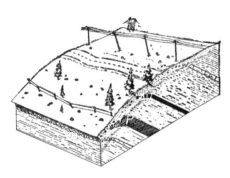

2.2

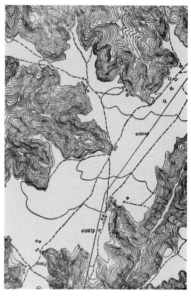

2.1

3

1   相机很难辨识熹微中山体浮现时的微妙光感。
    拍这张照片时，已天光大亮。
    摄：刘东洋
    An early morning view of the Back Hill.
    from my home's rear window.
    Photo: Liu Dongyang
2.1 测绘学里的地形图：关注的是对大地表面的投影。
    周顺麟《地貌测绘》，北京：中国工业出版社，1964 年，35 页。
    A typical topographical map in Land Survey whose main concern
    is the projection of the earth surface.
    Zhou Shunlin, *Introduction on Landform Surveying*, Beijing:
    China Industrial Press, 1964: 35.
2.2 地貌学里的地形图：关注的是大地从内到外、从上到下的构造与
    变化过程。
    严钦尚、曾昭璇《地貌学》，北京：高等教育出版社，2010 年，41 页。
    A typical topographic sectional view from Geomorphology whose
    main concern is the geotectonic transformation from inside the
    earth to the atmosphere.
    Yan Qinshang and Zeng Zhaoxuan, *Geomorphology*, Beijing:
    Advanced Educational Press, 2010: 41.
3   后山身上一处被凿开的断面。
    摄：刘东洋
    A facet of the Back Hill.
    Photo: Liu Dongyang

刘东洋
LIU DONGYANG

插图

4

5.1

5.2

4 一张 1984 年的航拍照片。照片捕捉的是这片土地走向城市化之前的农耕状态。
王国栋《马桥子·1984》，大连出版社，2009 年，16 —17 页。
An aerial-photo of Dalian Economic Development Zone in 1984 before its urbanisation.
Wang Guodong ed., *Maqiaozi 1984*, Dalian Press, 2009: 16—17.

5.1 从前身后地势起伏的"庙下屯"。
王国栋《马桥子·1984》，大连出版社，2009 年，80 页。
Miaoxia Village, spreading between the sea and the hill.
Wang Guodong ed., *Maqiaozi 1984*, Dalian Press, 2009: 80.

5.2 如今这里变成了平地上的一处高档温泉洗浴中心和一群高层。
摄：刘东洋
Now, this hilly landscape has been completely bulldozed to make space for a high-end spa and high-rises.
Photo: Liu Dongyang

FIGURES　　　　　　　　　　　插图

6

7

6　后山能被留作"森林公园用地",多源自它较陡的坡度。
　　摄:刘东洋
　　My home's Back Hill was zoned as a Forest Park mainly due to its steepness.
　　Photo: Liu Dongyang
7　后山树上的喜鹊窝
　　摄:刘东洋
　　The magpie nests on top of the tall elms at the Back Hill
　　Photo: Liu Dongyang

NOTES

1 The questions of 'What is *dixing*? & what is *dimiao*?' were posted by me on a website on 23 February 2011. Within 24 hours, this question had elicited 65 replies. Despite the variety of response, the 65 responders all have an advanced educational background. Their interpretations to *dixing* and *dimao* are largely related to their previous training and experiences.
See: ⌘ www.douban.com/note/141553786/
2 Ge Chuangui et al. ed., *A New English—Chinese Dictionary*, Shanghai Translation Press, 2002: 1475.
3 Joseph Rykwert, 'Topo-philia and -phobia', *Topophilia and Topophobia: Reflections on Twentieth Century Human Habitat*, Xing Ruan and Paul Hogben ed., London: Routledge, 2007: 12.
4 Ibid., 12.
5 Gu Zuyu, 'General Preface One', *Brief Notes upon Reading the Local History and Gazetteers*, Qin Yong ed., Beijing: Tianxia Press, 2000.
6 Ibid. 'Guides to the Book'.
7 Ibid.
8 Zhang Ziping, *Physical Geography*, Beijing: Business Press, 1923: 1.
9 Ibid., 2.
10 David N. Livingstone and Charles W. J. Withers ed., *Geography and Enlightenment*, University of Chicago Press, 1999: 236—275.
11 Zhou Shunlin, *Introduction on Landform Surveying*, Beijing: China Industrial Press, 1964: 3.
12 Yang Jingchun and Li Youli ed., *The Principles of Geomorphology*, Beijing University Press, 2009: 1.
13 Regarding the definition of Geomorphology, particularly of its relation to Geology and Geography, three key authors in the field of Geomorphology had once delivered a quite thorough discourse:
'There does not exist, as far as we can discover, a history of landform study, by which we mean the history of evolution of ideas relating to the development of the physical landscape. Before the early nineteenth century this study formed the bulk of geology, but thereafter, largely because of the rapid growth of stratugraphy and palaeontology, it occupied a decreasing share of that science, although its fundamental necessity in general geological knowledge was never in doubt. After about 1860 the study of landforms became part of both geology and physical geography and was later known as physiography or geomorphology. Today it forms the link between geology and geography and probably its association with the latter discipline is one of the main reasons why it has failed to develop along more strictly scientific lines. Even in 1963 it is still a 'pseudo-science' or a 'quasi-scientific art' and its historian is faced with the portrayal of its early rise both as an artistic perception and as a gradual transition towards its goal of truly scientific explanation which its has so far not attained.'
— Richard J. Chorley, Antony J. Dunn and Robert P. Beckinsale, *The History of the Study of Landforms or the Development of Geomorphology: Volume 1, Geomorphology Before Davis*, London: John Wiley & Sons Inc., 1964: xi.
Whereas in the Chinese context, despite the fact that *dixing* and *dimao* in Chinese are synonymies in the field of Physical Geography, Geomorphology or Geology, the adoption of the usage of *dimao* to replace *dixing* did happen in the mid-1950s when Geography Department in Beijing University began to re-organise its sub-disciplines. In a telephone interview, Emeritus Professor Wang Enyong of Beijing University had informed me, back then, the common Russian transliteration of *dixing* would be 'топография' (topography), which as a sub-discipline in the former Soviet Union belonged to School of Land Survey, for its main concern would be mapping out the earth surface. To match to the Russian 'геоморфология' (geomorphology) in their College of Geography, according to Professor Wang Enyong, his colleagues such as Han Mukang, coined the Chinese term Dimao Xue to differentiate topography in Geography from in Land Survey.

注解

1 我在2011年2月23日针对"何谓地形？何谓地貌？何谓地形学？"这三个问题，在豆瓣网上展开了意见征集。24小时内提供了有效回答者共65位。这些回答并不太具有普遍意义，但是代表了受过一定高等教育的非地貌学专业人士对于"地形、地貌"的看法。详见 ⌘ www.douban.com/note/141553786/
2 葛传椝等编撰《新英汉词典》，上海译文出版社，2002年，1475页。
4 同英注3，12页。
5 顾祖禹《读史方舆纪要》（秦勇编），北京：天下出版社，2000年，"总序之一"。
6 同注5，"凡例"。
7 同注5，"凡例"。
8 张资平《自然地理学》，北京：商务印书馆，1923年，1页。
9 同注8，2页。
11 周顺麟《地貌测绘》，北京：中国工业出版社，1964年，3页。
12 杨景春、李有利编著《地貌学原理》，北京大学出版社，2009年，1页。
13 有关"什么是地貌学"以及它跟地质学和地理学的学科关系，多年前，三位撰写地貌学学科史的作者曾经给出过如下简明介绍："在我们看来，并不真的独立存在着一种有关地貌研究的历史。我们说的是那种关乎自然地景发展过程的认识演化史。在19世纪初期之前，有关地貌的研究构成了地质学的主体内容。可是19世纪初期之后，主要由于地层学和古生物学的迅速崛起，地貌研究在地质学里的比重逐渐下降，虽说在一般性地质知识中地貌研究仍然有着不可或缺的重要性。到了1860年之后，地貌研究部分地属于地质学，部分地属于了自然地理学。后来，地貌研究就被称为地文学或者叫地貌学。今天，地貌研究构成了地质学与地理学之间的一架桥梁，或许，也正因为它与地理学的联系，才是它没能沿着更为严格科学路线继续发展下去的主要原因。即使在1963年，人们还是会把地貌学当成'伪科学'或是'准科学性艺术'。而研究地貌学学科史的史学家则经常看到人们会这么以为，就是把地貌学看成是始于一种艺术化感知、逐渐走向真正科学化解释可是仍然没有达到这一目标的学科。"
—《The History of the Study of Landforms or the Development of Geomorphology: Volume 1, Geomorphology Before Davis》。详见英注13。
而在中文语境里，尽管在自然物理学、地质学里"地形"和"地貌"可以被当成同义词来使用，到了1950年代中期北京大学地理学系进行专业调整时，北大的地理学者们还是采用了"地貌学"而不是"地形学"的称谓。在晚近的一次电话访谈中，已经退休的北大地理学教授王恩涌先生告诉我说，当时，字典上对应着"地形学"的俄语词汇是'топография'，这是前苏联测绘院校里研究大地测绘的专业。为了跟俄语里地理学系下的"геоморфология"准确对应，当时北大的老师们像韩慕康先生就把翻译成为"地貌学"，以便区别于测绘学里的地形学。

NOTES

14 Ning Jinsheng et al. ed., *General Introduction to Land Surveying*, Wuhan University Press, 2009: 2.

15 A.K. Lobeck, *Geomorphology: An Introduction to the Study of Landscapes*, New York: McGraw-Hill Book Company, Inc., 1939: 15.

16 China Coastline Survey Committee, *China's Bays: Volume 1, The Eastern Bay of Liaodong Peninsula*, Beijing: Haiyang Press, 1991: 272.

If this geological report sounded terribly dry, it is not so for a typical Geomorphology class. One of the 65 participants who replied my question about topography is currently studying Landscape Design in France. Her impression on her first Geomorphology class is anything but boring:

'We went for an excursion to the forest of Fontainebleau for a whole day. After lunch break, my teacher standing in front of a huge rock began to talk about the geomorphologic history of Fontainebleau, of its crust movements and climatic changes. My teacher would sketch a few sectional drawings to illustrate the morphoristic changes of the rock stratums at each stage. And then, we would take a hike. My teacher would explain a series of characteristic yet different types of landforms along the trip, or take soil samples every now and then to illustrate his previous points based upon the alterations of soil colours and acid-base properties. To add to our geomorphologic learning, another teacher specialised in Ecology would talk about the distribution and variation of vegetations that match the local topographical structures.'

As such, a geomorphologic understanding and an ecological reading of landscape extend into each other.

See: ⌘ www.douban.com/note/141553786/

17 Li Dehua, et al. ed., *City Planning Principles*, Beijing: China Architectural Industrial Press, 1991: 47.

18 In this textbook, by the 'social' aspect of urban land use, it really meant the difference between Stated-Ownership or Collective-Ownership of a parcel of land, whereas by the 'economical' aspect the buildable floor area ratio, the 'legal' aspect the planning permit application process—with no hints of public participation.

19 Dalian Municipal Gazetteer Publication Committee, *Dalian Municipal Gazetteer: Its Natural Environment and Hydraulic Facilities*, Dalian Press, 1993: 20.

20 In a similar tone, Kenneth Frampton and others had already called into question of topography back to the 1980s. For instance, in his seminal essay 'Towards a Critical Regionalism: Six Points for an Architecture of Resistance', Frampton wrote, "*It is evident that modern development favours the optimum use of earth-moving machinery so as to achieve a totally flat site which is almost regarded a precondition for the rationalisation of construction. Here again, one touches in concrete terms the fundamental opposition between universal civilisation and autochthonous culture. The bulldozing of an irregular topography into a flat site is clearly a technocratic gesture which aspires to a condition of absolute 'placelessness', whereas the terracing of the same site to receive the stepped form of a building is an engagement in the act of 'cultivating' the site*".

— Kenneth Frampton, 'Towards a Critical Regionalism: Six Points for an Architecture of Resistance', *Labour, Work and Architecture: Collected Essays on Architecture and Design*, London: Phaidon Press Limited, 2002: 86.

Also, in David Leatherbarrow's writing, 'topography' has become a mutual dialogue, hermeneutically, between designer and site conditions, between body and atmosphere, between past and future, and between potentiality of a place and architectural realisation

— David Leatherbarrow, *Topographical Stories: Studies in Landscape and Architecture*, Philadelphia: University of Pennsylvania Press, 2004.

For certain, thus, the Topography currently debated in the field of architecture, goes far beyond topography as it in Land Survey, or simply scientific landform study in Geomorphology, or site improvement, but at- tempts to involve them all and some sense of eco-philosophy, to return to daily world through building's making.

注解

14 宁津生、陈俊勇、李德仁、刘经南、长祖勋等编《测绘学概论》，武汉大学出版社，2009年，2页。

16 中国海湾志编纂委员会《中国海湾志》第一分册《辽东半岛东部海湾》，北京：海洋出版社，1991年，272页。

如果这种科考报告式的描述的确有些枯燥，那倒不是地貌学课堂真实的代表。在回答何谓"地形学"的问题时，有一位留学法国的网友这样跟我写到："关于'地形学'的第一节课"：

"我们去枫丹白露的森林徒步一天，中午野餐后老师就着一块大岩石开始讲解枫丹白露森林的形成过程。从地壳运动和气候演变开始，老师一边讲解一遍手画出各个阶段的地形变化地层结构，随后徒步过程中有几个典型的不同地貌，挖土观察颜色和酸碱性的变化来验证之前的讲解，生态学老师配合分析不同的植被情况讲解植物。"

没错，这样地貌学就和生态学对于地貌的解读相互交织在了一起。

见 ⌘ www.douban.com/note/141553786/

17 李德华主编《城市规划原理（第二版）》，北京：中国建筑工业出版社，1991年，47页。

18 那时土地的社会属性仅指"土地所有制"，经济属性仅限于"容积率"，法律属性主要指土地使用的审批过程而不是后来的产权——也跟公众参与毫无关联。

19 大连市城乡规划局法规编写组编撰《大连市城市规划管理条例：实施参考材料》，大连市城乡规划局，1992年，20页。

20 这个问题也正是肯尼思·弗兰姆普敦等人在1980年代开始询问的问题。在"面向一种批判的地域主义"一文中，弗兰姆普敦就设问过，

"将一块不规则的地貌铲平成为一块平坦的场地，明显就是一种追求绝对'无场所感'状态的技术官僚的做法。而将同一块基地处理成为台地，让建筑去错落，则是在从事一种'滋养'基地的行为。"

— "Towards a Critical Regionalism: Six Points for an Architecture of Resistance"

当然，在戴维·莱瑟巴罗那里，"地形学"已经更多地具有了阐释学意义上的设计与场地的对话，身体与环境的对话，历史和未来的对话，场地可能性和建筑的对话，等多重意义。

—《Topographical Stories: Studies in Landscape and Architecture》。

详见英注20。

所以，当下建筑学里的地形学，肯定不止是测绘学里的地形学，也不是简单的地貌学里的大地形态构造，更不是简单的地形美化，而是可能包涵了上述诸多思考的介乎生态哲学和强调建筑直接面向日常世界的努力。

TERRAIN & *XING SHENG*

A comparison between the landscape gardens
built in the Italian Renaissance and
the Chinese Ming and Qing Dynasties

地形和形胜

意大利文艺复兴时期台地园
和中国明清时期山水园比较

陈薇
Chen Wei

Figures in the previous page and
above are by Chen Wei

前页与上图为陈薇提供

Based on my observations, the historical charm of a garden can only materialise by being experienced in person.

When visiting the landscape gardens of Italy a dozen years ago, I was left with a strong impression that the best way to view Italian gardens is from a high vantage point, where one could take in their magnificent entirety—scenery comes alive with the downward expansion of the garden. The best example is Villa d'Este in Tivoli, Rome. Before learning the appeal of Italian gardens, I had studied Chinese gardens for quite a while, and had visited many of them. A Chinese garden, such as the Chengde Mountain Resort, guides one upwards and further and further away to experience its intricate layout. I was struck by the dramatic differences between Italian and Chinese gardens and have been intending to explore the essence of this. This forum provides me with a good opportunity to talk about gardens, allowing me to put together my old memories and the findings derived from my garden studies, and compile them into this paper I am presenting before you.

从真实现场体验园林很有必要，认识也许更接近古典意趣，这是我的体会。

十几年前考察意大利台地园时，有一印象特别深刻，即鸟瞰园林时最能感受到它的整体和壮观，而随着向下、向下地展开，景致才逐步丰润起来，如罗马蒂沃里爱斯特别墅园°；而在这之前也一直研究中国园林，倘徉于山水之间，却常引导我向上、再向上，或向远、再向远，这样才能得到山水园的整体认识，如承德避暑山庄。两相对照，给我很多鲜活和生动的感受，也促使我一直想探讨意大利台地园和中国山水园的本质差异。感谢这次会议的推动，令我重拾记忆和研究，遂成此文。

如果说，一脚一脚往下行走时，我更多关注到地形和景观；而一步一步向上攀登时，我眼中看到的是形胜和景色，这就是本文题目的来历。地形，是和地质、地貌相关的地志，和大地的解剖相关；* 形胜，则和山形走势以及风水环境相关，是自然的外部形态。* 也可以说，地形和形胜分别是这两种园林的关键概念和设计出发点。本文以意大利文艺复兴时期台地园和中国明清时期山水园作为研究对象，不仅试图在比较中探索不同园林的设计思想和方法论问题，也期望依此深刻了解中西方园林的异趣。

* 图1
* 图2

This essay explores the distinctive experience of upward and downward movements when visiting a garden. When stepping downward in a garden, I argue that one would notice the details in terrain and landscape; on the other hand, when climbing up in a garden, *Xing Sheng* and scenes would dominate one's experience. This is where the topic comes from. Topography, or terrain, refers to geological and geomorphological features and an anatomical examination of the landscape,* while the concept of *Xing Sheng* refers to the shape and ridgelines of mountains and its *Feng Shui*; it focuses on the external features of the earth.* The focus on terrain and *Xing Sheng* are essentially two different key concepts and approaches to landscape design. This essay makes a comparison between the Italian terrace gardens from the Renaissance and Chinese landscape gardens from the Ming and Qing dynasties. The contrast highlights conceptual differences in design approach and understanding; this also allows for a deepening appreciation of the unique flavours of European and Chinese gardens.

* FIGURE 1
* FIGURE 2

I.

CHARACTERISTICS & VIEWING VANTAGES
OF THE TWO TYPES OF GARDEN

I.I.

ITALIAN TERRACE GARDEN:
AN OVERLOOKING PLANE

The Italian terrace garden made its debut during the European Renaissance. To some extent, Italian terrace gardens qualify not only as the origin of, but also the most important form of European Renaissance gardens.

During the Renaissance, commerce and trade activities were conducted through merchant boats travelling between Venice, Genoa, Florence, Constantinople, North Africa, Asia Minor, and the Black Sea coasts among other locations, which generated large amounts of wealth for all cities along the trade route. In the attempt to dissociate themselves from the Church doctrine, emerging urban bourgeoisie began to champion a revival of Classical culture and a positive attitude to daily life and reality. This mind-set was reflected in literature, science, and the arts. People started to reincarnate the long lost ancient Greek and Roman forms of architecture that featured skilful utilisation of terrain and symmetry, such as at the theatre of Dodona,° the Temple of Jupiter Optiums Maximus, Juno, and Minerva.°

° third century BCE
° 509 BCE

陈薇
CHEN WEI

一、阅读两种园林的特点和视角
（1）意大利台地园：俯瞰的平面

意大利台地园诞生于欧洲文艺复兴时期。而欧洲文艺复兴发源于意大利，公园 14 和 15 世纪是早期，16 世纪由盛而衰。因此，意大利台地园从某种程度上讲是欧洲文艺复兴时期园林的重要代表和滥觞。

当时意大利威尼斯、热那亚、佛罗伦萨有商船和君士坦丁堡、北非、小亚细亚、黑海沿岸进行商业贸易，逐渐地，城市资本积累，城市新兴的资产阶级也要求在意识形态领域反对教会精神，开始提倡古典文化，肯定人生以及对现实采取积极的态度。这一指导思想反映在文学、科学、艺术等领域，而在园林和建筑中最直接地表现在古典的"再生"——擅用地形和对对称完美风格的追求——那种久已存在于诸如古希腊多多拉剧场。和古罗马朱庇特、朱诺和密涅瓦庙之中。°

○ Dodona, theatre, Totale third century BCE
○ Temple of Jupiter Optiums Maximus, Juno, and Minerva, 509 BCE

特别值得提出的是古希腊露天剧场创造，表现出古典建筑中对环境的化用和对大地的关怀，

> "它们从来没有丧失过与大地的密切联系，也从来不曾被当作独立的建筑物，尽管就一种场所而言它们已经相当古老了……最初，剧场用地只不过是一个供舞蹈和典礼合唱使用的水平空间，并没有什么建筑方面的意图；观众们聚集在周围的山坡上，或者也许会拥挤在某种临时搭建的木看台上。"•

• 1

到公元 4 世纪，出现了布置成半圆形的石阶座位，观众自上而下地俯瞰观看演出，合唱团的演出发展成为戏剧表演，较低的平台是演员演出之地，而戏剧由怀抱的宏大剧场背景产生气势非凡的效果——场景。在这里，"地形"是一关键，是和大地密切关联的地形处理，进而诞生出和竖直向相关的景观。如希腊的埃比道拉斯剧场，°依据地形将看台分为 34 排大理石座位，全场能容纳一万五千余名观众，中心舞台直径 20.4 米，而由中心次第升高，像一把展开的巨大折扇，建成后的剧场和自然融为一体，是十分壮阔的景观。*

○ Epidaurous Peloponnesos, Theatre, 350 BCE

* 图 3

It is worth mentioning that the design of ancient Greek amphitheatres shows careful utilisation of the surrounding environment and concern for the land:

> They never lost a close contact with the land, nor treated as separate structures, though they were undeniably very old... At the beginning, the site was just a horizontal space for dance and ritual chorus, not meant to be a structure. Audiences gathered on the surrounding hills, or crowded on a makeshift wooden platform.*

• 1

In the fourth century, some stone seats were arranged in a semicircle fashion, allowing audiences to have an overlooking view of the performance. Choir performance gradually evolved into drama. The lower level of the terrace served as a stage. The magnificent embracing backdrop generated a surreal effect for the theatre. Here, terrain is key, a feature closely associated with the land, depicting a vertical landscape scenery. Epidaurus Theatre in Peloponnese° is a good example of this. The theatre has 34 rows of marble seats for some 15,000 spectators. The centre stage is 20.4 metres across. The theatre steadily rises from the centre, like a huge folding fan. The theatre makes itself a magnificent landscape by being part of nature.*

○ 350 BCE

* FIGURE 3

The combination of natural and artificial scenes reflects the principle of utilising the terrain emphasised by Classical doctrines, a very important feature of terrace gardens built during the Italian Renaissance. Residential structures were placed at a higher level, allowing people to have a panoramic view of the entire site. At this point, the terrace is the liveliest feature of the garden.

Set in Tivoli, 40km east of Rome, the Villa d'Este, built in 1549, is a magnificent terrace garden constructed during the Italian Renaissance.* Covering an area of 200 by 265 metres, the garden sits on a gentle slope, overgrown with towering cypress trees, through which people may see the remaining old city walls. Looking westwards, one can see the sunset resting squarely on Rome. The site is apparently selected to show a logical connection between the terrace garden and its surrounding environment. The garden is built on the terraces at six levels, with a major sightseeing spot on the top of the terrace on the central axis, a structure named 'Casino', from which stretch out one longitudinal axis and three horizontal axes that define the terrace at different levels. Beautiful fountains are placed at the intersections between the horizontal and vertical axes, making them the focus of the villa, feasting the visitors' eyes with a neat, magnificent, and breathtaking terrace view. From a bird's-eye view, people are not only given the opportunity to see the layout of the garden, but also led to associate it with a Roman sunset. One has to visit each level of terrace to experience the changing scenes. When reaching the bottom level, a look back would surprise you with an unanticipated and splendid view of the landscapes in the garden.*

* FIGURE 4

* FIGURE 5

这种对地形的利用而形成的自然与人工结合的景观，是古典主义的一个方面，这在意大利文艺复兴时期台地园中是十分重要的特点，往往住宅在高处，或从建筑楼层观览，由此往下俯瞰，平面格局尽收眼底，而利用台地建成的景观恰为园林中最生动活泼的内容。

　　爱斯特别墅园，位于罗马东面40公里的蒂沃里城。始建于1549年，是意大利文艺复兴时期最雄伟壮丽的台地园。* 规模宏伟，占地为200米×265米，选址于罗马东郊缓坡地，柏树参天，柏树后可看到残留的老城城墙，向西望的日落点正好是罗马，这样的选址为台地园和周围环境取得了很好的联系。总体共有6层台地，重要观赏点位于最高台地，是中轴线上的房屋，° 自此下瞰，纵向有中轴线，横向依不同台地有三条轴线，居中的横轴与纵轴交叉处，设精美龙喷泉，是全园中心。台地壮阔，整齐有序，十分气势。下瞰时，不仅了然总体布局，还和日落中的罗马呈现整体上的关联和意向上的联系。而只有当你逐层观览时，才可以看到每一层的景观变化和丰富性，至达到最底层，暮然回首，整个园林景观和场景才丰富地表现出来。*

　　罗马美第奇别墅园，建于16世纪，别墅3层，局部望楼6层，自别墅楼层下瞰是观赏园林的最好视角，园基本保留16世纪格局——由木本植物为主体，用冬青、月桂和黄杨围成绿篱，用树木分割空间，形成规则的几何图案。原先的花园结构简单，建筑物前种植有甘菊、月季、罂粟和牡丹，北部用十字形小路形成十六个分割空间，这些分隔空间用荚迷镶边，中间种植低矮的果树。1576年红衣主教美第奇购买了这座园林，兴趣是建一个能展示他收藏的古代雕塑和大理石雕刻品的花园，在其入口的轴线上布置有水池喷泉，但总体布局仍由整齐有序、图案井然的植物构成。另一佛罗伦萨布拉托里诺美第奇别墅园俯瞰的平面是园林最重要的场景。°

* 图4

° 术名"Casino"，意大利文，意为房屋。

* 图5

° the Medici villa at Pratolino, 1590s

The Villa Medici, built in the 16th century in Rome, is an architectural complex mainly made up of three storied structures, though partially six-storied for a panoramic view of the garden. The villa retains the basic layout conceived in the 16th century, sheltered within woody plants, such as pines, cypresses and oaks, and fenced by holly, laurel, and boxwood hedges. The garden is cut into regular geometric patterns with trees. The original garden has a simple layout, with chamomile, rose, poppy and peony in the front of the structures. In the north, there were 16 small individual spaces divided by cross paths and trimmed by viburnum, with dwarf fruit trees grown in the middle. The Cardinal's father Cosimo I de'Medici bought the villa in 1576, and used the garden to show his antique marble sculpture collection. There is a fountain in the foreground. The garden remains in a neat and orderly pattern made up of plants even with additions. Another example is the Medici villa built in the 1590s at Pratolino, where an overlooking view of the garden depicts the most impressive scene.

1.2

CHINESE SHANSHUI GARDEN:
A FAÇADE TO LOOK UP

The Chinese *shanshui* garden originated in the Wei and Jin Dynasties,° although it developed fully in the Tang and Song Dynasties,° and became popular in the Ming and Qing Dynasties.°

In the Wei and Jin Dynasties, China was ripped apart by civil wars and political unrest. Some scholars chose hermitage to shun the confused social reality. They revelled in *shanshui*,° making poems, paintings and gardens of *shanshui*. Sun Chuo• depicted a disillusioned scholar venting his frustration or forlornness through natural landscapes in a postscript to a poem collection:

○ 220—420
○ 618—1279
○ 1368—1911

○ literally mountain and river
• 2

> Disillusioned scholars would enjoy themselves in natural mountains and streams, listening to birds signing, appreciating the beauty of bamboos, and fishing in the streams. They would write down their humble thoughts in a realisation that good times had gone and what they could is cherish these old memories.

Tao Yuanming illustrated the connection between natural landscapes and his life and home in the countryside in a prose named 'Return to the Countryside'.•

• 3

> I was no part of an ordinary folk's life when I was young, though I was born to love the mountains. I mistakenly chose a wrong way of life that killed my past thirty years. A restrained bird would have the nostalgia for trees, and a fish in the pond would have a wish to return to the

一（2）中国山水园：仰视的立面

中国山水园，源起于魏晋时期，兴盛于唐宋，普及于明清。

中国魏晋时期，世道混乱，战争频繁，政治多变，一些士大夫由于对现实不满，愤然归隐，陶醉于山水，一时，山水诗、山水画、山水园兴起。孙绰的《兰亭集诗》：• 　　　　　　• 2

"流风拂枉渚，停云荫九皋。
莺语吟修竹，游鳞戏澜涛。
携笔落云藻，微言剖纤毫。
时珍岂不甘，忘味在闻《韶》。"

表达的是其时文人游牧山水的思想感情。
陶渊明的《归园田居》：• 　　　　　　　　　　　• 3

"少无适俗韵，性本爱丘山。
误落尘埃中，一去三十年。
羁鸟恋旧林，池鱼思故渊。
开荒南野际，守拙归园田。
方宅十余亩，草屋八九间。
榆柳荫后檐，桃李罗堂前。
暧暧远人村，依依墟里烟。
狗吠深巷中，鸡鸣桑树颠。
户庭无尘杂，虚室有余闲。
久在樊笼里，复得返自然。"

说的是田园风光、田园生活和宅院的关系，自然山水园滥觞的缘由也跃然纸上。值得注意的是，宅屋、桃李罗堂前和榆柳荫后檐、远人村与墟里烟，是由里而外一层层晕染的关系，是一种仰视的立面的层叠关系。

不过魏晋时期的山水画"虽有尺山片水，亦只画中衬贴，而无专学。"• 但唐代以后山水画学形成，当和士大夫山水诗 • 4
画联盟有关，其根源还是思想情操上通过自然而逸然相关。
李唐渐后山水画成为国画主流，是因为• 　　　　　　• 5

"厚重不迁之山，最足以代表静；
周流无滞之水，最足以代表动，
故画家根据二元之思想，
以山水画象征整个之自然。"

地形和形胜
TERRAIN & XING SHENG

> deep waters. So I decided to cultivate the wilderness in the south, and become a farmer. I built my residence and thatched cottages over a large piece of land, with peach and plum trees in the foreground, and large elm and willow trees in the rear. I can feel the warmness of the distant villages, and can see the bellowing smoke there, though from afar. I can hear dogs barking, and cocks crowing on the top of mulberry trees. The house and courtyard are clean from dust, and I have extra rooms to spare. Now I'm feeling relaxed in a natural environment after so long a stay in a confined one.

It is worth noting that Tao depicted a cascading relationship from inside to outside, and one has to look through the scenes with peach and plum trees in the foreground, large elm and willow trees in the background, and the villages and bellowing smoke in the distance.

Landscape paintings prevailed in the Wei and Jin Dynasties and depicted 'mountains and waters only as a background, without a dedicated theme',[4] while the formation of the School of Landscape Painting in the Tang Dynasty advocated the close association of poetry and painting, in which imagery expressed a philosophy of finding personal peace within nature. Gradually, Landscape painting became the mainstream in Chinese painting, for 'the mountains on the painting are heavy enough to define the tranquillity, and the flowing water in the painting is sufficient to represent a dynamic movement. Consequently, a painter is able to cover natural landscapes simply using two elements: mountain and water.'[5] That is to say, the practice of combining landscape painting, poetry and gardening is an affirmation of one's love for nature.

A love of nature fuelled the literati's enthusiasm in gardening. One can see such tendencies in landscape gardens built in the Ming and Qing Dynasties. The late Ming and Qing Dynasties that mark the end of Chinese feudal society, featured obscure politics and implicit art expressions. The gardens built in the Ming and Qing Dynasties, both royal and private, were designed as an extension of the residence, as a space for living and also a place for spiritual fulfilment. The dual functionalities of Ming and Qing gardens provide a landscape to ease one's frustration and forlornness, and a canvas for making symbolic statements relating ideals to reality. This association makes for a rich landscape with multiple layers of meaning, that entertain people's eyes and leaves room for imagination.

The Qing Yi Garden, or 'Summer Palace' by modern translation, is a good example. Located in west Beijing, it was initially built only as a holiday palace in the Jin Dynasty,[○ 1115—1234] and was expanded into a holiday garden palace during the Yuan Dynasty[○ 1206—1368] due to its beautiful location. The original Jinshan Mountain was altered and made into an encircled mountain, and the waters in front modified into an encircled lake. In the Ming Dynasty, the lake was renamed the West

如此我们可以理解山水画、诗、园的关系，乃均关乎自然之情。

这种从对自然之情出发而反诸于造园的发展，至明清的突出特征乃小中见大。明清是中国封建社会晚期，政治不够开明清朗，艺术不够酣畅简明，明清的私家和皇家山水园林，一方面是住宅的延续、生活的场所，另一方面又是精神寄托、理想追求之地。这种双重性就使得明清山水园具有这样的特征：注重创造自然、抒心中块垒，用象征的手法将理想的彼岸性和现实的此岸性连结起来。这种联系在直观上表现为景色丰富，层次叠韵；而在心理上则形成想象空间的扩大，产生意境。

以清漪园°为例。园位于北京西郊，金代建成行宫，是因这里风景优美；元代扩建，原金山改为瓮山，山前汇水为瓮山泊；明代湖称西湖；清乾隆十五年°大兴土木，改瓮山为万寿山，湖为昆明湖，称清漪园，后光绪十二年°西太后重建改名颐和园。当时西太后乘船自京城宫出，一路西行，一年中多数时间住在颐和园，最好看和最壮观的景色就是自水上观赏万寿山前山的壮观形胜。*

° 即现在的颐和园

° 1750年

° 1886年

* 图6

万寿山为东西长南北窄的山体，南有开阔湖面，北有悠长后湖，山体浑厚高耸，有势但形并不起伏生动，经过前山和后山建筑的建造，则形成变化的形势。前山中心有南北轴线的牌坊、排云殿建筑群和佛香阁，结合台地层层展开，位于高处的大体量佛香阁加强了山的稳重感。山脚沿山水交接面，水平向布置长廊，既解决"步移景异"的观览，也增加了山体和水体之间的层次。自水上而登山，有框景、有隔景、有对景；至山顶，东可览东宫院落鳞次栉比，西可借玉泉山岚色美景，南可瞰水烟迷茫，北可接佛光塔影。因为有距离，有层次，也就有了想象和期盼，是中国明清时期山水园的真谛，也是自然的属性赋予的特色。后山以"须弥灵境"为立意，结合山体形成错落的山形，加上后湖幽深静谧和环绕，产生山水的流动感。颐和园和严谨的北京城、壁垒的宫城是反差，是互补，是生活的延伸，是精神的需求。

Lake. In 1750 when Emperor Qianlong was in power, a massive reconstruction was launched, and the mountain redesigned and renamed Longevity Hill, and the lake—Kunming Lake. The garden itself was renamed the Qing Yi Garden. In 1886, Empress Dowager Cixi had the garden renovated, and renamed it as Summer Palace. The Empress Dowager took a westbound boat trip to the Summer Palace, several times a year, from her official palace in the capital. The best and most spectacular landscape in the garden is the magnificent scene of the front section of Longevity Hill viewed from the lake, a spectacular *Xing Sheng* view.*

* FIGURE 6

Longevity Hill has a long ridge running from east to west, narrowing down from north to south. It has a vast open lake in the south, and a long rear lake in the north. The round and towering hill is imposing but lacks undulating features. The reconstructed front and rear contribute a changing curve to the appearance of the hill. At the front, a range of structures, including an arched gate, a group of buildings named Pai Yundian, and a Buddhist Scripture Library, sit at different levels along the north-south axis line in the centre. The massive Buddhist Scripture Library on the top adds weight to the hill. The hill is interspersed with a long horizontal corridor and a vast lake at the foot. The layout presents multiple walking scenes, allowing more layers to be added to the hill and water. When climbing the hill from the lake, one may enjoy framed scenes, decorative scenes, and side scenes along the way to the top. In the east, one has an overlooking view of the structures and associated courtyards at different levels, or enjoys the beauty of the mountain ridges in the west, the distant waters and bellowing smoke in the south, and the reflection of the Buddhist Temple in the north. This variety of distances and levels defines the essence of a landscape garden in the Ming and Qing Dynasties, as it provides more room for imagination and expectation, and reveals more of Mother Nature. At the rear, a mysterious secluded aura implies the spiritual movement of both mountains and waters, taking advantage of the tranquillity presented by the peaceful long lake and terrains at different levels. The relaxing Summer Palace complements the severe Forbidden City as a place for the extension of life and pursuit of spiritual fulfilment.

Other examples which can demonstrate *Xing Sheng* are the Lion Grove Garden and the Art Garden in Suzhou. The Lion Grove Garden is built with large and labyrinthine grottos of rocks over an area of 1,153 square metres. All rocks found in the garden were collected from Taihu Lake. The major part of the rockery was completed in the Song Dynasty. Numerous notable visitors adored the garden with poems and paintings. For example, a scroll of Lion Grove painted by Ni Yunlin° depicts beautiful rock scenes. During his visit to the garden, Emperor Qianlong toured around the rockery comparing it to a rockery painting in his hand. The rockery is arranged in three levels, with steep rocks on the top, a rock path in the middle, and rocks in the pond. A stretch of vertical rocks constitutes

° 1301–1374

另一充分体现仰视的立面是园中主景的园林，是苏州狮子林和苏州艺圃。狮子林是以大假山为中心建造的，大假山总面积约 1153 平方米，几乎全用太湖石堆叠，其中大部分为宋代朱勔花石纲遗物，狮子林自元代建园以来，文人墨客以狮子林假山为题的诗词题咏和图卷图跋层出不穷，倪云林的《狮子图》绘出假山大成，乾隆皇帝南巡时展卷对照而游，就是比照假山的立面。狮子林假山可分为上中下三层，高者立峰突兀于山顶，低者石矶沉浸于池中，中者盘旋有山道，加上绵延的难以尽数的立峰，便形成起伏有势和层峦叠嶂的山林，《狮子林联句》赞叹曰：

• 6

"有峰有岫，有碉有湫
　降观深窟，忽焉崇丘。"

周围的亭台楼阁设置均和观览这大假山有关。艺圃大假山是明代叠石掇山的佳品，最重要的是它的位置经营十分讲究看它的层次，大假山位于大池南面，从立面看，由水面向上，峰石错落重叠，越后越高，呈现出巉岩交错、峰峦纷出、洞穴幽入、横梁飞架的一派自然山峦景象，*《园冶》说：

* 图 7

"池上理山，园中第一胜也。
　若大若小，更存妙境。
　就水点其步石，从巅架以飞梁；
　洞穴潜藏，穿岩经水；
　峰峦飘渺，漏月招云；
　莫言世上无仙，斯住世之瀛壶也。"•

• 7

艺圃大假山背衬白墙，山上林木参天，收有这样的效果。其简练开陈，宛若大自然剪得片山。

an impressive overlapping peak view, running along a rippling curve. An ode to the rockery states that the rockery has:•

• 6

> Both mountains and peaks, and streams and cascades. Here you dive into a deep grotto, and there you look up at a hill.

The surrounding pavilions and structures are defined by the positions of the rocks.

The Art Garden, on the other hand, is a typical exemplification of the rock craftsmanship of the Ming Dynasty. The designer positioned the rocks very carefully. Massive rocks were manoeuvred to stand on the water surface of a large pond in the south. The rocks in different shapes are intricately placed with the one in the rear standing higher than the one in the front, depicting a scene of natural landscape made up of rocks and cliffs, peaks and ranges, grottos and hidden paths.* *Yuan Ye*,° an ancient writing on gardens, put it vividly:

∗ FIGURE 7
○ 1634

> The rocks erected from the pond make the most impressive scene in the garden. Rocks, small or large, were borrowed to build a wonderland. Stone steps were rested in the water for walking, and a flying beam on the top bridges the ranges. One can hide in the grottos, or travel through the rocks or waters. The looming ranges and peaks shade the moon but attract the clouds. All these make one feel like living in a wonderland, though people say there is no wonderland at all.•

• 7

In the garden, massive rocks stand against a white wall, with towering trees sticking out on the rocky mountains. The skilful arrangement of the rockery mimics mountains in the wilderness.

II.

IMAGERY & MEANINGS
OF THE TWO TYPES OF GARDEN

II.I.

ITALIAN TERRACE GARDEN:
NEAT, GEOMETRIC & ORDERLY

Terrace gardens, the most important form of landscape expression during the Renaissance, evolved from a simple decoration in the beginning, to an extremely enriched landscape expression in their heyday, and further developed into a faded beautiful memory. In addition to the Villa d'Este, many other villas shared the same fate, including the Boboli Garden. These Italian Renaissance terrace gardens feature orderly, geometric, and artificial scenes, denoting the pursuit of order and humanitarian spirit. They provided a kind

二、分析两种园林的图像特征和意义
(1) 意大利台地园：规整、几何与秩序

意大利台地园作为文艺复兴时期的重要园林，依时间发展有对应的初期简洁、极盛丰富和衰落期装饰化的过程。和上述爱斯特别墅园基本同时期的台地园，还有佛罗伦萨的波波里花园等，° 它们作为台地园代表，也反映文艺复兴时期意大利台地园兴起的原由：用规整、几何、人工的图像特征传递对秩序和真实人文精神的追求，以在时序上抗衡和跨越中世纪的繁复和长期战争留下的混乱，重塑经典。

° Boboli Garden

波波里花园，位于佛罗伦萨西南隅，原是彼蒂宫° 的庭园，1550年开始扩建而成。整体布局井然有序，* 包括东西两部分和连接部分：东部为南北轴线，由旧宫° 和台地园构成；西部为东西轴线，为平地园；衔接处另有一条轴线斜向交接，为柏树林荫大道，东高西低，坡道连贯，其东端为观景平台花园，从这里可以俯瞰全园的灿烂景观和佛罗伦萨的城市景观，此高地处，1592年建有长条碉堡，在当时围城十一个月中，它帮助保卫了城市。

° Pitti Palace
* 图8
° 即彼蒂宫

园林和城市其他部分既对比又承接的关系，是解读园林的重要方面。波波里花园南高北低，和城市成环抱状，东部台地园自北而南依次为花园洞屋、露台剧场、尼普顿神° 塑像喷泉水池和转而斜向轴线最高处的雕像与观景平台。值得一提的是在洞屋南面的露天剧场，吸收了古典剧场依山就势而造的做法，设有6层座凳，不高，但座凳外围用园林的手法均匀立设放有雕塑的壁龛，再后背景为月桂树篱，就增加了垂直面的高度和山林的气氛，* 在剧场北面的二层平台上，布置有壮丽的八角形喷泉，在剧场中心放有大水池和罗马带来的埃及方尖碑，更加突出了古典主义的对称与中心感以及园林的独特性。

° Neptune

* 图9

波波里花园地处城边，地形不规则而有高差，但通过有组织的布局，尤其是几何图形的有机组织，整合为大而不乱、主次分明、景有特色的园林。也是文艺复兴时期园林追求秩序的精神体现，它和中世纪留下的不规则街巷的佛罗伦萨老城形成鲜明对比，也和佛罗伦萨文艺复兴时期创造的古典美的建筑互相呼应。

罗马阿尔多布兰迪尼别墅园，° 始建于1598年，1603年完成。它位于罗马东南的一个山腰上，为红衣主教彼埃特罗·阿尔多布兰迪尼所有，花园依山就势做成有层次的台地，

° Villa Aldobrandini

of order after years of rampant war and chaos throughout the Middle Ages.

The Boboli Garden in the southwest of Florence was originally built in 1550 on a court at the Pitti Palace. The garden has a well-organised layout,* separated into east and west sections connected by a transition. The east section runs on a north-to-south axis, covering the older palace, Pitti Palace, and the terrace gardens. The west section follows an east-to-west axis, sitting on the ground garden. The two sections of the garden are connected by a sloped boulevard lined with cypress from east down to the west. The boulevard leads into a terrace garden in the east, where one can have a splendid panoramic view of both the terrace gardens and the city of Florence. At the high land is a long bunker built in 1592 to defend the city amid an 11 month long siege.

* FIGURE 8

Landscape gardens are not only decorative but also a part of a city. This relationship reaffirms the underlying meaning of landscape gardens. For example, the Boboli Garden which is high in the south and low in the north embraces the city. The eastern part of the garden is dotted with a range of structures from north to south, including houses, an open theatre, the statue of Neptune, and a fountain pool. From that point, the terrain line turns to a statue and a viewing platform on the top level of terrace. The amphitheatre in the south of the grotto house was built to take advantage of the magnificent mountains, similar to other Classical amphitheatres. The theatre has six levels of spectator seats, proportionally surrounded by sculpture niches against a background of laurel hedges,° which enhances the height of the vertical façade, and enriches the aura of mountain forests.* At a two-story platform north of the theatre sits a spectacular octagonal fountain, and in the centre a large pool and Egyptian obelisks brought from Rome, highlighting Classical symmetry and centralism and the uniqueness of the garden.

° a traditional way to decorate a garden.
* FIGURE 9

Sitting on the edge of the town, the Boboli Garden is irregular in terrain and elevation. The well-planned layout, especially the thoughtfully connected geometric patchwork, makes the garden large yet not confusing, allowing its visitors to enjoy splendid views in a prioritised and tiered manner. The garden is a spiritual reflection of the pursuit of order during the Renaissance. It makes a sharp contrast against the irregular medieval streets in the old town of Florence, and echoes of the beauty of Classical architecture created by Florentines during the Renaissance.

The construction for Villa Aldobrandini started in 1598 and the building was completed in 1603. It sits on a hillside southeast of Rome and is owned by the Aldobrandini family. The villa is built on terraces against an imposing forested mountain, aligned with an axial avenue starting from the forests. The villa is designed with strict symmetry and is very well balanced.* Water features and sculptures along the axis are the main scenes to feast the visitor's eye. The neat greeneries were trimmed in regular patterns. Both the

* FIGURE 10

背景为丛林。自丛林始便自上而下为一条轴线，整体布局为规则式，严整对称，方圆均衡。*水景和雕塑相结合，依台地跌落成为轴线的主要内容，绿化是构成规则图案的主体，修剪齐整。建筑和雕塑装饰有巴洛克风格，是文艺复兴后期的特色。

* 图10

另一台地园兰特别墅园，°位于罗马西北巴格内亚村，°1560—1580年红衣主教°修建花园，依山坡建4层台地，从低往上为壮观的方形花坛和方形水池、扁长方形的的梧桐树群、圆形喷泉池、纵长方形的瀑布和平台，4层台地有中轴线，贯穿起有节奏、有空间、有变化的几何台地。这是一个文艺复兴早期的作品，其规整、几何和秩序，与后来延续的台地园图形风格一样，表达了文艺复兴时期台地园对完美秩序的追求。

○ Villa Lante
○ Bagnaia
○ Cardinal

二 (2) 中国山水园：层叠、自然与想象

和意大利文艺复兴时期台地园相反，中国明清时期的山水园，最突出的成就是宛若自然。无论是皇家园林还是私家园林，其特征是层叠的、自然的，予人有想象空间的。其中，最著者当属江南园林。江南的湿润气候、山水环境、人文气质成就了层次丰富的山水园，同时由于清代康熙和乾隆二帝的南巡，推动了这种自下而上的江南园林对中国北方皇家园林的深远影响。

如承德避暑山庄，位于河北承德市北部、武烈河西岸，西北为狮子岭、狮子沟和广仁岭西沟，占地5.6平方公里，造园部分呈南北向长型，南面设门，进门为前朝后寝格局，为皇帝朝政和住区，实为夏宫的宫殿部分，前宫后苑，宫后湖光山色，风光旖旎。首先进入眼帘的是湖面，此为山上流下的热河泉水汇成，是园林的第一层次；水上桥亭起伏、水中岛屿层出，形成园林的主要内容，*并且主要景致为江南胜景移植；再北部为成片的万树园，背倚狮子岭，形成自然和人工的自然过渡；再北为自然山脉；再远处可借景园外棒槌山的棒槌石。这种层叠的园景、内外交融的自然景色、山水呼应的风光，确实是宛若自然的，并且和京城的严谨宫殿、等级分明的城市有着霄壤之别，于此，还可以享受江南景色，想象或萧瑟迷离或风和日雨的自然柔美。

* 图11

architecture and landscape are Baroque styled, a major signature of late Renaissance architecture.

Another good example is Villa Lante, located at Bagnaia northwest of Rome, built by the Cardinal Gianfrancesco Gambara during 1560–1580. The villa was built on the four-level terraces against a hill. The gardens of the Villa Lante are designed with spectacular square flowerbeds, square pools, plane trees in rectangular form, circular fountains, waterfalls in rectangular shapes, and vertical platforms from a lower terrace to a higher one. The four-level terraces run through a central axis linking the terraces in a rhythmic and fluid manner. As a masterpiece of the early Renaissance, the villa's neat, geometric, and orderly layout mirrors the pursuit of a perfect order, like other Renaissance terrace gardens do.

II.2.

CHINESE GARDEN:
A CASCADING VIEW OF NATURE & IMAGINATION

Contrary to Italian Renaissance gardens, Ming and Qing *shanshui* gardens are known for their refined natural beauty. Cascading and natural features are important characteristics of both royal and private Chinese landscape gardens, which leave room for imagination. The gardens of southern China are most well-known. The humid climate, landscape environment, and culture nurtured a rich *shanshui* garden culture. Qing Emperors Kangxi and Qianlong's tours of southern China had a tremendous influence on the design of royal landscape gardens in the north, which incorporated the imageries of southern *shanshui* gardens.

A good example is the Chengde Mountain Resort. The Mountain Resort, in the north of Chengde City, Hebei Province, is bordered by the Wulie River in the west, and the Lion Ridge, Lion Grove, and Guangren Range in the northwest, covering an area of 5.6 square kilometres. The Resort is set in a long form running from north to south, with a large gate in the south, which opens to an administrative hall where the emperor attends to state affairs. Behind the offices is a large sleeping quarter. The administrative hall and sleeping quarter constitute one palace. Behind the palace stand beautiful mountains and waters. When visiting, one would be greeted first by the lake that gathers hot spring water from the mountain above. In the lake, there are a number of bridges connecting dotted small islands. The attractive scenery is borrowed from typical southern Chinese *shanshui* gardens.* In the north, patches of forest stand against the Lion Ridge, forming a natural transition between artificial and natural views. Further north rests a natural mountain, from which one may see the looming Club Stone in the distance. The garden is layered and blended with man-made and natural scenes corresponding to the natural beauty of the mountains and rivers within the vicinity, which makes the resort a seamless part of the natural

* FIGURE 11

扬州瘦西湖是从水上观览景致的江南园林，狭长的水面曲折位于扬州城西北，和唐宋时期的城市护城河相传承，1765年前后，扬州盐商为取悦南巡的乾隆皇帝，密集在河两岸造景植树、叠石建屋，建成二十四景，如西园曲水、长堤春柳、四桥烟雨、梅岭春深、白塔晴云、蜀岗晓照、万松叠翠等，其实这些景致从南而北、自水而山，便是形成层叠景色的组成部分，同时，水渐宽、离城市愈远，景致也更近自然，也更需要仰视而观，如蜀岗晓照、万松叠翠，便是最北端的景色。另外，沿湖岸而观，也有石景树木、建筑造景、远山遐迩的不同景深和层次。* 自然本身是变化丰富、层出不穷的，园林中或人工或自然的景色变幻，因为有了空间和气息，就令人产生遐想了。这种真假自然的衔接和转换、感受和体悟，便是明清山水园的重要图像意向了。 * 图12

苏州环秀山庄的叠石假山，位于园林北侧，假山面积不大，但占园中主要地位，面积也达全园三分之一。从其南侧的平台望去，水萦如带，青石如画，* 是戈裕良继承清代山水画家石涛的笔意叠石而成。 * 图13

假山是层叠的：
主峰突起于前；次山相衬其后；后有远山之姿。

假山是自然的：
主山以东北方的平冈短阜作起伏，似山之连绵不尽；而在西南角，山成崖峦，动势雄奇，向外突出，直壁临水；临水山石以大块竖石为骨，秀峰平地拔起，峰态倾劈，山根和池水相接，水面蜿蜒，犹如江水直出磅礴。

假山是想象的：
整个假山立面如山水画面，大胆落墨，小心收拾，卷云自如，峰自皴生。行走其间，内部洞穴空灵，外部涧谷虚空，有无限山林感受。

而登上山顶，山上覆土植树，极具真自然意趣，尤其是主山山尾由西向东在东侧截于墙外，给人以深山老岭、云探远处的感受。

wilderness. It makes a sharp contrast to the rigid style of palaces and hierarchical divisions of the city. In this garden, one enjoys a landscape that could only be found in the southern part of China, though physically in the north, indulged in the natural tenderness of breezes, sunshine and rains.

Slender West Lake in Yangzhou, in southern China, is a *shanshui* garden which should be experienced from the water surface. Connected to the city moats built during the Tang and Song dynasties, the long and narrow waters twist and turn throughout the northwest part of the city. To please Emperor Qianlong on his southern tour in 1765, the salt merchants in the city planted trees and built stone houses at 24 scenes along the lake banks, including twisting waters, willow banks, misty drawbridges, deep spring peaks, a white pagoda, sunny hills, and pine forests among others. The scenes automatically became layered landscapes running from south to north, and from waters to mountains. Meanwhile, the wider the waters became, the further away the scenery travelled from town, though closer to the wilderness. The landscape scenes created further away from the town, such as the sunny hills and the pine forests that sit on the northern tip of the town, have to be observed from a raised angle. Additionally, from the lake banks one can see scenes of rocks, trees, buildings, and distant mountains at different levels and depths.* Nature is changing, and the change is endless. The scenes in a garden, either artificial or natural, would change with one's imagination, credit to the space and aura of the garden. The combination of natural elements and craftsmanship allows people to experience the natural wilderness in a man-made *shanshui* garden, which is the intention of Ming and Qing *shanshui* gardens.

* FIGURE 12

The rockery at the Huanxiu Villa in Suzhou stands on the northern side of the garden. The rocks, though not very large in size, take up an impressive part, one third of the garden. Viewing from a stand in the south, the water runs like a belt, and the rocks look a painting.* Ge Yueliang arranged the rock piles based on a Shi Tao painting in the Qing Dynasty. The rocks were arranged in layers, allowing the main rocky peak to stand out in the front, followed by smaller rocks against a looming mountain background in the distance. The rocks were made to look natural and wild, with the flat northeast part of the rockery being cut into a waving curve, suggesting the endless rolling hills afar. On the southwestern corner, a magnificent cliff juts out from the rocky mountain, with a façade facing the waters. The large rocks bordering the waters stand vertically. The beautiful rocky mountain stands out from the ground, with multiple peaks. The mountain connects to the water at its foot. The zigzagging waterway flows like a natural river. The rocks in the garden were set to create room for imagination. The entire façade of the rockery looks like a landscape painting by a bold yet careful brush. Walking in-between the stones and water, one feels the ethereal air emitted from the grottos inside, which engulfs the emptiness of gullies outside,

* FIGURE 13

三、研究两种园林对地形的理解和设计
(1) 意大利台地园：台地、界面与整体的立面

如果说意大利台地园和中国山水园在阅读视角和图形呈现方面表现出不同的特征外，那么，从根本上说是对地形或自然的不同认识以及价值观念差异所致。也因此设计方法迥异。我相信麦克哈格在他的著作《设计结合自然》○中分析"东西方对人与自然关系的态度"的差异十分中肯。他认为：

○ Design with Nature by Ian Lennox McHarg

> "不管西方人对待自然的态度最早源于哪里，但是他们的态度得到犹太教的肯定，这是很清楚的。一神教出现的结果必然是对自然的排斥。"

> "基督教毫无改变地吸收犹太教有关创世的故事。它强调了人的绝对神圣性，上帝特许他征服地球和支配一切的权力。"

> "了解了这一点，就可以理解人类对自然的征服、压迫和掠夺，同样就可以理解这种不完善的价值体系。"●

● 8

不过他说的这个时期应该不包括古希腊，当时的泛神论其实隐含着对自然的尊重，而文艺复兴时期从整体上说是强调人本主义的时代。有意味的是，意大利台地园既传承了古希腊强调的和环境协调，同时又张扬了人工创造的伟力，也因此，它的影响是深远的。

具体到意大利文艺复兴时期台地园的设计手法和思维，就是利用台地、强调界面、形成立面。整体的场景恢宏和壮观是结果，重视地形和人工结合是手段。兰特园这个意大利16世纪60年代建造的台地园，是最重要和宏伟的意大利花园之一，由建筑师维尼奥拉○和工程师托马索济努齐○为红衣主教卡纳迪尔·冈伯拉设计。○园林分为四层台地，整体中轴对称，贯穿全轴的是水景，与其他别墅园不同，建筑不在构图中心，而是对称于两侧。园林与原红衣主教里阿里奥○于1514年修建的围猎公园旁接。*

○ Giacomo Barozzi da Vignola
○ Tommaso Ghinueei
○ Cardinal Gianfrancesco Gambara
○ Cardinal Raffaele Riario
* 图 14

当你拾级而下，每个台地的主题尽显眼前。最高层台地的尽端为德鲁格洞窟，是花园主要水源，该台地的主题是海

while wondering at the infiniteness of mountains and forests. The rocky mountain is covered with earth on top for planting trees. It is interesting to point out that the designer aligned the west-to-east mountain range to a completion outside of the garden wall in the east, creating a delusion of deep mountains with clouds drifting afar.

III.
INTERPRETATION OF THE TERRAIN & THE DESIGN OF THE TWO TYPES OF GARDEN

III.1.
ITALIAN TERRACE GARDEN:
A FAÇADE OF TERRAIN, INTERFACE & INTEGRITY

If we agree that Italian terrace gardens and Chinese *shanshui* gardens differ in perspective and in their presentation, the difference is the result of different understandings of terrain, nature and value. This would, of course, lead to completely different methods of design.

Ian Lennox McHarg made a pertinent observation of the oriental and western attitudes towards the relationship between man and nature in his book *Design with Nature*. He stated Judaism affirms Westerners' attitudes toward nature, no matter from where they originated. Monotheism would automatically result in the expulsion of nature. He added that Christianity absorbed the story of creation told by Judaism without reservation, emphasising man's absolute sanctity, and his power enabled by God to conquer the Earth and everything on it. Understanding this opens a door to understanding the conquest, oppression, and exploitation of nature by man. In the same context, it helps to understand the imperfect value system.* McHarg indicated, however, that this period does not include that of ancient Greece, as the pantheism that prevailed at the time implied a respect for nature. The Renaissance as a whole emphasises humanism. Italian terrace gardens retain the touch of their ancient Greek heritage, advocating harmony with the surrounding environment, while boasting the greatness of man-made creation. This allows Italian terrace gardens to cast a far-reaching influence on garden design.

• 8

Specific to the technique and philosophy involved in designing a terrace garden, the utilisation of terrace, emphasis on interface, and façade formation are key elements. A grand and spectacular landscape is the result, while the skilful combination of terrain and man-made creations makes the means to achieve this result. The Villa Lante, created in the 1660s, is one of the most important and grandest terrace gardens built in Italy. Designed by architect Giacomo Barozzi da Vignola and engineer Tommaso Ghinucci for Cardinal Gianfrancesco Gambara, the garden is made up of four terraces, symmetrically distributed along the central axis, with water features running through the entire garden. Unlike other gardens, the struc-

豚喷泉；第二级台地通过餐桌喷泉强化轴线，在上级和这级台地之间是河神喷泉，并由上层天鹅水渠引导水和联系；第三级台地呈横向窄地，与上级台地界面为卢瑟尼喷泉；最低层台地最大，被分为十八方格，中央方格是水池，四座桥通往中间的小岛，岛上有圆形的喷泉，中央有四个裸体男子举着红衣主教蒙达多°的冠冕。因为是台地，每层之间的界面很重要，界面之间的总和既成为整个园林的立面，界面之间的差异也拉开了园林的景深。兰特园最底层和第二级台地之间为大的斜坡，布局有菱形植物图案，两侧为台阶，再外侧为对称的别墅，形成第一个立面；其上的台地与再上一层台地间为四层跌落的水池喷泉形成立面主景，两侧为挡土墙和侧上台阶；最上一层台地和第二级台地间为河神喷泉，跌落的贝壳形的水池和羊角装饰加上斜向的楼梯栏杆装饰构成又一立面；在最上层的台地上，流淌着水的天鹅渠道和台阶形成有层次的界面，花园高处，则是洞窟形成最具焦点的立面，加上两侧的亭子和环绕的高大悬铃木，使得整个园林郁郁葱葱和充满生气。

○ Cardinal Peretti Montalto

这种有俯瞰的形式美，而下到最低层蓦然回首又有整体的立面美和纵深感的设计，是台地园重视台地、界面和立面而形成场景的突出之处，也是人工结合自然的一种成熟的设计方法。具体而言，就是大多利用缓坡做成台地，而利用高差做成水景，喷泉和雕塑是重要的立面及层叠的组成，绿化和丛林是形成界面及背景的重要要素，如此形成平面规则几何的、景深悠长的、立面有高度层次的景观。

tures in the garden were not built at the centre, but arranged symmetrically on the two flanks. The garden was connected to a hunting park built by Cardinal Raffaele Riario in 1514.*

* FIGURE 14

When walking down the steps, you notice all the different themes of each terrace. On the top terrace sits a grotto at one end, creating a major water source for the garden. The terrace was designed to show the dolphin themed fountain. The second level of terrace has a dinner table fountain highlighting the axis. A river god fountain runs between the terrace above and the terrace at this level, linked with a swan water channel above it. The third level of terrace becomes narrower horizontally, interfaced with a fountain above it. The bottom terrace, split into 18 squares, is the largest in size. In the central square sits a pond with four bridges leading to a small island in the middle. The island has a circular fountain, with four naked men holding the crown of Cardinal Peretti di Montalto in the centre. The interface connecting each level of terrace is very important, as it creates the façades of the garden, and the difference between the connections deepens the depth of the landscape. There is a large slope between the bottom and second levels of terrace at the Villa Lante. It is designed with a plant pattern in diamond shape with stairs aligning both flanks. Together with the structures symmetrically distributed at the external flanks, it creates the first façade. The terraces above are the source of a stream running through the four levels, creating a major façade scene supported by earth walls and side steps. Between the top and second levels of terrace is the river god fountain, with a dropping shell-shaped pond, crescent decorations, and oblique staircases, making up another façade. The top level of terrace is built with a tiered interface for a water flowing channel and steps. The grotto sitting on the high land of the garden makes a most impressive façade. With the pavilions at the flanks and towering plane trees in the circle, the garden stands in a lush green and vibrant landscape.

The beauty of this garden, enhanced by the panoramic views and by the intricate façade at each level, highlights the effect of terrain, interface, and façade on shaping a landscape. It mirrors a matured design philosophy combining man-made creations with nature. For example, slopes were turned into terraces, water features were built taking advantage of differences in elevation; making fountains and sculptures major façades, and the greenery and foliage part of the interfaces and background. This skilful use of the elements resulted in a geometric landscape with intricate scenes and layered façades.

The Villa di Castello near Florence has a similar design plan.* It is located northwest of Florence, and was owned by the Lorenzo de'Medici family. The villa, when first built in 1537, positioned the structures at a lower level of terrace in the south, with the gardens on the northern slope. The gardens were built on three-level terraces, borrowing the mountain contours in the background. On the first level of terrace stands an open flowerbed fountain with statues. The

* FIGURE 15

佛罗伦萨卡斯特洛别墅园°亦如此设计。*该园位于佛罗伦萨西北部，它是美狄奇家族的别墅园，初建于1537年。建筑在南部低处，园林在建筑北面坡地上。园林依山地形成3层台地园：

○ Villa Castello
* 图15

一层为开阔的花坛喷泉雕塑园，
格网状的花坛中心为两神角力喷泉雕塑，景深大；

  第二层为柑橘、柠檬、洞穴园，
  芳香的树木后是凉爽的洞室，
  雕塑下为水池，寓意水源；

    第三层是丛林，
    环抱大水池，是全园的真正水源，
    有水库作用，水中小岛上为象征亚平宁
    山°的老人雕像，胡子水滴下流。
    三层台地有中轴串联，
    形成规则的图案。

○ Apennines

在一层和二层的台地界面为槛墙，
中间有踏步拾级而上，
人穿过时秘密喷泉会喷出许多细柱水；

  在二、三层台地之间为壁墙，
  有雕塑和盆栽，人工而不乏趣味。

从别墅楼层观览，台地园和山地背景融为一体，又秩序井然；
从高处下瞰，建筑是终点，又延伸到人工的广场和城镇中。

girded flowerbeds have two statues of wrestling gods in the centre. The second level of terrace is planted with citrus and lemon trees. Behind aromatic trees sits a cool grotto. A water pond appears at the foot of the statues, implying the source of the water. On the third level of terrace, a large pool is surrounded by waterlogged undergrowth, making it a real water source and reservoir for the garden. In the middle of the water stands the statue of an old man with a water-dripping beard, symbolising the Apennine Mountains. This level of terrace is set in regular patterns connected through a central axis. The first and second levels of terrace are connected by an enclosed wall with steps in the middle. Walking up the steps, one would be met with fine water columns springing from hidden fountains. A cliff wall decorated with interesting statues and potted plants connects the second and third levels of terrace. Viewing the structures from within the villa, the terrace gardens and mountains in the background appear in an integrated and orderly form. Overlooking from above, the structures herald the end of the villa, though seemingly an extension to the man-made squares and town.

III.2.

CHINESE GARDEN:
A PLANE OF XING SHENG, DEPTH & INTEGRITY

Ancient Chinese had a long history of nature worshipping originating from their inherent love, respect, fear and desire for nature. They developed a unique understanding of the natural landscape in *Xing Sheng*, which creates a groundwork for building a *shanshui* garden as such. In the context of mountains, *Xing Sheng* is a vague word applied to indicate something imposing and massive, or 'a thousand feet means *Shi*, and a hundred feet implies *Xing*'. Specific to the combination of nature and man-made creations, people say 'watching a dragon to see the *Shi*, and probing its cave to see the details'. That means '*Shi* is empowered by God, while *Xing* represents the best part of the details'. As a result, 'one would not see the spirit of a dragon without *Shi*, nor would he see the details of its cave without *Xing*'.• It became a rule to observe *Xing Sheng*, when conceiving a scene or mimicking a natural landscape in a garden. One builds a *shanshui* garden as the terrain allows, making the garden a natural landscape with appropriate elements. Meanwhile, it is believed that the mountains and waters are inseparable from one another. The water flows around the mountain, and the mountain becomes a landscape because of the water. The two combine to create a nature full of vitality.

• 9

The Jichang Garden in Wuxi is a good example of a garden that takes advantage of natural mountains and rivers. The garden sits in the Huishan Mountain on the western suburb of Wuxi. It was originally named as Fenggu Xingwo, meaning 'a nest in the phoenix valley', before being renamed as the Jichang Garden. The wording

三（2）中国山水园：形胜、景深与整体的平面

中国古人对自然有着天然的喜好、敬畏和研究，自然崇拜历史悠久并生生不息。其中对自然山水形胜有独特的理解，是山水园建造的根本依据。这个形胜就山而言，有"千尺为势，百尺为形"的定量规定，这是指在大自然中。而具体到如何将自然和人工结合，则有"观龙以势，察穴以情"的看法，

> "势者，神之显也；形者，情之著也。"
> "非势无以见龙之神，非形无以察穴之情。"●   ● 9

所以在山水间的营造或造园中的模拟山水，古人注重脉络形胜已成为一种传统。因地构筑，值景而生，方可宛若天然。同时，山水是一体的，水绕山转，山因水活，这才是有生命力的自然。

  无锡寄畅园是在山水间进行造园的优秀典范。寄畅园位于江苏无锡西郊惠山山麓，初名"凤谷行窝"，后改称寄畅园。"寄畅"者，取王内史《寄畅山水阴》诗句：

> "取欢仁智乐，寄畅山水阴，
> 　清泠涧下濑，历落松竹林。"●   ● 10

以山水比君子之仁德，由寄托而畅心，解脱中岁罢归之心结。这样的追求必注重山水形势，《鸿雪因缘图记》所载寄畅园与今日园址规模相似，面积14.8亩。园门临惠山街，入园后是一片竹林，名为"清响"，出"清响"，即可看到迎面耸立的惠山和位于园中心数亩之广的水陂"锦江漪"，一园山水悉呈眼前。有了竹林、锦江漪，就有了层次，惠山的绵延遂成大势。后来改筑，更臻完美。尤尊重"寄畅"之意，皆顾锡、惠之背景来处理园中每一局部，并且行叠石引水，独具匠心。涧、峰、谷、泉、林俱全，且巧于因借：锦江漪倒映东南方向锡山龙光寺塔于漾波；八音涧背倚惠山引山涧如天成，从而寄畅园成为内外宛若自然的园林。在这里，惠山有势，锡山有形，通过整体的平面组织，产生景深和距离，便有了大的寄畅山水的意境。

  值得强调的是要感受形势和山水，景深是必要的，需要在整体布局上进行流线和视线设计。首先是景深，基本可以概括为近景要遮掩，如苏州拙政园进门乃黄石叠山，* 绕过山   * 图16

of Jichang was taken from a poem written by an appointed historian official, 'enjoying a wise man's fun, Jichang° in the mountains and rivers, and watching cold streams flowing downhill and hitting the pine forests'.*

° relaxing

• 10

In the poem, mountains and rivers were borrowed to imply a place where a gentleman's virtue shall rest, in an attempt to free the poet from the gloomy sadness of being fired and asked to go home. Such mentality naturally leads to the pursuit of mountains and rivers as a comfort. The Jichang Garden recorded in ancient literature is similar in size as what we see today: it occupied an area of 14.8 mu. The gate of the garden opened to a street named Huishan. Entering the garden, one can see a patch of bamboo forest named 'crispy clear', before catching the view of the towering Huishan Mountain and a vast pond in the middle. The presence of Huishan Mountain defines the tone of the garden, while the bamboo forest and large pond create the layering effect. Reconstruction and renovation in the following dynasties added more picturesque elements to the garden, though under the same philosophy of resting one's virtue within the mountains and rivers. Every detail of the garden was placed in relation to the pond and the mountain. Craftsmen piled rocks and diverted water into the garden in a unique manner, equipping the garden with gullies, peaks, valleys, springs, and forest. The Jichang garden is also a good sample of landscape borrowing. It borrowed the pond to reflect the image of the Longguang Temple in the southeast. In addition, an eight-sound grotto was built along the mountain, making it resemble a natural cave. Consequently, the garden was designed in a seemingly natural environment, both inside and outside. The garden borrowed the *Shi*° of the Huishan Mountain, and the *Xing* of the Xishan Mountain. A well-arranged layout made the garden both deep and distant, creating an ideal landscape for resting one's virtue within the mountains and rivers.

° massiveness

Depth is another important element used to generate the *Xing* and *Shi* in a *shanshui* garden by defining the right flow and view of the layout. As far as the depth is concerned, a close-range scene shall be obscured. For example, the piled up yellow rock when entering the gate of the Humble Administrator's Garden in Suzhou* would block one's view, while a detour would expose the visitor unexpectedly to the Distant Fragrance Hall. Additional layers create medium-range scenes, such as in the Humble Administrator's Garden, where a small bridge was built to serve as a decorative or framing scene.* As for the distant scene, more skill is needed, such as the Bamboo House in the Humble Administrator's Garden, where the scene of the Northern Pagoda is borrowed as a background.* This technique of arranging a view operates in a similar manner to the principle of 'three farness' in traditional Chinese painting. A mountain view should have three distances: a high distances when viewing the mountain from bottom to top, or a deep distances when viewing through the mountain, or a level distance when viewing a distant mountain from a nearby mountain.* Organising, or putting

* FIGURE 16

* FIGURE 17

* FIGURE 18

• 11

屏后乃远香堂豁然开朗；中景增加层次，如拙政园运用小飞虹桥廊隔景又透景；* 远景要巧借，如拙政园于悟竹幽居借景北寺塔等。* 这同于中国山水画中强调的"三远"：

* 图17
* 图18

"山有三远：
　自山下而仰山颠，谓之'高远'；
　自山前而窥山后，谓之'深远'；
　自近山而望远山，谓之'平远'。"•

• 11

其次是组织，童寯先生说过：

"园之妙处，在虚实互映，
　大小对比，高下相称。"•

• 12

这和谢赫的作画六法同理：

"气韵生动、骨法用笔、应物象形；
　随类赋彩、经营位置、传移模写。"•

• 13

最后，整体布局其实是为层层展开山水画卷进行设计和经营后的结果，平面的曲折变化，实乃为感受自然形胜尤其是山水意境而为之。

在这个过程中，形胜是核心也是出发点。如果将山水形胜和园林的人工建造进行联系理解的话，就是势全形顺、形全势就和巧形展势。形可能就是中观层面或曰尺度可以把握的造园，而目的是要成就大自然的势、气和境界。以北京圆明园为例，由于选址在北京西郊大自然的山水环境中，多数景点都不破坏形胜，造园在山麓或山前，并和自然山水有一定的景深以便欣赏，这就是势全形顺。如《九洲清晏》：•

• 14

"周围支汊，纵横旁达。
　诸胜仿佛浔阳九派。"

——是势；

"念彼沟壑，曷其饱诸？
　水榭山亭，天然画图。"

——是形。

地形和形胜
TERRAIN & XING SHENG

elements together, is also important. Tong Jun said, 'The beauty of a garden mirrors natural elements in a man-made setting, proportioned in both height and size'.* This statement echoes the six principles applied in painting: vivid spirit, bold brushing, essence catching, colouring the kind, position oriented, and copying when needed.* The final layout should be the result of careful design and planning, allowing the scenes to be unveiled one layer after another, and the plane to show the twists and turns. This approach allows people to experience the natural *Xing Sheng* in a *shanshui* garden.

Xing Sheng is the starting point as well as the guiding principle for designing a *shanshui* garden. The best marriage between natural landscape and landscape gardening depends on the meticulous coordination of the beauty of nature and a man-made aesthetic, such that *Xing* and *Sheng* complement each other boosting the overall aesthetic experience. A full *Shi* would generate a smooth *Xing*, a full *Xing* would make *Shi* a success, and a skilful *Xing* would boost the *Shi*. *Xing* could be interpreted as a scale you have to weigh in when creating a medium-range view, in a bid to mirror the *Shi*, air, and aura of the natural wilderness. Take the Yuanming Garden Palace in Beijing as an example. Mountains and streams in the west of Beijing embrace the Yuanming Garden palace, which is most desirable for retaining the *Xing Sheng* of many scenes. In the Yuanming Garden Palace, some gardens were built either at a foothill or in front of the mountain, allowing deep connection with the natural landscape, which agrees with: 'a full *Shi* would generate a smooth *Xing*'. An example is the Jiuzhou Qingyan Garden,* where 'the garden is surrounded by tributary waters, reaching in all directions, like a large river' — *Shi*. 'Gullies, streams, waterside pavilions, and rocks make the garden look like a natural painting' — *Xing*. In the case of Danbo Ningjing Garden* in the same palace, 'the garden has green mountains for tranquillity, and clean waters beautify the garden' — *Shi*. 'Structure mimics the form of the Chinese character *Tian*,° and secret chambers were skilfully hidden to shun away even the dust' — *Xing*. The simple composition 'leaves room to natural elements to come in, such as wind, water, and plants, allowing people to feel the tranquillity and listen to distant sounds. Ruru Hanjin Garden* is described as 'surrounded by parapets, and decorated with encircling corridors. The guest house has clear glass panels with numerous books'. It is apparent that the structure's wavy curves° refer to the nearby hills, and form a *Shi* with the towering and deep mountains where the loyal mausoleums rest in the distance. This is a *Shi*, or 'a full *Xing* would make *Shi* a success'. Jia Jing Ming Qin,* another garden in the Yuanming Garden Palace illustrates that 'a skilful *Xing* would boost the *Shi*'.* In the garden, Guang Yu Hall sits on top of a small hill, highlighting the massive-ness° of the towering mountains, and a bridge above the waters between the two mountains. This design leads one to looking up and down, and forges a scene as Chinese poet Li Bai put it, 'migrant birds have no

- 12
- 13
- 14
- 15
- 田
- 16
- Xing
- 17
- FIGURE 19
- Shi

又如"澹泊宁静": • 15

　　"青山本来宁静体，绿水如斯澹泊容。"

——是势；

　　"仿田字为房，密室周遮，尘氛不到。"

——是形。

由于建造简单，于是

　　"风水沦涟，蒹葭苍瑟，澹泊相遭，
　　　洵矣视之既静，其听始远。"
而

　　"茹古涵今"• • 16

　　"缭以曲垣，缀以周廊，
　　　邃馆明窗，牙签万轴。"

——是用建筑群体形成起伏变化的形，宛若近山，它和其后深远和高远的山，乃成

　　"今人不薄古人爱，我爱古人杜少陵。"

——长安汉杜陵之势，这就是形全势就。圆明园另一处"夹镜鸣琴"，•* 是巧形展势的做法——将广育宫建在小山头上提升山势之耸，将两山之间的水面上架设虹桥一道又上构杰阁，一上一下的建筑营造，俯仰之间， • 17
＊ 图19

　　"移情远，不在弦，付与成连。"

获得李白两水夹明镜的诗意。

intention to stay but depart afar; standing before a pond having no intention to fish; and the sky and water aligned in one'.

IV.
WHAT STEMS FROM THE COMPARISON?

IV.1.
THE GAP IN UNDERSTANDING NATURE

All landscape gardens are beautiful, and are associated with nature. However, Chinese and Western gardens differ in their aesthetic and methods of expressing nature. I believe the focus on terrain versus *Xing Sheng* when creating the two types of garden plays an important role.

Terrain unveils the geological and geomorphologic traits of a piece of land in an objective and anatomic manner. An Italian terrace garden was built to show the designer's philosophy through both terrain and landscape, from the terrace to the skilful handling of the terrain.

Xing Sheng is a philosophy applied in ancient China to mirror the external shape and state of the environment where a landscape is created. A Chinese landscape garden is created to express an objective landscape in subjective consciousness, from the mountains and rivers in one's imagination to the physical scenes in the garden, and further back to the landscape in hallucination.

The difference between Italian terrace gardens and Chinese landscape gardens lies in their unique understanding of nature, with the former from terrain to landscape, and the latter from *Xing Sheng* to framed scenes. In the development of Western history, Monotheism advocates changing and conquering nature or mythologising nature as gods, developing an ontological philosophy that emphasises materiality. The world doesn't exist within 'emptiness' or 'void'. The value of a Western aesthetic is built upon material reality. This approach explains the extensive use of fountains and trees in a Western garden. The sound of water is created by artificial mechanism, and the lushness of mountains constructed by extensive planting of trees. Terrace is an easy approach to handle sloped terrains. Italian terrace gardens achieved an artistic effect by creating magnificent scenes and beautiful landscapes, in pursuit of order and rules during the Renaissance period.

Chinese culture traditionally advocates the belief of 'integration between heaven and man'. Nature originates from heaven, and consciousness is an integral part of man, or 'born by heaven, fed by land, and became by man'. This mentality minimises the gap between the subjective and objective world, melting man into nature. Artistically speaking, the objective world is not the goal to represent, but rather a medium to imply one's consciousness and value at a deeper level. As a result, Chinese landscape gardens attach great importance to

四、理解两种园林比较的意义
（1）了解自然认识观的差异

园林无疑是美的，园林无疑和自然有关，但如何寻美，如何真实理解和表达自然，中西方园林有显著差异。其中，地形和形胜是一组关键切入词语。

地形，含有更多对土地相关的地质和地貌的客观认识及构造解剖的理解，意大利台地园从台地出发，乃借助地形客体处理，来表达主体意识——从地形到景观。

形胜，主要反映山水环境的外部形态，中国山水园以此为意向，通过主体认识，叠合客体进行表达——从胸有丘壑到山水景色再到意念自然。

从本质上讲，从地形到景观、从形胜到景色，反映出中西方关于自然认识观的差异。

在西方漫长的历史发展过程，由于一神论的影响，自然常作为可以改造和征服的对象，或者将自然和具体的神联系在一起，在哲学层面则形成重视物质可感的本体论。在他们的眼中，世界不存在"虚空"和"无"。同时也在审美上建立以表现客观的真实性为参照系作为衡量艺术的价值尺度。如此也可以理解喷泉和树木在园林中普遍运用的原由：用人工的机械装置生产出自然的流水声，树木则是山林葱郁的表征，而台地则是最接近用人工处理坡地的做法。意大利台地园不仅合理地传承了这样的传统，更结合文艺复兴时期对秩序和规则的追求，达到了创造完美的宏大场景和景观的艺术效果。

中国传统强调"天人合一"，自然之源在于天，意识却内在于人性，即所谓"天生之，地养之，人成之"。实际上就是泯除一切主体和客体的显著区别，达到个人与自然相融的状态。在艺术上，往往客体不是表现的目的，只是一个中介物，力图表现的是深层的人的意识和价值取向。中国山水园林对形胜的重视，乃将它作为自然赋予人类的恩惠，其动静起伏表达的是生命的状态，园林中创造自然的层次、云蒸霞蔚的气氛、四季的变幻，均为宛若自然而能够因寄所托。

Xing Sheng, reckoning it a gift endowed by nature to humans. Static or fluid scenes both imply the flow of life. In the garden, scenes were intricately dressed up in a naturally overlapping manner, creating a splendid aura and changing seasons, allowing people to rest their souls in a seemingly natural world.

IV.2.

DIFFERENCE IN METHODOLOGY

The philosophy, ideas, techniques, and aesthetics applied in landscape design are inseparable. The techniques employed in designing a Western terrace garden or a Chinese landscape garden cannot be compared because they are different in terms of design method. What we have to understand is the differentiation.

The design of an Italian terrace garden would start from terrain, or specifically from the terrace, before creating detailed scenes, the interface between different levels of terrace, and the entire landscape supported by the façades. A Chinese *shanshui* garden would start with *Xing Sheng* as the foundation, before to the placement of physical elements, such as rocks, streams, distance, position, texture, and elevation, and their proportion in the garden, in an attempt to show the charms of nature on a plane through the manipulation of distance and techniques. The philosophy, ideas, techniques, and aesthetics initially applied in the two types of garden design schools are different, though sharing the similar principle of the unified whole being greater than the sum of separate elements. However, we see completely different approaches in the creation of the unified whole in Italian terrace gardens and Chinese *shanshui* gardens.

Taking a step back, the methodological differences also apply to the understanding of Classical architectures and cities. The terrace garden in the Italian Renaissance focuses its design on the plane and interface, while the Chinese *shanshui* garden during the Ming and Qing Dynasties on the design of layers and depth. Such emphasis also found its expression in the design of other structures. For example, the tombs, palaces, and temples built during the Ming and Qing Dynasties were very particular in the design of layer and depth. In the same context, ancient Chinese cities paid great attention to the *Xing Sheng*. This is completely different from Western Classical architecture which focuses on the functionality of surface and façade proportion, city layout, and the depiction of terrain and the skyline of cities. The comparison between Western and Chinese gardens can be enlightening in the study of traditional designs of Western and Chinese architecture and cities.

四 (2) 探索设计方法论的区别

观念、思维、手法、审美是一体的，中西方园林在设计方法论上的不同，没有高下之分，但有区别。

意大利台地园从地形的台地平面出发，进而设计景观、台地之间的界面和形成立面的整体场景。中国的山水园以自然形胜为大意，再关注山水、远近、前后、软硬、上下等自然要素和人工建造的关系，转而在平面上实现体悟自然魅力的距离和途径。其切入点的差异，包含着由观念、思维、手法、审美构成的设计方法论上的不同。但有规律可循，对于园林来说，整体大于局部之和的格式塔原理，° 是比较通用的，但如何从整体到局部，以意大利台地园和中国山水园为代表，方法迥异。

° Gestalt

放大一点看，这种设计方法论也适用于我们认识不同文化类型的古典建筑和城市，意大利文艺复兴时期台地园重视的平面和界面的设计，中国明清山水园关注的层次和景深的设计，也反映在其他建筑设计中。譬如中国明清陵墓、宫殿或寺庙建筑群体，均对层次和景深有特别的要求和成就卓然，而中国古代城市的图像表达也十分注重形胜。这与西方古典建筑重视平面功能和立面比例，以及西方古典城市以平面图表达见长和在图侧示意地形及城市轮廓完全不同，也相应成趣。从这点出发，园林的比较可以对深入研究中西方建筑和城市的设计传统有启示之功。

IV.3.
TAPPING THE HISTORICAL VALUE OF
THE TWO TYPES OF GARDEN CULTURE

In the late 15th century, Italian gardens started to influence gardens built in France, Spain, Britain and Germany. This development led to designs that combine man-made elements with natural landscape, enhanced by clearly defined axes and terraces. Important examples are the Château de Fontainebleau and the Château d'Anet in France, Powis Castle in Powys, Mid Wales, the Alcázar of Seville in Spain, and Sanssouci in Potsdam, near Berlin.

Starting from the 17th century, the Chinese *shanshui* garden design found its way into Europe in the form of Chinoiserie, which became popular in the 18th century and lost its momentum in the 1920s and 1930s. However one still can find traces of Chinese-styled gardens in many Western countries, including France, Great Britain, Germany, and later in Hungary, Russia, and Sweden.

Because of this, we can find terrace design as well as the Chinese-styled tea pavilion in Sanssouci in Potsdam, Germany. Inverting this, there are also traces of Italian terrace garden in the use of Western-styled structures in the Changchun Garden on the northeast tip of the Yuanming Garden Palace built in the 18th century, as the palace had embraced all the proven garden styles in the world. These interesting stories are worth further study.

May, 2011
Nanjing

四（3）挖掘两种园林的史学价值

意大利文艺复兴时期的台地园在 15 世纪末以后影响到法国、西班牙、英国以及德国等国家的园林发展，其突出特征是将人工和自然进行整体设计，并有明显的轴线和台地。如法国枫丹白露园和安尼特园，° 英国波维斯城堡台地园，° 西班牙赛维鲁阿尔卡扎尔园，° 德国柏林波茨坦无忧宫苑等。°

○ Fontainbleau & Anet
○ Powis Castle
○ The Alcazarin
○ San Souci
○ Chinoiserie

而中国山水园随着 17 世纪强大和独特的中国风° 也影响到欧洲，始于 17 世纪、盛于 18 世纪，约 19 世纪 20、30 年代趋于平淡，影响的国家有法国、英国、德国，又由德国传入匈牙利、俄国、瑞典等国。

因此在德国波茨坦无忧宫，° 既可以看到台地的格局，又可以看到中国式茶亭及亭子上的中国人物。而中国 18 世纪建造的圆明园，虽然是"谁道江南风景佳，移天缩地在君怀"，• 但在东北端的长春园西洋建筑群中，甚至可以寻觅到意大利台地园的踪迹。这些都是值得深究和玩味的。

○ Sans Sous

• 18

辛卯年端午前夕于金陵
修改于是年立秋

FIGURES

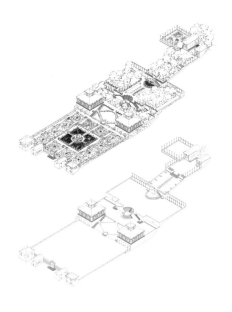

1

2

3

1   地形与园林示意，罗马兰特别墅园。
  Diagram of Terrain and Garden, the Villa Lante.
  Charles W. Moore, William J. Mitchell, William Turnbull, Jr.,
  the Poetics of Gardens, The MIT Press, 1993: 145.
2   形胜。
  董其昌（明）《烟云供养》。
  Xing Sheng.
  Dong Qichang (Ming Dynasty),
  Yan Yun Gong Yang.
3   希腊的埃比道拉斯剧场。
  The Little Theatre of Epidaurus, 350 BCE.

插图

4

5

6

7

4 罗马爱斯特别墅园台地景观。
  A terrace view of the Villa d'Este.
5 罗马爱斯特别墅园整体场景。
  An overall view of the Villa d'Este.
6 北京颐和园（前身为清漪园）万寿山前山。
  The front view of the Longevity Hill, Summer Palace, Beijing.
7 苏州艺圃大假山。
  Rock scenes at the Art Garden, Suzhou.

地形和形胜
TERRAIN & XING SHENG

8

9

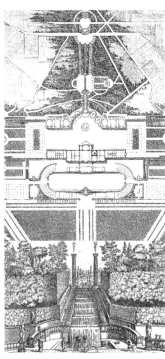

10

8		佛罗伦萨波波里花园整体布局。 The layout of the Boboli Gardens, Florenz.
9		佛罗伦萨波波里花园露天剧场及其树篱背景。 The open-air theatre and hedge background at the Boboli Gardens, Florenz.
10		罗马阿尔多布兰迪尼别墅园。 张祖刚《世界园林发展概论》，北京：中国建筑工业出版社，2003年，85页。 The Villa Aldobrandini, Rome. Zhang Zugang, *An Overview of Landscape Gardens in the World*, Beijing: China Architecture and Building Press, 2003: 85.
11		冷枚（清）《承德避暑山庄》。 Leng Mei (Qing Dynasty), *Chengde* Mountain Resort.
12		扬州瘦西湖景色。 Scenes at Slender West Lake in Yangzhou.
13		苏州环秀山庄假山。 Rock scenes at the Villa Huanxiu, Suzhou.
14		罗马兰特别墅园。 作者整理。图片来源：[意]佩内洛佩·霍布豪斯著，于晓楠译，意大利园林，北京：中国建筑工业出版社，2004：96；张祖刚，世界园林发展概论，北京：中国建筑工业出版社，2003：56。 The Villa Lante Garden, Rome. The Villa Lante Garden Composed by Author. Source from: Penelope Hobhouse, trans. Yu Xiaonan, *Gardens of Italy*. Beijing: China Architecture and Building Press, 2004:96. Zhang Zugang, An overview of landscape. Beijing: China Architecture and Building Press, 2003:56.

11

12

13

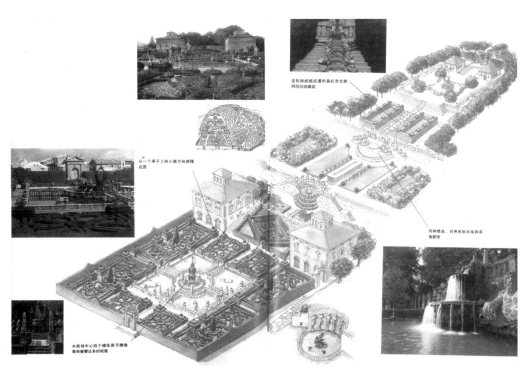

14

地形和形胜
TERRAIN & XING SHENG

15

16

17

18

15 佛罗伦萨卡斯特洛别墅园。
张祖刚《世界园林发展概论》，北京：中国建筑工业出版社，2003 年，53 页。
The Villa Castello, Florenz.
Zhang Zugang, *An Overview of Landscape Gardens in the World*, Beijing: China Architecture and Building Press, 2003: 53.

16 苏州拙政园近景黄石叠山。
A close view of the yellow rocks at the Humble Administrator's Garden, Suzhou.

17 苏州拙政园中景小飞虹桥廊。
A mid-range view of a small bridge at the Humble Administrator's Garden, Suzhou.

18 苏州拙政园远景北寺塔。
A distant view of the North Pagoda at the Humble Administrator's Garden, Suzhou.

19　北京圆明园。
　　"势全形顺"
　　I.　北京圆明园"九洲清晏"。
　　II.　北京圆明园"澹泊宁静"。
　　"形全势就"
　　III. 北京圆明园"茹古涵今"。
　　"巧形展势"
　　IV. 北京圆明园"夹镜鸣琴"。
《圆明沧桑》，北京：文化艺术出版社，1991年，22、56、38、86页。
Gardens in Yuanming Garden Palace, Beijing.
'A full *Shi* would produce a smooth *Xing*'.
I.　'*Jiu Zhou Qing Yan*' Garden.
II.　'*Dan Bo Ning Jing*' Garden.
'A full *Xing* would make *Shi* a success'
III.　'*Ru Gu Han Jin*' Garden.
'A skilful *Xing* would boost the *Shi*'
IV.　'*Jia Jing Ming Qin*' Garden.
The Vicissitudes of the Old Summer Palace, Beijing: the Culture and Arts Publishing House, 1991: 22, 56, 38, 86.

NOTES

1 Roland Martin, *Greek Architecture*, Zhang Sizan and Zhang Junying tr., Beijing: China Architectural Industrial Press, 1999: 130.
2 Sun Chuo (314–371, Eastern Jin Dynasty), or Xing Gong, born in Zhongdu, Taiyuan (now Pingyao in Shanxi). He lived a secluded life in his early years at Huiji, traveling across the mountains and rivers for more than ten years. In 353, he and other celebrities, including Wang Xizhi, Xie An, Sun Chuo, gathered at a mountain pavilion called Orchid Pavilion in Kuaiji for a ritual to eliminate disasters and seek for happiness. They wrote poems in front of the waters, venting their feelings. The gathering became a renowned event in the history of literature. The poems they wrote at the gathering were later compiled into a collection named the Orchid Pavilion Poems. Wang Xizhi wrote a prelude for the collection, and Sun Chuo a postscript.
3 Tao Yuanming (365–427, Eastern Jin Dynasty), or Yuan Liang, born in Xunyang Chaisang (now Jiujiang in Jiangxi). He was an official at several different posts. Tao relinquished his post as the head of Pengze when he was 41 years old, for not willing to see an inspector with his official belt on. He retreated to his countryside home, leading a secluded rural life on his own, until death. Tao was honoured as the father of 'hermit poet'. His poems are plain, natural, and deep in flavor.
4 Bu Yantu (Qing Dynasty), 'Essence of Painting: Q & A'. Qtd. in Li Xiyu, *On Chinese Painting*, Zhengzhou University Office of Academic Affairs, 1982: 152.
5 Li Xiyu, *On Chinese Painting*, Zhengzhou University Office of Academic Affairs, 1982: 156.
6 Pan Mo, Zhu Yizun, Zhang Shijun et al., 'Lion Grove Poems'. Qtd. in Zhang Chenhua, *Lion Grove Garden*, Suzhou: Guwuxuan Publishing House, 1998: 5.
7 Ji Cheng (Ming Dynasty), 'Section 6: Ponds and Rocks', *Notes on Garden Making (second edition), volume 39*, Beijing: China Architectural Industrial Press, 2009: 11.
8 Ian Lennox McHarg, *Design with Nature*, Tianjin University Press, 2006: 33.
9 Meng Hao, *Xingshi bian (About Shi)*.
10 Qin Yujun (Qing Dynasty), *Story of Jichang Garden*.
11 Guo Si (Northern Song Dynasty), *Lin Quan Gao Zhi*.
12 Tong Jun, *Gardens in South China*, Beijing: China Architectural Industrial Press, 1984: 7.
13 Xie He (Southern Qi Dyansty), *Catalog of Ancient Paintings*.
14 The 'Jiuzhou Qingyan' Garden is the largest structure group in the Yuanming Garden Palace. The lines cited in the paper are the one quoted from a prose and poem collection on the Yuanming Garden Palace. (Originally published in 1744; re-published in *The Vicissitudes of the Yuanming Garden Palace*, Beijing: the Culture and Arts Publishing House, 1991: 23.)
15 The 'Danbo Ningjing' Garden, quoted from a prose and poem collection on the Yuanming Garden Palace. (*ibid.*, 57.)
16 The 'Rugu Hanjing' Garden, quoted from a prose and poem collection on the Yuanming Garden Palace. (*ibid.* 14, 39.)
17 The 'Jiajing Mingqin' Garden, quoted from a prose and poem collection on the Yuanming Garden Palace. (*ibid.* 14, 87.)

注解

1 罗兰·马丁《希腊建筑》（张似赞、张军英译），北京：中国建筑工业出版社，1999 年，130 页。
2 孙绰（314—371），字兴公，东晋太原中都（今山西平遥）人。早年隐居会稽，游牧山水十余年。公元 353 年上巳节（永和九年三月三日），王羲之、谢安、孙绰等文士名流在会稽山阴的兰亭集会修禊（一种在水边祛灾求福的活动），临流赋诗，借景抒情，成为文学史上的一次有名的盛会。他们在兰亭所写的诗歌称为《兰亭集诗》。王羲之还写了《兰亭集序》，孙绰写了《兰亭集跋》。
3 陶渊明（365—427），字元亮，东晋浔阳柴桑（今江西九江）人。曾做过几任小官，41 岁任彭泽令时，因不愿束带接见督邮，毅然解职而归。从此，隐居田园，躬耕自资，终老一生。陶渊明被誉为"隐逸诗人之宗"，诗风平淡自然而韵味深长。
4 布颜图（清）《画学心法问答》。自李戏鱼《中国画论》，郑州大学教务处教材科，1982 年，152 页。
5 李戏鱼《中国画论》，郑州大学教务处教材科，1982 年，156 页。
6 潘耒、朱彝尊、张士俊等《师子林联句》。自张橙华《狮子林》，苏州：古吴轩出版社，1998 年，5 页。
7 计成原（明）《园冶注释（第二版）》园冶卷三九《掇山（六）池山》（杨伯超校订，陈植注释，陈从周校阅），北京：中国建筑工业出版社，2009 年，11 页。
8 伊恩·伦诺克斯·麦克哈格《设计结合自然》（黄经纬），天津大学出版社，2006 年，33 页。
9 孟浩《形势辨》。
10 秦毓钧《寄畅园考》。
11 郭思（北宋）编《林泉高致》，记载了其父郭熙之说。
12 童寯《江南园林志》，北京：中国建筑工业出版社，1984 年，7 页。
13 谢赫（南齐）《古画品录》。
14 "九洲青晏"是圆明园中最大的建筑群之一。"九洲青晏"引文出自乾隆九年（1744）"九洲青晏"诗序及诗文。《圆明沧桑》编辑委员会编《圆明沧桑》，北京：文化艺术出版社，1991 年，23 页。
15 "澹泊宁静"引文出自乾隆九年（1744）"澹泊宁静"诗序及诗文。同上，57 页。
16 "茹古涵今"引文出自乾隆九年（1744）"茹古涵今"诗序及诗文。同注 14，39 页。
17 "夹镜鸣琴"引文出自乾隆九年（1744）"夹镜鸣琴·调寄水仙子"诗序及诗文。同注 14，87 页。
18 王闿运（清）《圆明园宫词》。

MISTY RIVER, LAYERED PEAKS: A CONCEPTION OF CHINESE SHANSHUI[1]

从《烟江叠嶂图》看中国山水画

林海钟
Lin Haizhong

TRANSLATORS 翻译
Liu Ming 刘 明
Li Hua 李 华
Yu Changhui 于长会

I am not an architect. I am a landscape painter. I am honoured to talk to my architect friends about landscape painting. Let me start with one painting of my choice: *Misty River, Layered Peaks** painted by Wang Shen.•°

The reason I picked this painting is that it is one of the most important paintings in Chinese art history, although the size of it is small, only 26 centimetres in height and about one metre in width. Besides its artistic quality, the story behind the painting is extraordinary. It is a present to Wang Dingguo from Wang Shen. When Su Dongpo° saw this painting in Wang Dingguo's house, he was completely moved by its utmost beauty, and wrote down a famous poem about it, named *Inscribing Misty River, Layered Peaks in the collection of Wang Dingguo*.° From then onwards, a great deal of paintings and poems inspired by Su's poem were created from one generation to another. I will show some of them to you in the hope that they will help you understand more about Chinese landscape painting.

* FIGURE I
• 2
○ c. 1048—1122

○ 1037—1101

○ *Shu Wang Dingguo cang Yanjiang diezhang tu*

我不是建筑界的，我是一个从事山水画创作实践的。这次很荣幸受约给建筑界的朋友讲山水画。让我从一幅画讲起。

我选了上海博物馆馆藏的《烟江叠嶂图》* 作为讲演的内容。这张《烟江叠嶂图》并不大，高度 26 厘米，宽度 1 米多一点。但这张小小的画却是中国美术史上非常重要的作品，被称之为"画中之兰亭"，这个评价一点不为过。这件作品除了画得精彩之外，还有它发生的故事：《烟江叠嶂图》是王诜°画给王定国的，苏东坡°在王定国家里看到的这张画非常感动，觉得人间居然有那么美妙的作品，于是写了一首千古绝唱，也是苏东坡非常重要的作品，叫《书王定国藏王诜烟江叠嶂图》。历代因为这首诗而创作的书法作品和绘画作品非常之多。我把其中部分作品聚起来请大家看看，希望大家通过这些作品，对中国山水画有一些了解。

* 图 1

○ c. 1048—1122 年
○ 1037—1101 年

What is intriguing is that Wang Shen's seal is absent from the painting, but Su Dongpo's inscription is present. Moreover, a great deal of other paintings were entitled 'Misty River, Layered Peaks'. For instance, another painting with colour by Wang Shen should not be the *Misty River, Layered Peaks*, but its meaning is very close to what Su Dongpo's poem depicts. That painting was then named 'Misty River, Layered Peaks' at a later date. Another example is a hand scroll painting. It begins with Zhao Mengfu's° calligraphy of Su Dongpo's poem in the Yuan Dynasty; followed by Shen Zhou's° painting *Misty River, Layered Peaks* in the middle, and Wen Zhengming's° painting *Misty River, Layered Peaks* in the last section.* Shen Zhou is a famous painter of the Ming Dynasty and one of the artists of the Wu School. Wen Zhengming was Shen Zhou's student. That is to say, the whole painting was not completed by one person at one time. It is usual in China that painting or calligraphy is embedded with and inherits stories from one generation to another. The last one I want to show is Dong Qichang's° *Misty River, Layered Peaks* from the late Ming Dynasty.

○ 1254–1322
○ 1427–1509
○ 1470–1599
* FIGURE 2

○ 1555–1640

The emergence of landscape painting was closely connected with Chinese ancient philosophy. In his work *Lofty Ambitions in Forests and Streams*, Guo Xi,° a famous painter of the Northern Song Dynasty, makes a clear statement: 'Why does a nobleman love mountains and water? What is that for? *Qiuyuan* can cultivate and elevate one's personality and mentality.'° *Qiuyuan*, a garden with hills in its literal meaning, means that people take mountains and water as their home for reflection. Home in the Chinese sense involves a concept of sky, earth and man. As man positions himself between sky and earth, he lives between mountains and water, which is *Qiuyuan*. As shanshui• could help people cultivate themselves, so it is called *Qiuyuan yangsu*. In ancient China, people believed that they could achieve a comprehension of life through *xiuwei*, which is advocated by Confucius as 'the cultivation of character' both mentally and physically; aligning with nature in Taoism, and emptiness in Zen teachings. *Yangsu* means calming down to peace and returning to nature. Mountains and water could help people enter into this kind of state. Stones in streams flowing from springs, the simple happiness of fishermen, and a closeness to birds and other animals, would enable them to stay far away from the mundanities of life, and pursue a godlike spirit. Where are the gods? In the woods and mountains. Mountains and water would ease people's minds from the anxiety and sorrow of everyday life, from ordinary people to emperors. *Wangchuan Tu* by Guo Zhongsu° is a manifestation of this philosophy. The scene it depicts looks like an emperor's garden, such as a summer palace, rather than Wang Wei's° retreat for seclusion.* That is to say, emperors also need easiness and rest. So it is inevitable that Chinese people pay great attention to mountains and water.

○ after 1000–c.1090

○ *Qiuyuan yangsu*

• 1

○ ?–977

○ 699–759
• 3

林海钟
LIN HAIZHONG

王诜的这件作品没有印章，后面有苏东坡的书题。另外，与这张画相同的被定为《烟江叠嶂图》的也有多卷。比如，其中一张也是王诜画的，是设色的，应该不是《烟江叠嶂图》，但描绘的意境与苏东坡的诗意非常接近，后人也把它称之为《烟江叠嶂图》。°这是一个手卷，前面是元代赵孟頫的书法，°书的是苏东坡的这首诗，中间是明代吴门四家沈周°创作的《烟江叠嶂图》，后段是沈周的学生文徵明°画的《烟江叠嶂图》，* 流传有绪。中国有历史传承的书画都是很有故事的。而接下来看到的这张是明代末期董其昌°画的《烟江叠嶂图》。*

○ 设色本

○ 1254—1322 年
○ 1427—1509 年
○ 1470—1599 年
* 图 2.1

○ 1555—1640
* 图 2.2

　　山水画的形成，与中国人的哲学观是密切相联的。北宋著名画家郭熙°在《林泉高致》里，比较清楚地讲了为什么中国人要亲近山林，以下录用这段文字加以说明："君子之所以爱夫山水者，其旨安在？丘园养素"。"丘园"，是以山水作为自己的家园来观照的意思。中国人的家有天地人的概念，人在山水之间。山水就是丘园。人需要陶冶性情，在山水中就能得到陶冶，叫丘园养素。人生的体悟，在古代是通过修为来实现的：儒家称之为修身养性，在道家那里就是自然，禅家是在空性。养素的观念就是让人沉静下来，返璞归真。山水恰能使人进入这样的状态。山水里的泉石，渔樵的隐逸，鸟兽虫鱼，并能与之亲近，远离尘嚣世俗，追求神仙的境地。神仙在哪里？神仙就在山林之中。世俗的烦躁，心灵的慰籍，就会从山水中得到清洗、涤刷。从平民百姓到帝王都是这样。郭忠恕°的《辋川图》，应该是误传的画，图中画的并不是王维°隐居的地方，看起来像皇家的园林，应该是避暑宫图一类的作品，可见皇帝也需要逍遥，也要休息。中国人最终关注山水，是必然的趋势。

○ 1000 后
—c.1090 年

○ ?—997 年
○ 699—759 年

Confucius, Buddhism and Taoism are the essentials of Chinese culture and constitute the philosophical roots of landscape painting. As in Chinese culture, Chinese landscape painting has three directions: the emptiness of Zen in Buddhism; the pavilions of the mountains of heaven in Taoism; and the utopian world in Confucianism. Taoists usually build their houses at the top of mountains. From the bottom to the top, the buildings are layered up as in heaven. In the world of Confucius, everything is arranged in order.

Mount Lu Shan painted by the monk Yu Jian° in the Song Dynasty depicts the perfect emptiness of the Zen world Yu Jian lived around the North Peak behind Lingyin Temple in Hangzhou. This painting projects an image of Lu Shan through depicting the landscape of the North Peak. Yu Jian had probably never been to Lu Shan, but intended to draw an image of it through the shape of the North Peak. The inscription on the painting alludes to Monk Hui Yuan, Tao Yuanming and Lu Xiujing of the Wei and Jin period. The imagery of this painting is extraordinary in terms of presenting a Zen world of emptiness. The skill of it is simple, and the brush follows the painter's wishes, which perfectly reveals the aesthetics of Zen. Yu Jian himself believed in Zen. And the inscription of the painting also indicates a mix of the three traditional philosophies.

○ in the mid-13th century

My painting *Ode to the Autumn Sound* depicts a view of autumn.* The imagery of autumn is deep in its meanings: every beautiful thing is falling down, and will come to an end after all. The world is transient. When the wind blows, we feel the suddenness of death revealing. Autumn has a special beauty and a very deep richness.

* FIGURE 3

The *Pavilion of Buddhist Voice** is a garden I built upon the paper. It is a Buddhist temple. In contrast to Taoist temples, Buddhist temples are usually built in remote mountains. Nothing could be seen from the outside. But when you get into the mountains, you may suddenly find a courtyard there.

* FIGURE 4

Here is an example of my *Guoqing Temple of Tiantai*.* This temple is located at Tiantai in Zhejiang. It is built in the Sui Dynasty, and is kept in a very good condition. Guoqing Temple of Tiantai is located in deep forests. From the outside, you cannot see it at all. What is interesting about this landscape painting is that people can 'wander' in it. This is what Chinese landscape painting is about: viewers wander in the painting as painters move with ink.

* FIGURE 5

The other painting is about the *Imperial Palace in the Forbidden City*.* It is my graduate work of my Master studies. This painting treats architecture as a landscape. The high walls of the Palace are like mountains, and the roads are like rivers in a deep valley.

* FIGURE 6

林海钟
LIN HAIZHONG

儒释道是中国文化的根本,山水画也不离其根。

山水画的审美有如中国的文化,也有三大去向:虚空禅境,是佛家的境界;仙山楼阁,是道家的境象,道家往往把房子筑在高高的山崖上,由下往上,一层一层的楼台建筑,犹如天界一般,楼阁仙山的意味;儒家是世外桃源,在那里一切井然有序。

这幅是宋代僧人玉涧°所画的《庐山图》,好一派虚空禅境。玉涧住在杭州,在灵隐北高峰一带。《庐山图》画的就是北高峰的山水,画的意思却是庐山,玉涧可能没有去过庐山,也许是借北高峰的形画庐山的意,题跋上的诗写的是对魏晋时期慧远虎溪三笑的追思。这幅画特别的空灵,如虚空直接化现出的清凉世界,笔墨技巧简单,并没有很复杂的笔法呈现,而是很随意的墨戏。禅家思想的审美表现得一览无余。玉涧本身也是禅家之人。"过溪一笑意何疏,千载风流入画图。回首社贤无觅处,炉峰香冷水云孤。"从诗文来看,有浓厚的三教合一思想。

° 13世纪中叶

我的作品《秋声赋》,*画的是秋天的景象。秋意是深刻的,美好的事物终究会结束,可以说是无常,秋风飒起时,给人的感受是突然,以前美好的东西突然停止,一片肃杀,让人警醒。秋有一种特殊的美,意味极为丰富。

* 图3

《梵音殿阁图》*是我在纸上造的园,是一个佛家寺院。佛家寺院和道家不同,喜欢藏在深山里,可能在外面什么都看不见,走到里面忽然有个院子。

* 图4

我画的《天台国清寺》就是一个例子。*这个寺院在浙江天台,是一个隋代寺院,现在还保存得非常好。天台国清寺是在密林之中,在外面什么都看不见。这也是山水画有意思的地方,人可以在画里游走。这是中国山水画的高妙之处,观画的人在画里游,画者在笔墨中畅游。

* 图5

这是我研究生时期的毕业作品,画的是故宫。*这幅画是把建筑当作山水来画。高耸的墙就像是大山一样,这些路就像是河流、深壑里的溪流。

* 图6

'THREE TYPES OF DISTANCE'

Distance is a key point in understanding Chinese paintings and writings. With distance there comes space. Space in distance comforts people psychologically. To the ancient Chinese, a mountain is big, so it is far or distant.

People in the Northern Song Dynasty proposed that mountains impressed upon people 'three types of distance'. In *Lofty Ambitions in Forests and Streams*, Guo Xi writes, 'Looking up from the lower slope of a mountain is called high distance, looking straight ahead at background hills is called deep distance; looking into the distance from a hilltop is called level distance.' The view of high distance is bright, and its shape is lofty and imposing. This impression evokes the imagination of the peak of the mountain. *Travellers among Mountains and Streams* by Fan Kuan° is a typical articulation of this.* Deep distance is another important characteristic of mountains. This feeling is formed by springs flowing out from valleys and paths leading into mountains. So 'the view of deep distance is dim', and it 'gives an overlapping impression'. Such a mysterious sense leads us to yearn. The painting *Glen* painted during the Northern Song Dynasty* in the collection of the Shanghai Museum depicts the conception of deep distance very well; a secluded place often gives us more room for imagination, which can be described as 'deep distance'. The third one is 'level distance'. 'The view of level distance is at once bright and dim', and it 'gives a misty impression'. This is also explained in the poem by Du fu, 'Surmount extreme summit, Single-glance many-mountains little.'• Due to the shielding of clouds, the scene is bright, dim and misty at the same time. Many Chinese landscape paintings are enabled by high distance, deep distance and level distance. *Twin Pines, Level Distance* painted by Zhao Mengfu is a good example of this.*

Later on, 'the new three types of distance' is put forward in the *Chunquan's Compilation on Landscape*° written by Han Zhuo. In fact, I think Han refers to water instead of the mountain, which was the focus of the old 'three types of distance'. To Han, the 'new three types of distance' are broad distance, hidden distance and obscure distance. 'Broad distance can be explained from the sight of inshore, vast waters, and a distant mountain; hidden distance from the sight blocked out by smoke and fog, and obscure distance from the sight of far distance and blurred views'. The image depicted by 'a fishing boat like a leaf where the river swallows the sky'• in Shen Zhou's *Misty River, Layered Peaks* shows the artistic conception of broad distance. It is hard to separate water from sky. The painting *Water** by Ma Yuan° is a representation of hidden distance. It depicts a scene of the Dongting Lake, which is hidden by fog and gives a feeling of mistiness and distance. This is one of the 12 pieces of Ma Yuan's *Water* in the Beijing Museum.

° c.990—1020
* FIGURE 7

* FIGURE 8

• 4

* FIGURE 9

° in the late 11th century

• 5

* FIGURE 10
° c.1160—1225

林海钟
LIN HAIZHONG

"三远法"

"三远法"远是中国人图像和文字的着力处,有远才会有空间。远的空间给人以心理上的释然。古人云山乃大物故有远意。

宋人提出山有"三远"是对山的一种印象。《林泉高致》说道:"自山下而仰山颠,谓之高远","高远之色清明","高远之形突兀",高远之巅即山之巅给人一种想象。范宽°的《溪山行旅图》,是"高远"很典型的作品。*深远也是山很重要的特点。很幽深。所以说"深远之色重晦","深远之意重叠",这种神秘感使人有向往之意,上海博物馆的《幽谷图》,*有深远之意;幽深之处常常给人较多的想象空间,这就是深远。第三是平远,"从近山而望远山,谓之平远","平远之色有明有晦","平远之意冲融而飘飘缈缈"。登临望出去,一览众山小,云的遮挡是有明有晦同时又是飘飘缈缈的。很多中国山水画都是以高远、深远、平远来命名的。赵孟頫的《双松平远图》是其中例子。*

○ c.990—1020 年

* 图7

* 图8

* 图9

之后,宋代韩拙在《山水纯全集》°论山一篇中,又论"新三远"。事实上我感觉他在论水,老"三远"是论山而"新三远"是论水。"所谓有近岸广水旷阔遥山者,谓之阔远。有烟雾暝漠野水隔而髣远不见者,谓之迷远。景物至远而微茫缥渺者,谓之幽远。"所谓新三远也就是阔远、迷远和幽远。沈周的《烟江叠嶂图》画的阔远,"渔舟一叶江吞天"的意境。水天一色。马远°的《水图》,*是迷远法。画的是洞庭湖,有野雾相隔,是迷迷蒙蒙的远。马远的《水图》收藏在北京故宫博物院,共有十二幅。

○ 11世纪后期

○ c.1160—1225 年
* 图10

'MISTY RIVER, LAYERED PEAKS'

'A painting in a poem, a poem in a painting',° this is Su Dongpo's praise to Wang Wei.° Chinese painting is very much concerned with meaning. The meanings of poetry and painting are consistent. That is why *Misty River, Layered Peaks* is regarded as a masterpiece in the history of Chinese painting. It has Wang Shen's painting at first, followed by Su Dongpo's poem *Inscribing Misty River, Layered Peaks in the collection of Wang Dingguo*, and then Wang Shen's postscript. Poetry, calligraphy and painting are perfectly integrated.

○ *Shi zhong you hua, hua zhong you shi*

○ 701–761

Through Su Dongpo's poem, we can get the feeling of how the painting is constructed and the meaning is articulated.

Firstly, let us look at the scenery of high distance. The poem says,

> On the river, anxious mind,
>   a thousand layered mountains,
> Patches of green float in space like clouds and mist;
>
> Are they mountains?
>   Are they clouds?
>   Too far to tell;
>
> When mist parts and clouds scatter,
>   the mountains are just as always.•

• 6

These are the images of high distance explained by literature. Regarding the expression of overlapping mountains, different painters have different ideas. We can find these differences in the paintings of Shen Zhou, Wen Zhengming and Dong Qichang.

The next two sentences express the meaning of deep distance. Su wrote,

> We only see two rugged green cliffs shading a deep valley,
> In it hundreds of springs cascade in waterfalls,
>
> Winding through woods, wrapping around rocks,
>   lost and seen again,
> Falling to form swift streams at the valley's mouth.•

• 7

Shen Zhou, Wen Zhengming and Dong Qichang each display an individual understanding and expression of 'hundreds of springs cascade in waterfalls'. However, Wang Shen does not paint the scene in his painting. The scene is only depicted in Su Dongpo's poem as it can only be seen and heard very faintly. Thus I interpret that: what Wang Shen draws is a dark valley, deep and far away. There are no springs cascading in the waterfalls of the painting, but the sound of the springs can be heard. What Su Dongpo depicts is

《烟江叠嶂图卷》

"诗中有画,画中有诗。"这是苏东坡赞扬王维°的话。中国画讲究意境,诗的意境和画的意境是相合的。王诜《烟江叠嶂图》后有苏东坡唱和的诗《书王定国藏王诜烟江叠嶂图》,加上王诜的跋。诗、书、画被完美地结合在一起,成为中国绘画史上的杰作。

° 701—761年

通过苏东坡的诗来感受画面的营造和意境。

首先看高远之意。诗云:

"江上愁心千叠山,浮空积翠如云烟,
　山耶云耶远莫知,烟空云散山依然。"

即高远,这是文字的表现。对千叠山的表现:对于高远的理解,不同的画家又有这不同的理解。沈周、文征明、董其昌画的都不一样。

接下来两句,讲到深远之意。诗云:

"但见两崖苍苍暗绝谷,中有百道飞来泉。
　萦林络石隐复见,下赴谷口为奔川。"

沈周、文徵明、董其昌三人又各有理解和表达所谓的"百道飞来泉"。而实际上,王诜原画没有画百道泉,是东坡诗中的描写,是隐隐看到或听到,后面这三人便画出来了。所以此处我加了评注:王诜原图所画绝谷、幽壑深远,不画飞泉,但泉流之声在其间,东坡先生诗中所谓"中有百道飞来泉"句,正是观想中的所见。由此又想到王献之°的那句话:"人在山阴道上走,千岩竞秀,万壑争流。"很精彩。王献之用八个字已经表达清楚,而苏东坡却要花这么多诗句才说清楚。这样的比较太有意思了。

° 344—386年

《烟江叠嶂图》是一幅手卷,画卷是慢慢移动展开的,诗文也是一样:

"川平山开林麓断,小桥野店依山前,
　行人稍度乔木外,渔舟一叶江吞天。"

his imagination. This reminds me of Wang Xianzhi's° words, 'traveling along the Shanyin road, a rapid succession of peaks and rocks, *[and a rapid succession of springs and water]*, enfolding each other in all kinds of postures.'• It is so splendid. Wang Xianzhi expresses the imagery very clearly in eight words while Su Dongpo has to use several sentences. Such comparison is very interesting.

° 344–386

Wang Shen's *Misty River, Layered Peaks* is a hand scroll painting. As the painting is unfolded slowly, so is the poem,

> The river calms and mountains part,
>   the foothill forest ends,
>
>   A small bridge and country shops nestle before
>   the mountain;
>
> Few travellers cross beyond the tall trees;
>
> A fishing boat like a leaf
>   where the river swallows the sky.•

• 9

From this poem, we feel that we are wandering around the painting, moving and feeling the landscape step by step. The ideas of the Chinese regarding landscape painting are different from the Western view of landscape painting. Western paintings depict a perspective—we are outside the painting and see the scenery within it, while in Chinese paintings, we are engaged inside it. We are inside the paintings, and are integrated directly with the sky and earth. We are part of the mountains and water of the work, whether it is a poem or a painting. The scenes depicted by a Chinese painting and poem and its viewers are integrated as a whole.

The Peach Blossom Spring or the realm of Spirit Mountain is the pursuit of Chinese scholars in ancient times.• It is also reflected in the poem of Su Dongpo. When Su Dongpo sees Wang Shen's *Misty River, Layered Peaks*, he says with emotion:

• 10

> Tell me sir, what was your source?
>   Delicately detailed with brush tip,
>   painted with purity and beauty;
>
> I don't know where in the world where is such a scene,
>   Or I would go at once and buy two fields,
>
> You, Sir, have not seen Wuchang and Fankou,
> the remote and inaccessible places
>   Where Master Dongpo lived for five years.
>
> Spring wind shook the river, the sky was boundless,
> Evening clouds rolled up rain, the mountains were lovely,

我们感觉到在画中游走，慢慢移动看到所呈现的景致，步步在移。中国人对山水画的认识，与西洋风景画不同，西画是画透视的，在看风景，我们是融进去，我们就在画里面，直接与天地相融，是山水中的一份子，诗也好画也好，人完全融入。我们再看唱和中"渔舟一叶江吞天"这句。沈周和文征明都画了渔舟一叶江吞天。而董其昌没有画，因为他有一个毛病，他的画从来没有点景人物。如果画人物的董其昌就是假画，所以这里面他连一叶扁舟都不画。

桃花源或仙山境界是古代士大夫追求的理想。在苏东坡的诗上也有体现。苏东坡看到这幅画很有感慨："使君何从得此本，点缀毫末分清妍。"这是说笔墨的精妍的程度。而这样的山水，"不知人间何处有此境，径欲往买二顷田。"我想去画里买个二顷田，呆在里面。同时又感慨自己"君不见武昌樊口幽绝处，东坡先生留五年。"在武昌樊口如此美丽的地方，留了五年，很不错。他是怎么度过的呢？对四季的描述作为他的写照。"春风摇江天漠漠，暮云卷雨山娟娟。丹枫翻鸦伴水宿，长松落雪惊醉眠。"春夏秋冬的变化，感慨人世。"桃花流水在人世，武陵岂必皆神仙。"接下来两句特别棒，"江山清空我尘土，虽有去路寻无缘。"山水是如此清明的世界，而我是俗人、是尘土，感慨人生无奈。不过觉得自己终究是完美的，是神仙种，所以诗的结局很完美。"还君此画三叹息，山中故人应有招我归来篇。"我还是山中的故人，本是神仙，我一定还会回到如此美好的境界。苏东坡人生的经历和理想都包含在这首诗中。

> "使君何从得此本，点缀毫末分清妍，
> 　不知人间何处有此境，径欲往买二顷田，
> 　君不见武昌樊口幽绝处，东坡先生留五年，
> 　春风摇江天漠漠，暮云卷雨山娟娟，
> 　丹枫翻鸦伴水宿，长松落雪惊醉眠。
>
> 　桃花流水在人世，武陵岂必皆神仙，
> 　江山清空我尘土，虽有去路寻无缘，
> 　还君此画三叹息，山中故人应有招我归来篇。"

宋代时，画有"可行可望可游可居，画凡至此，皆入妙品"，"可行可望不如可居可游之为得"。为什么呢？"观今山川，地占数百里，可游可居之处十无三四，而必取可居可

> In scarlet maples crows fluttered,
> > companions to a waterside dwelling,
> From tall pines snow fell, startling my drunken sleep.
>
> Peach Blossom stream is in the world of men,
> > How could Wuling be only for immortals?
>
> River and mountains are pure and empty; I am in the dust,
> > Although roads lead there,
> > > it is not my fate to follow them;
>
> Returning your painting, I sigh three times,
> > Friends of the mountains should summon me
> > with poems of return.*

• 11

In terms of the ideas of the Song Dynasty, a good painting should have a quality of being viewable, traversable, wanderable, and inhabitable, and the activities of wandering and dwelling are better than viewing and traversing. Why? Because there are not so many places for wandering and dwelling in the world, and those places must satisfy us with viewing and traversing at first.

I made a hand scroll painting, entitled *Clear Distance of Xishan*.°* It depicts the scenery of a riverside in the early morning. Unfolding the painting, it will show an understanding of distance indicated by the scene of wild geese landing on the beach.° There are clean rivers and distant mountains in the painting, within which people walk, view, wander and dwell within the jungles and reed grass, and all that along the riverbank in early spring, with refined scholars singing in the pavilion by the river. Another painting of mine, *Broad Distance of Xiangjiang*,°* presents a scene where people look down mountains with water on top, which abides by the style of broad distance. Opening up the painting, you can see courtyards and pavilions built in the mountain. The refined scholars gather there. When looking outside, there are scenes of clouds, water, sky and earth for reflection.

I think I should stop my introduction of landscape painting here. Chinese landscape painting is extensive and profound. What I have said above is very brief and limited.

o *Xishan qingyuan*
* FIGURE 11

o *Pingsha luoyan*

o *Xiangjiang kuoyuan tu*
* FIGURE 12

游之品。"因为这是真正的佳处。可行可望没有问题，但真正可以居住的地方确实少之又少。这是宋人对山水画的精辟见解。

这是我画的一张手卷，题目是《溪山清远》，*是早晨清旷的江边景象。手卷展开，是平沙落雁的意境，画的溪山清远，人在画中，有密林、芦草、岸边诸境。可行可望，可游可居，又有文人雅士在岸边茅亭之中唱和，是早春的景。我的这幅《湘江阔远图》，*人站在高处看山水，是阔远法。画卷开卷，山中建有院落和亭台，文人雅士在其中雅集，望出去，是云水天地的观照。

＊ 图11

＊ 图12

对山水画的介绍我就讲到这里，中国山水画博大精深，这只是其中的管窥，谢谢大家。

1 *Shanshui*, literally means mountains and water or rivers, and in translation normally refers to scenery or landscape. But in the Chinese sense, *shanshui* does not only refer to physical existence, but also to a mental imagination and ideal. As it is hard to find a corresponding English word, we use Chinese Pinyin for this particular Chinese concept.
2 This painting is now in the Shanghai Museum.
3 Wang Wei, active during the Tang Dynasty, is one of the most famous poets and an important painter in Chinese history. Wangchuan is his estate at Lantian, southeast of Chang'an, the capital of the Tang Dynasty.
4 Translation see David Hawkes, 'Gaze Mountain', *A Little Primer of Tu Fu*, Oxford: Clarendon Press, 1967. Du (Tu) Fu (712–770) is one of the most famous poets in Chinese history.
5 Alfreda Murck, *Poetry and Painting in Song China: The Subtle Art of Dissent*, Harvard University Asia Centre, 2000: 131.
6 Su Dongpo is quoted in *ibid.*, 130.
7 *Ibid.*
8 Yang Liu, Edmund Capon and Stephen Little, *Fantastic mountains: Chinese landscape painting from the Shanghai Museum*, Sydney: Art Gallery of New South Welse, 2004: 224. *Shanyin* here means the paths in the mountains.
9 Su Dongpo is quoted in *ibid.* 5.
10 The Peach Blossom Spring is a utopian society depicted by Tao Yuanming (365–427) in his famous poem *Notes on the Peach Blossom Spring*.
11 Su Dongpo quoted in Alfreda Murck, 2000, p.131

FIGURES

1

2.1

2.2

1   王诜（北宋）《烟江叠嶂图》。
Wang Shen (Northern Song Dynasty),
Misty River, Layered Peaks.
2.1   明代沈周、文徵明《烟江叠嶂图》补图卷。
Misty River, Layered Peaks.
Shen Zhou & Wen Zhengming, Ming Dynasty.
2.2   董其昌（明代）《烟江叠嶂图》。
Dong Qichang (Ming Dynasty),
Misty River, Layered Peaks.

插图

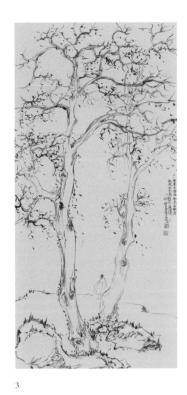

3

4

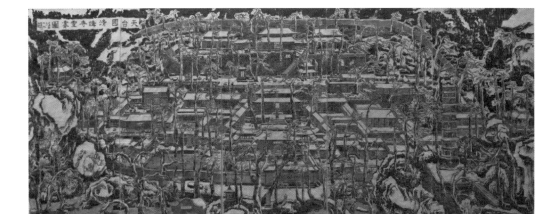

5

3 林海钟《秋声赋》。
　Lin Haizhong, *Ode to the Autumn Sound*.
4 林海钟《梵音殿阁图（局部）》，2007。
　Lin Haizhong, Pavilion of *Buddhist Voice* (detail), 2007.
5 林海钟《天台国清寺》，2002。
　Lin Haizhong, *Guoqing Temple of Tiantai*, 2002.

6

7

8

6　林海钟《故宫》。
　　Lin Haizhong, *The Imperial Palace in the Forbidden City.*
8　佚名（北宋）《幽谷图》。
　　Name Unknown (Northern Song Dynasty), *Glen.*
7　范宽（北宋）《溪山行旅图》。
　　Fan Kuan (Northern Song Dynasty),
　　Travellers Among Mountains and Streams.

9

10

9 赵孟頫（元）《双松平远图》。
　 Zhao Mengfu (Yuan Dynasty), *Twin Pines, Level Distance*.
10 马远（南宋）《水图》之一。
　 Ma Yuan (Southern Song Dynasty), *Water*.

11 林海钟《溪山清远》, 2005。
   Lin Haizhong, *Clear Distance of Xishan*, 2005.
12 林海钟《湘江阔远图》。
   Lin Haizhong, *Broad Distance of Xiangjiang*.

BUILD A WORLD TO RESEMBLE NATURE

建造一个与自然相似的世界

王澍
Wang Shu

建造一个与自然相似的世界
BUILD A WORLD TO RESEMBLE NATURE

TRANSLATOR
Jiinyi Hwang

翻译
黄锦宜

Today my lecture will start from the phrase: 'a picture-like river and mountain'. I once had the experience of traveling from Suzhou to Nanjing. I saw the road sign of Tianchi Mountain and continuous mountains, which were blasted out of a row of pits and holes. Tianchi Mountain is depicted in the famous painting *The Stone Cliff of Tianchi** by Huang Gongwang.° The riddled mountain made me think about this topic: 'a picture-like river and mountain'.

  I think everybody in China is familiar with the phrase 'a picture-like river and mountain'. We often say that 'the scenery of our country is like a picture'. Say it like a slip of the tongue and deprive this concept of any specific referents. You can say it in any way, so long as referring to some landscape pictures. Today what I want to say is that: 'a picture-like river and mountain' is not just a vocabulary of description. It actually conceals an existing truth and very important key power regarding the intention or image of Chinese landscape across the whole country. Meanwhile, it is an idea and reality that has a major influence on Chinese architecture and architectural practice. 'River and Mountain' is an interesting concept because we use 'River and Mountain' to indicate our country. We do not use 'Cities and Rivers' or 'Cities and Mountains'. Why don't we say that? When we say 'River and Mountain', you see no trace of a city, or in other words, the city has no significance in such an expression. So there appears the implication of such when we say the phase.

* FIGURE 1
○ *Huang Gongwang*: 1269—1354

我今天讲的题目从"江山如画"一词谈起。曾经坐车从苏州到南京，路上看到了天池山的路牌，看到了连绵不断的山，山上被开山炸石炸出了一排的坑和窟窿。因为天池山是黄公望°著名的《天池石壁图》*所描绘的对象，这座山被炸得千疮百孔，让我想起这个题目"江山如画"。

  "江山如画"这个词大家都很熟悉，我们经常会说：我们的祖国江山如画。说得太溜嘴了，导致这个概念好像成了没有具体所指物，怎么说都可以的一个词，配几张风景图片就可以说江山如画。我今天想说的是，"江山如画"这个词不只是一个形容和比较的词汇，而是在它的背后其实藏着一个真实存在的非常重要的关于中国整个国家的景观意向或景象的一个关键性的力量，它同时是对中国建筑学和建筑活动产生过巨大影响的一种思想和现实。"江山"这两个字很有趣，因为我们用"江山"指中国，我们没有用"城江"或者说"城山"。为什么不会这样说？我们说"江山"，这里你看不到城市的踪影，或者说在这个描绘里城市是没有地位的。所以当我们说"江山如画"这样一个词时，似乎有这样一个含义。

○ 1269—1354 年
* 图1

In fact, in the recent past there existed a system of landscape architecture which covered the whole nation. Just like the urbanisation that we are talking about nowadays, if we do not guess and only look into the reality, we will find our urbanisation started from the Ming Dynasty under another kind of concept. Every village is not simple scattered houses. It involves a complete landscape system, with a space configuration like a small city. Considering the scale of European cities, every village resembles a small city. And such urbanisation not only refers to villages, but the whole landscaped system enclosing towns, and road systems. This system has been depicted repeatedly in Chinese landscape painting, and it really exists. I remember when I visited Panan in the deepest centre of Zhejiang last summer and 90% of Panan County was covered by ridges and mountains. It was very difficult to access. Local people showed me the ancient post road which was built with very high quality stone and runs continuously along mountain cliffs. There were thousands of pine trees along the road in the past, but only twelve to thirteen trees are left now and are infected by pine caterpillar disease. Basically all those trees should have died by this year. We just caught the last scene of it. I remember my visit to Xinchang in Zhejiang Province two years ago to follow the footsteps of poet Li Bai° on his way to Tianmu Mountain. On the way, we saw very high quality ruins of bridges, guesthouses, pavilions, to villages which are still there. Therefore, we could probably realise that this system was only abandoned in the last thirty years. Now, the traces of the system can still be found, but in general it has been forgotten in our minds. Therefore, I feel frightened sometimes. For a nation with a culture of several thousand years history - not a simple culture, but a culture accumulated by activities of writing, planning and architecture, it is actually very easily forgotten within thirty to fifty years without any consciousness. On the other hand, to the architecture we're talking about, I always think that the system of traditional Chinese architecture is actually a system in which literati, or someone a bit like a philosopher or poet, collaborates with craftsmen. What we call today planners, landscape designers or architects do not exist in Chinese history. There are no such figures. China once had a system of architecture different from the system that we have got used to nowadays. Next we are going to see how this concept occurred and the discussions that followed this concept.

○ 701–762

我认为实际上在不太远的过去我们曾经见过在这个国家存在过一个覆盖了整个国家的景观建筑体系。就像我们今天说的城市化一样，如果我们只要看下实物，就会发现中国在另外一种观念下从明朝开始就全面开始了城市化。我们的每一个村落都不是简单散落的几个民宅，它是有严整的景观体系，是一个小城市一样的空间系统。如果按欧洲的规模，每一个这样的村落就是一个小城。而且这个城市化不仅仅是指村落，而是从大的城镇到小的村落之间的整个的一个联系系统，包括它的道路系统都全面被景观化了，它们在我们的山水绘画中被反复描绘，这个系统是真实存在的。我记得我去年夏天去浙江最中心的深处磐安寻访，磐安县百分之九十以上都是崇山峻岭，道路非常难走。在看村落的时候，当地人带我去看当年的驿道，非常高质量的石阶砌筑，在深山绝壁中连绵不绝，曾经在道的两边有几千棵以上的参天松树，而现在只剩下十二三棵，还染了松毛虫病，基本上今年应该全部死完了，我们看到了某种东西最后的尾巴。我记得前两年去浙江的新昌，在深山里重新去走李白○当年走过的去天姥山的那条道。一路上从桥、驿、亭、村落可以看到非常高质量的遗迹，还仍然存在在那里。所以我们会意识到，大概在刚过去的三十年时间里这个系统被放弃了。这个系统现在还能找到它的踪迹，但基本上在我们的头脑中被忘却了。所以有时候我觉得挺可怕的，对于一个国家，对于一个有几千年历史的文化，而且不是简单的文化，有大量的书写文字活动、规划活动、包括建筑活动累积起来的一个文化，要忘记它其实挺容易的，只需要三五十年就可以到完全不知道、完全没有意识的程度。反过来说，对于我们现在讨论的建筑学的范畴，我一直有一个观点，就是中国原来的建筑学系统实际上是一个文人或者说有点像哲学家或者诗人的人和工匠一起工作的一个系统。我们所说的规划师、景观设计师包括建筑师在中国历史上并不存在，没有这样的人。中国曾经存在一个和今天习常的建筑学不同的建筑活动系统，那么接下来我们就看一下，这个观念的发生和它中间一些观念性的讨论是怎么样的。

○ 701—762 年

There is a very important term 'imagined view'° in Chinese classical landscape painting. We, architects, often use words of ideas, thoughts, designs and methods which can be implemented as site plans, elevations, sections and perspective drawings etc. In fact, the most important term to the Chinese system is 'imagined view'. 'Imagined view' is not only about the things that you can see on-site, but also at the same time it includes the whole experience of entering and leaving your imagination. It is not a word about seeing directly and making analyses.

○ *guan xiang*

So first of all, let's look at the category it refers to.

The first case is *A Thousand Miles of Rivers and Mountains*,* painted by Wang Ximeng° in Northern Song Dynasty. If we look at the painting in a normal manner, we cannot see the whole, because it is 30—40 centimetres in height and about 11 metres in length. This is a scroll. When you view the painting, the process requires you to roll it out. When we roll out the whole painting, we will find out that the entire drawing was painted with a consciousness of integral control. It is not segmented into disconnected parts. It is continuous. Besides, the painting itself brings out another topic of 'a picture—like river and mountain' in Chinese landscape painting. Many painters depicted this theme and it has become not only a painting type, but also something like a philosophical discussion, or a research on the horizon from generation to generation. That is to say, such consciousness and a way of looking at things, including what I call 'The Law of Nature', the laws of precision and depth behind nature, could be passed from one generation to another by the method of drawing.

* FIGURE 2
○ 1096—1119

This is an early painting, falling into a primary category called landscape painting. Chinese landscape painting originated in the Wei and Jin Dynasties. It is a special type of Chinese painting. Why is it special? For example, we don't know when figure painting originated, but it was very early on. However, there was no landscape painting before the Wei and Jin Dynasties. Therefore I call it a theory-led genre of painting. It was preceded by theoretical discussions, and it exists in the atmosphere of a large number of discussions that such a kind of painting has generated. This kind of painting embodies an effort, not simply to depict the river and mountain, but an attempt to use the form of painting to capture the mutual understanding and harmonious relationship between humanity and nature, and an activity of how to make sense of it. It clearly formulated at that time a new ideology—i.e. interests in human society were lost, and an alternative idea, belief or interest in war, politics and the city was established by a strong cultural notion. It has maintained a crucial influence in China ever since, and this is a very important moment for Chinese painting. You can imagine how extraordinary it was for this painting to be created from scratch—which never happens in other cultures. The geographic scale that it depicts is

中国古典山水画论讲到了一个很重要的词，就是"观想"，我们建筑师经常会说想法、思考、设计、方法这些词汇，落实到建筑上说总平面、平面、立面、剖面、透视图等等。那么实际上对中国这套系统最重要的词之一就是"观想"，"观想"不止是在场地的现场直接可以看到的东西，同时它叠加着你的整个进去出来的经验和你的想象，它不是简单的直接看到和做分析的一个词。

首先我们来看一下"观想"这个词所涉及的广度。

第一个例子是北宋王希孟°的《千里江山图》。*如果以我们正常看的方式是看不全的，因为这张画的比例是高三四十厘米，长度大概十一米。这就是手卷，看的时候是要卷着展开。但是当我们以把它全部展开的方式看的时候，你会发现它的整个描绘是带有整体性控制的意识去做的，而不是说既然是手卷就切成很多段，段与段之间不相关联。它整个是连续性的。《千里江山图》还带出了另一个话题，就是中国山水画里江山图这样一个主题。因为这个主题被很多画家无数次地描绘，它已经变成了不只是一个简单的画的类型的创作了，它更像是一种哲学的讨论或者视野的琢磨，一代一代之间的传承——就是说这样的一个意识和一种看事情的方法，包括所能掌握到的我称之为"自然之道"的东西，即自然背后那个法则的精度和深度，是不是可以通过这样的描绘方法一代一代地传承下去？

° 1096—1119 年
* 图 2

　　这张是比较早期的，按大的类别叫山水画。中国的山水画起源于魏晋时期，它是我们中国绘画里面一个特殊的画种。怎么特殊呢？比如说人物画，我们不太知道其最早的源流是什么时候，但很早就有。而山水画在魏晋以前没有，我称之为是一个理论在先的绘画。它是先有理论的讨论，我们称之为雅集也好清谈也好，在这种大量讨论的氛围之中产生的绘画。那么这种绘画就有一个努力，不是简单地去描绘山水，而是试图通过绘画的方式，捕捉到人可以参照的人和自然之间意会和唱和的关系，以及到底怎么去把握的一种活动。它在那个时期比较明确地树立起了一种新的意识形态的思考——就是对人的社会失去了兴趣，就是面对战争和政治包括城市的另外一种理想、信仰或者说兴趣所在怎样以强有力的文化观念建立起来。它对后代的中国产生重大的影响。这是绘画很重要的时期。因为你可以想象，当没有人能够画

huge, from a territory of 10 kilometres to 20, or even 100. Today I will discuss a very specific issue of what kind of consciousness can help us to manage the making of large-scale city planning and architecture. This is still a difficult issue in today's architecture. As we can see, if you can draw, it means in some sense that you have understood, and understanding means there is a method which could lead to design and construction. So it is very important to understand this kind of painting.

We can look at this painting in four parallel sections. It is not a simple conceptual drawing. But what kind of drawing is it actually? In the painting, level distance, high distance, and deep distance° all exist, including at a later stage the three new types of distance. This kind of painting cannot rely on a single method, because the integrated depiction of the large scale needs to use all devices available to control it, including both the three new types of distance and three types of distance. When Wang Ximeng painted this drawing, he was only seventeen years old, very young. The Prime Minister, Cai Jing recognised his talent and kept him in his court. The colour of the painting is also very good, which used the royal pigments, the best paint at that time. This man also died very young, around twenty-three years old. We can see that from the beginning of the Wei and Jin Dynasties, this kind of painting has been profoundly developed in the Five Dynasties and early Song.

○ three types of distance

Another case is *Pure and Remote View of Streams and Mountains** by Xia Gui.° It is a particularly long painting. The characteristic of it is the scenery of the South. We can see that clouds and a blank section in the middle, which we call space, occupy a very large proportion and it is very different from previous works. In the previous paintings, there is a sense — very materialistic, very solid, which we call a sense of the North. Let us take a close look at *The Autumn Scenery of Rivers and Mountains** painted by Zhao Boju.°

* FIGURE 3
○ 1195–1224

* FIGURE 4
○ 1120–1182

One key point of Chinese painting is to make the painter himself live in the mountains. You can find he draws himself inside a pavilion. Meanwhile, he also stands outside. It is the same situation when viewing these paintings. When we look at one, we do not just stand outside looking at the painting. What we call 'marchable, viewable, livable, navigable', means that you really enter it. This whole method of painting strives to make you realise that you are really in the scene. So it is not a painting merely to be viewed from outside. I will come back to discuss this issue later on.

出这样的图的时候，山水画的出现是了不得的，在其他的文化中没有。它面对于这么大广度的东西，十公里二十公里甚至一百公里范围的描绘。以这种绘画的观想广度，今天我会说一个很具体的问题，即城市规划和建筑学中大范围地域尺度的建造到底以什么样的意识来控制，这对今天的建筑学仍然是一个难以掌握的问题。那么我们可以看到，既然可以画得出，在某种意义上就意味着理解了，理解了就意味着这是有办法的，有办法的就意味着可以设计和建造。所以对这种画的理解非常重要。

我们把这张画切成四段并列地看。它不是简单的概念性的描绘。那它到底是一种什么样的描绘呢？这里面平远、高远、深远°都存在，包括后面的新三远。这种绘画不是简单地用一种方法，因为这种大尺度的整体性的描绘需要动用新三远老三远可以说是全部的力量才能够控制。王希孟画这张画的时候好像只有十七岁，非常年轻，当时的丞相蔡京发现他太有才了就把他收在府里，像是一个门客。这幅画的颜色也非常好，用的都是宫里的颜料，当时最好的颜料。这人死的也早，二十岁左右就死了。可以看到，从魏晋发端，这类绘画在五代和北宋初期已经很成熟了。

○ 老三远

另外一张是南宋夏圭°的《溪山清远》，* 也是特别长的一幅画。这张的特点是南方的景象，大家可以看到中间的云雾和空白，我们称之为空间的部分占了特别大的比例，和前面的画法非常不一样。前面的那种非常物质感、非常实体感，我们称之为北方的那样一种感觉。我们可以仔细看一下赵伯驹°的《江山秋色图》。*

○ 1195—1224 年
* 图 3

中国画的一个要点就是画的那个人要能住在山里面，你可以发现他把他自己画在某个亭子里面，同时他也站在外面。我们看的时候也一样，绝对不会只是站在画外看那张画，我们说的可观可望可居可游指的是你真的进去了，它整个绘画的方法都是为了让你意识到你真的进去了。所以说它不是一个简单的从外面看的绘画。我后面还会讲到这个问题。

○ 1120—1182 年
* 图 4

We just talked about the issue of 'breadth', the horizontal scale. Now I will discuss the issue of 'height'. Let's look at another genre of painting, *The Creek Bank** by Dong Yuan° of the Five Dynasties. We called this kind of painting a 'vertical scroll'. In the Song Dynasty, the main hall of every residence hung such a kind of painting on the building axis. This kind of painting has difficulty depicting level distance, instead it normally expresses high distance and deep distance. If we take a closer look, it is interesting when we talk about the issue of primary propensity/composition, we cannot ignore the other side. The large formation and the small compositions, as well as the very subtle details, are created at the same time. Its thinking and method is different from what architects normally do in a rationalised procedure: drawing a site plan first, then a rough plan, some vague sketches, and gradually refining them. It contains a consciousness of control on the largest scale, and the greatest sensitivity to subtle details at the same time. These two things have to be done at the same time. Such capacity is very important for architects.

* FIGURE 5
° c. 934 – c. 962

There is a big problem. If we say that these kinds of things have a very significant impact on the history of ideas and notions, who are those people performing the thinking, in which society, and where? This is very important.

Let's now look at the lower part of Dong Yuan's *The Creek Bank*.* Standing in a pavilion, we see a typical Chinese literati wearing a tall hat. Although situated at home, his eyes look towards the outside, apparently he is absentminded. In the background is a happy family scene, but he is absent metaphysically. There is a horizon of thought, on which his eyes look out. Many people think this is a literati in reclusion, but in the Chinese system, reclusion is not easy, as Su Dongpo said: 'reclusive life is very tough, it needs a strong determination to leave the hustle and bustle of city life behind and reside in the mountains'. Reclusive life also has real and fake incarnations. We know that a real hermit would move his residence immediately when people became aware of his habitation. He does not want anyone to know his location. He is completely segregated from social affiliations. In fact, for Chinese people, this is a kind of ideological criticism of faith. When he does not like the world, he will leave for another world. Of course, it evolved into an opportunistic solution in garden making, i.e. build a garden next to the house, ready for an escape. It is not necessary to go out of the city and enter the mountains. Compared to this, mountain reclusion for the Chinese is more lofty and pure.

* FIGURE 6

So who does the thinking? In the painting *Hermits** by Wei Xian° in the Five Dynasties, a hermit is the one doing the thinking. There is a very steep mountain painted in the higher part of the painting. Under it there is a place to live. We call that inhabitable and wanderable. In fact, generally the space under a mountain cliff is uninhabitable. A place like this is very dangerous. Mudslides and floods can happen at any time.

* FIGURE 7
° date unknown

我们刚才讲到了广度，这种横向水平的尺度。现在看一下高度。五代董源°的《溪岸图》*是另一种绘画。我们称为立轴。在宋朝的时候每一家的中堂上都有一幅这样的画，摆在建筑的中轴线上。这种画很难表现平远之意，它一般是表达高远、深远。我们再仔细看一下，有意思的是当我们讲到大的形势这种问题的时候，不能够忽略的是，不是简单的讲大的形势，大的形势和小的形势包括非常精微的细节，它们是在同一个时刻一起出来的。所以就像我们建筑一样，不是先画一个总平面，画一个粗略的平面，再画些不清晰的草图，逐渐推敲把一个带有理性秩序的建筑做出来，不是这样一种思想和方法。它是在同一时刻既包含了最大尺度的控制的意识，同时包含了直接在手上的非常敏感的细微的细节。这两件事情是要同时进行的。这样一种能力对建筑师非常的重要。

○ c.934—c.962 年
* 图 5

这里有一个很大的问题，如果我们说这些在思想史或观念史上有非常重大的影响，那么这些事是谁在想？什么人在想，在这个社会里？在哪想？这个问题很重要。

让我们看董源《溪岸图》的下半部分。*一个亭子里，戴着高帽子的就是典型的中国的文士。尽管在家中，但他的眼睛看着画外，显然他心不在焉，后面是家庭的天伦之乐，而他有形而上的游离，有一个思想的视野，他的眼睛这样看出去。很多人认为这是一个文人在隐居，其实在中国的系统里，隐居并不容易，如苏东坡所说的，哪那么容易啊，隐居的生活非常艰苦，在深山之中隐居要下非常大的决心，才可能做到离开那么热闹的城市。隐居也还分真隐居和假隐居。我们知道真正的隐居是只要别人知道他住在这里，立刻举家搬迁，不希望有任何人再知道，彻底断绝与社会的关联。其实对于中国人来说，这是一种带有意识形态批判色彩的信仰的力量。当他不喜欢这个世界的时候，会去另外一个世界。当然最后演化成了在现世中有投机色彩的作为，解决的办法就是园林，就是在院子旁边造一个园林，随时可以进去，也用不着出城了，用不着进山了。山中归隐和这个相比，对中国人来讲，品格更高、更纯粹。

* 图 6

那么谁在想呢，五代卫贤°的《高士图》*非常典型，就是高士在想。画的上面是一个非常陡峭的山，下面是一个居住的地方。我们叫可居可游。其实这种山崖之下一般不适合居住，这种地方非常危险，随时有泥石流和洪水的爆发。那么，接下来的问题就是这种绘画的意识里，人到底在哪里？人只是站在画外去想么？

○ 生卒年不详
* 图 7

Then, the next question: where are the people in the consciousness of this kind of painting? Do they just stand outside the painting and think?

Let's have a look at how a literati prepares for painting from the painting *Pure Dawn on Lake and Mountain** by the Five Dynasties Taoist Monk Liu. We can see that he is followed by a boy servant carrying the Four Treasures of the Study and references. This is a very important element in the life of Chinese literati. Therefore, there are road systems, landscape systems etc. to form a complete set. A whole set is related to life. From that you can see a basic consciousness: Liu did not paint elevations but painted a space for experience. So the intention of his whole painting is to invite you to come in, to ask you to participate in the whole experience of the process. You will not only need to have this consciousness in several kilometres to tens of kilometres on a site plan scale, but at the same time you must be ready to have the same consciousness in a cave or under a cliff. This is the most difficult and important issue for city planning and architectural design. Physical models or 3D animations cannot replace it.

* FIGURE 8

So what to look at? What are they imagining viewing? We'll look back at Dong Yuan's painting.* No matter if this work is genuine or a replica, it is still a good painting. Let's look at the structure of the painting and divide it into the upper and the lower sections at the middle. The upper section is what I call 'imagined view'. This may be five or ten or twenty kilometres away, or even it is not here at all, but at another place. The high mountains in the painting of the hermit represent a place hard to live a reclusive life. It also symbolises the determination to depart from politics and society in pursuit of a quiet and liberal life. The lower part can be called the subtlety of imagined view. He did not simply paint a symbolic painting. He was interested in the whole world around him and never let it go. Therefore, there is a strange feeling about this painting. If we discuss this painting by separating the upper and lower sections, it is almost like a painting of theory. It has very abstract meanings and meanwhile gives us such vivid depictions of everything he has seen. And there is a standard precision in this drawing. He never makes one part the main subject, and the others secondary — painted vaguely. This is the most difficult issue in understanding Chinese thinking and painting. This is what we call the Chinese metaphysical way. It is very difficult to understand but it does exist. If we enlarge the drawing to check the details of the bamboo fences, they are almost as accurate as an architect's drawings. The drawing of the bamboo fence can be used as an architectural drawing directly. Traditionally, we would say this is the understanding of Tao, because Tao and implement° are related. The detailed observation of the implement links to the depth of understanding of Tao directly. This is why we think that detailed arrangements in architecture and the consciousness of total control are of equal importance.

* FIGURE 5

° *qi*

从五代刘道士的《湖山清晓图》,*我们可以看到文士是如何做绘画的准备的。他后面跟着一个书童,带着文房四宝和参考资料。这是中国文人阶层几千年以来生活里非常重要的一个内容。所以它有非常配套的道路系统、景观系统等等,整个和这种生活是相关的。从中你可以看到基本的意识:他在画这张画的时候,不是在画立面,他是在一个可进去的空间体验当中。所以他整个绘画的意图就是请你进来,请你参与整个的体验过程,而且你不仅在几公里几十公里的总平面的尺度上要能有这个意识,同时你随时要有在某个山洞里、某个山崖之下同时存在的这个意识。这个实际上对从城市规划到建筑设计来说,是最难培养的、最重要的问题,模型、三维动画都代替不了。

＊ 图8

　　那么是在看什么呢?观想什么呢?我们再回头看一下董源的这张画。*这张画不管真假画的很好。我们看一下这幅画的结构。将这幅画分成了上下两段。上面一段我称之为观想的东西,这段可能在五公里十公里二十公里之外,甚至是不在这里在另一处,是他所观想的东西。《高士图》里高山代表着很难隐居的地方,也代表着想脱离政治和世俗社会,过一种安静的自由的生活的决心到底有多大。我们再看下面一段,我们叫观想之精微。他画这幅画的时候不是简单的象征,他对这个世界的所有事情都感兴趣,这是他身边的事情,他是绝不放过的。所以这幅画有一种很奇怪的感觉,如果我们用上下两段的方式讨论的话,它几乎像一种理论绘画。它表达的意思是非常抽象的,但他对所看到的东西又是如此细微的描绘,而且整幅画描绘的精确度是非常的均匀。不会说这边是主体,那边虚掉了,是次要的。这是中国思想包括绘画里面最难理解的,我们称之为中国式的形而上的方式。非常难以理解但它是真实存在的。如果我们再放大去看一下对竹篱笆描绘的精度,绝对赶得上建筑师的图纸,这个篱笆完全可以作为设计图纸使用。我们在传统上会说这是个人对道的理解,因为道、器这两者是相关的,对器的细微的观察和对道的体会的深刻度是直接相关的。我们讲做建筑对细节的配置的程度和他对整体控制的意识完全是等价的。

＊ 图5

Now we are going to talk about the *precision* of the imagined view. Let us look again at the eleven metres long *A Thousand Miles of Rivers and Mountains* by Wang Ximeng. When you look at its different sections, observe the degree of precision in his depiction of detail. You can see the structure of the boats, trees and mountains and the relation between the roads and mountains including the manner of treating the riverbanks, which are just like what we are talking about nowadays in landscape design. But I do not think there has ever been a system in China of applying architectural design to landscapes. Our basic idea is that there is only a landscape architecture system as a whole since ancient times. And it is a kind of system in which landscape precedes buildings, a system in which landscape is always more important than architecture. Because for us landscape has an alternative meaning of faith, it represents spirit and quality of life, rejects all vulgar things, and keeps one's character independent and mind pure. At the same time it is united with art directly. After painters have created drawings, artisans will follow these with buildings. Of course, there are great artisans and normal craftsmen. Artisans are close to architects since they also make models and drawings, a sort of combination. Do today's architects have the opportunity to make contact with artists? Do architects have opportunities to make contact with philosophers? No, they do not. But this system enables those contacts.

  * FIGURE 2

Let's look at another section of this painting. Because the long scroll is very long, Wang Ximeng painted the picture section by section. You can see that each section has different locations. Some have a village on the side, some have a village on top, and further up there is a temple. Each object has an appropriate location. If you put those objects in the wrong positions, the understanding of Tao is incorrect, because this involves the discussion of the habitable and the inhabitable. You can see the establishment of the whole system. Now you can find fragments of this system in the provinces of Jiangsu, Anhui and Zhejiang, these relatively developed areas in the Jiangnan Region. This system exists not only in painting, but in reality as well.

It is very interesting that you can see Zhao Boju's method of thinking in his painting, as in *The Autumn Scenery of Rivers and Mountains* discussed before. It is painted in a style depicting extreme distance. We can say that this is a scene viewed from five kilometres or ten kilometres away, but when you look at the details of this painting, the figure seems only one hundred metres away. He drew this picture by overlapping far and near together. Therefore many people argue that this is a realistic painting. Somebody said Song Dynasty is the most realistic and naturalistic epoch in China. On the contrary I think it is actually an age of enormous metaphysical temperament. It used such special means to overcome difficulties. It is very difficult for us to get a handle on large-scale cultural geography today. When we see distant figures we cannot see small

* FIGURE 4

我们接下来就说一下观想的<u>精度</u>。我们来看一下十一米左右长的王希孟的《千里江山图》。*当你放到一个局部的时候，看一下他所描绘的东西的细微程度。你看包括船、树、山体的结构，完全可以看到道路和山体之间的关系，包括岸边应该如何处理，似乎如我们今天讲景观时所讨论的内容。但我从不认为中国有一个建筑学加上景观的这样一个关系。我们所有的基本观念，如果套今天的词汇，从古至今就是一个景观建筑的体系，而且都是景观在先建筑在后的一个体系，永远都是景观比建筑重要的一个体系。因为山水对我们来说带有一种替代信仰的意义，它代表的生活精神和品质是拒绝所有俗气的东西，保持人格的独立和清净的心灵。同时它是和艺术直接结合在一起的，因为画家画完之后工匠接下来建造，当然这里面有大工匠小工匠，大工匠接近于建筑师，也做模型画图，它是这样一个相结合的体系。我们今天的建筑师有机会接触到艺术家吗？有机会接触到哲学家吗？接触不到，而这个体系是可以接触到的。

  我们可以看另一段。因为长卷很长，所以他一段一段地描绘，你可以看到一段一段所在位置的不同，有的在边上是乡村，有的在顶上是山村，再往上是寺庙，每一个都有合适自己的位置，位置摆错了对道的理解是不对的。因为这里面包括可居不可居这样的一些讨论，你可以看到这里整个系统的建立。现在我估计江苏、安徽、浙江，整个江南比较发达的区域如果去找的话，这个系统断章残简，但大体上还是找得到的。这个系统不只是绘画中出现，它是真实存在的。

  非常有趣，你可以看到他绘画中想问题的方法。这张图是赵伯驹的那张图，*是一种非常远的画法，我们可以说是五公里十公里之外画的这样一张景象。而你仔细看这个画的细节的时候，那个人离你只有一百米远。他是远和近叠在一起画。所以很多人说这是写实，说北宋是中国最自然主义、最写实的时代。我认为恰恰相反，这实际上是极有形而上气质的一个时代。它用这样一种特殊的手段来克服这个难题。今天我们想对一个大的人文地理进行控制，其实这很难，看到远的看不到小的，看到近的又看不到大的，真正在山里左一个弯右一个弯每处又是不同的，而我觉得中国的绘画提供了非常好的方式，尤其对我们今天的建筑师。当我们用这个方式理解的时候，你会觉得非常好。因为能画得出应该就能做得出，一定有一个可解决的方案在里面。

* 图2

* 图4

objects; when we view closely we cannot see the whole picture. In view of one's real experience of mountains, it is different at every turn of the road. Each turn has different views. But I think Chinese painting provides very good methods to make sense of this, especially for today's architects. Because an object can be drawn, it should be able to be built. There must be a solution in it.

This is like the Chinese garden. The fact that gardens can be made illustrates the issue. The reason why I always like to discuss Chinese gardens and admire Chinese gardens is because they are clearly created by literati and artists by a building process similar to that of today's architects. Their entire operation is the closest methodology to today's architecture. I often say that was one of the most pioneering eras, it expressed ideology and belief in traditional times.

There is another aspect of Chinese belief that is difficult to understand—the Chinese possess a strong consciousness of the abundant changes of nature. The real capacity is the handling of change. That is why when some people ask me about what Chinese people believe, my answer is that we only believe in 'change'. Many things change, but what matters is—in what ways. This involves a discussion of values, including all views upon the horizon, including images in our mind, which influence our actions and practices. Look at the painting's accuracy: in such a large-scale of painting, when you observe fragments of it, you see how water is drawn, the structure under the roof of the building, even the details of windows. Basically this is a drawing that is ready to be used for construction directly. But at the same time it expresses control over a large-scale landscape. How can this be done with a control of ideas?

This is considered a very freehand brushwork painting, because this painting is drawn in very narrow and small scale. Now, of course, we are going into another topic, which is the control of the brush, the so-called 'idea of brush'. Although the lines look more 'free' and 'lose' compared with previous paintings, in fact all positions are still depicted highly accurately. That is why I often use mathematical language to discuss traditional Chinese paintings, architecture, and landscape systems with my students. In order to understand this clarity of taxonomy and mathematical accuracy applied to the control of position and movement, the simple word of 'feeling' is not enough.

The issue of distance arises when discussing viewing and imagination. In the painting of Dong Yuan's *The Creek Bank*,* we can see a very serene and deep scene. The best word used in describing Chinese painting and gardening activities, is actually the word of 'distant'. I had no clear sense regarding it before. It is not only a matter of physical distance. At the same time we say that everything has a tradition. Its basic awareness is the pursuing of the past, or in another word, of the height and depth of culture. It exists before us. We are trying to get to it with very imaginative ways. You cannot inherit it automatically. If we do not use imaginative methods to discuss it, it will die like a corpse. In such consciousness, things exist in two temporalities.

* FIGURE 5

这像园林一样，园林能造出就说明了这个问题。而中国的园林为什么我一直特别喜欢讨论它和推崇它，因为它是清楚地由文人和艺术家以类似于今天建筑师的角色主动介入的建造活动，所以它整个的运作方式与我们今天的建筑学是最接近的。在那个传统的时代，我经常会说那是一个最先锋的时代，它表达了那种意识形态和信仰。

中国人信仰里的另外一个东西又很难理解——中国人意识到自然之间的那种丰富的变化。真正的能力是对变化的把握，所以有人问我中国人信仰什么，我回答说我们只信一个词就是"变化"。很多东西都会变化，问题是以什么样的方式变化。这里面就有一个价值观讨论的问题，包括整个视野包括你头脑中的景象，它影响着我们的行动和做法。你看看它的精度，那么大一张画当画到局部的时候，你看看水的画法、房子下面的结构，包括那个窗的每一个细节，基本上这个图你拿了之后可以进入施工图阶段，可以建造了。但同时它又是对大尺度景观的控制，它怎么做到这样一种观念的控制呢？

这张已经算是很写意的一张画了，因为这张画画得很窄很小。这时候当然进入另外一个话题，就是对笔的控制即我们说笔意的那个东西。尽管线条看着比前面画的要"放"和"松"，但实际上看所有的位置等等仍然是高度精确的。所以我经常会用一种类似于数学的语言去跟我的学生讨论中国传统的绘画和建筑、景观系统。要体会到这里面看似感性的东西背后所包含的分类学的清晰和它所有的位置、运动的控制像数学那样的高度准确，而不是简单地用"感觉"这样的词汇可以搪塞过去。

有观有想它就有距离的问题。这张还是董源的那张《溪岸图》，*我们可以看到那种很幽深的很深远的景象。这个实际上是中国绘画和园林活动里面评价一个东西用的最高的一个词，就是这个"远"字。我以前也没有明显的这种意识，它不光是一个物理距离的问题。同时我们说什么东西都是有传承的，它基本的意识是对过去或者说对高度和深度的一种追索，在我们之前已经有的东西是存在的，然后我们在它后面用非常有想象力的方式获得，不是你自动就可以传承，这个东西如果不用非常有想象力的方式讨论的话就死在那儿了，像一具僵尸一样。在这样一种意识中，事物以两种时间并存的状态存在着。

* 图5

Let's have a look at this painting again. Even in this degree of painting, you can still feel an 'easiness' that matches the 'distance'. It indicates the coexistence of times past and the present. I remember a poem by Tang Yin, which describes that he saw a buffalo approaching him from a village with a shepherd who sat on its back. Then he saw a Book of Han hanging on a horn of the buffalo. After that, he sighed emotionally and said: 'Nowadays people do not know the real sense of easiness'. You can see the width and the depth of thinking with a changed scale, including the depth of profound historical thoughts of the entirety of time and history. Today we can still talk about Han Dynasty as if it is a matter of today. This is very interesting because among world civilisations, only the Chinese civilisation is able to read writings from thousands of years ago. We can still discuss it. We still understand it and learn from it. This is very important. No other civilisation is capable of doing that. Even though China is in such status, I feel this kind of ability still remains.

Autumn Mountains and the Green at Dusk* by Guan Tong° of the Five Dynasties is an interesting painting. Some people said that there is no way to deal with this painting, because 'distance' is not different in the upper section or lower section of the painting. The upper left corner of the painting should be faraway in high mountains. In today's drawing techniques, it should be drawn vaguely because it is so far from us. We cannot see those things by our own eyes. We can see that the railings of small bridges are drawn as if they are close by, including the delicate details and structure of the stone on the river. They all exist in that space. This means that different descriptions time can be converted into depictions of different geographical locations and different depictions which can all be compressed into a two-dimensional plan.

* FIGURE 9
○ c. 907–960

Let's talk again about whether this is abstract or realistic. Look at the scroll painting *West Lake** by Li Song° of the Southern Song Dynasty. Many people put forward that this is an early example of idea-writing freehand brushwork painting in Chinese landscape painting. I suggest another point of view. In fact, this painting is a realistic one. He painted the *West Lake* landscape from such a far distance away. In reality you can see only the haze in the distance. If we view the painting from this perspective, the fine painting that we have seen previously is actually very abstract.

* FIGURE 10
○ 1166–1243

We have seen two other concepts about imagined view. One called temporal imagined view, another internal and external imagined view. Viewing the painting *Lush Forest with Distance Peaks** by Li Cheng° of the Northern Song Dynasty, I draw a line in the middle, and separate it into left and right halves. On the left half, it depicts the mountains, a land we say of 'reclusion'—*Lofty Spirit among Groves and Springs* by Guo Xi° refers to such place. The interesting point is that, first of all, everything is depicted so accurately. It is not like a barren mountain, which people cannot enter. Instead a very

* FIGURE 11
○ 919–967

○ 1023–c.1085

我们再看一下，即使在这样的程度的画，你仍然能感觉到"悠然"，和"远"配在一起，它表示时间过去的那种状态和现在的暂时状态的同存。我记得曾经看过一首唐寅的诗，就是说他看到乡村有个水牛过来，有个牧童坐在水牛的背上，然后他看到水牛的犄角上挂的是《汉书》，他就感慨：今人不知悠然意。你可以看到换了尺度的宽广和想的问题的深度，包括整个时间、历史的思考的深远程度。我们把汉代的事情当作今天的事情来讨论，这非常有意思。因为这个世界上的文明可能只有中国的文明现在还有这个能力直接阅读几千年前的文字，我们还可以对着它讨论，我们还认得，可以揣摩，这个非常重要，没有一个文明能够这样，即使中国到了现在这个程度，我觉得仍然还残存着这种能力。

　　这张图有趣，这是五代关仝°《秋山晚翠图》，*有人说你拿这张图没有办法，因为上下画得一样"远"，没有上半段和下半段的区别。我们简单地看一下图上方偏左的那个位置，那个位置应该是在高山的很远的地方，按今天的画法应该是虚掉的，因为它太远了，以我们的肉眼根本不可能看到的东西。我们可以看到画中小桥的栏杆这些东西都以你近在咫尺的感觉进行描绘，包括溪流上石头的精巧和结构，全部存在在那个地方。这意味着对不同时间的描绘可以转化为对不同地理位置的描绘，而不同的描绘可以被同时压缩在一张二维的平面上。

○ c.907—c.960 年
＊ 图 9

　　我们再来说一下到底是抽象还是写实。看这张南宋李嵩°的《西湖图卷》，*很多人把它说成是中国山水画里比较有写意精神的早期的一幅画。我倒可以提出另外一个看法，其实这张画恰恰是比较写实的。他画了在这么远的距离看西湖的山水，大概朦朦胧胧就只能看清到这个程度。如果从这样一个观念的角度去看，我们前面看的描绘得非常精细的绘画恰恰是非常的抽象。

○ c.1166—c.1243 年
＊ 图 10

我们看了和观想有关的另外两个概念，一个称之为时间的观想，一个是内外的观想。这张图是北宋李成°的《茂林远岫图》，*我从中间画了一条线，这张图是很典型的左右两半。我们看一下它的左边，画的是这种深山之中，就我们说的可以"归隐"之地——"林泉高致"，就指的这个地方。它有意思的是，首先所有的东西画得都如此精确，它不像是一个荒山之中，人不可以进入的地方，而是有一个很高质量的景观建筑的系统蔓延在其中。同时，我觉得对中国人的观念来

○ 919—967 年
＊ 图 11

high quality landscape architectural system spreads within it. At the same time, I think in terms of a Chinese point of view, three basic discussions—the past, the present and the future—it represents the past and future. Often we have interest in the idea of antiquity and always have a feeling that the past is better than the present. Then, we think about it to grasp the direction of how the path of the future should be. On the right is a field, what we call 'suburban land' considered suitable for living. There is almost no detail depicting a city. A city is only shown with a little edge at the right corner. Of course, this is not an exception, as in Chinese landscape painting there are rarely cases of painting a city, because in such a powerful literati ideology system, cities were not worthy of drawing. You can imagine that, if the Chinese followed such ideas today, it would give rise to a different scene. So you could say, since the Wei and Jin Dynasties, the powerful mainstream movement in the culture of Chinese cities is that of anti-urbanism. Chinese people used to say, I came to the city temporarily to fish for fame and money, or to achieve some personal struggle or goals, and after that I will go back. As Su Dongpo said, finally someone will call me back.

There is another important concept. In fact we call it the movement, i.e. these paintings cannot be viewed in a static way. Let's have a look at another picture, *Cottages*,* a legendary but debatable drawing by Li Gonglin.° In this painting, you can see the distance in his depiction is changing. If you do not pay careful attention, you cannot see the trick of it, because the structure of the mountains remains similar. When you look at it carefully, you will find that, in fact, the village on the left of the painting is almost like a site plan. It is very small. Its right side is in a real scale with a human figure in it. The previous drawing is somehow in a 1:2000 scale, and the right one is basically in a scale of 1:100. These two pictures are joined together directly without any transition. The height of that person exposes the real proposition of it. The red person joins the picture to real scale. Then what does he tell us? I think it is very important. It is what I call 'Distant View' and 'Close View'. Why cannot we simply use 'elevation' to describe it? It is actually a relation between a distanced view and close view. It is not simply divided into distanced views and close views. They coexist at the same time. If they do not exist at the same time, the meaning of 'distance' cannot be created. Of course, there are a number of other factors. As you can see, these things do not exist in a static way. They are in non-stop and continuous movement. It implies a continuous movement. We call that the manner of water and the description of the continuous spirit.° In fact, it implies that it is always moving. People are walking in it continuously.

* FIGURE 12
° 1049–1106

° *spirit: qi*

说，如果我们讲过去、现在、未来这样三个基本的讨论，这个景象，它代表的既是过去，也是未来。因为我们经常说古意，总是觉得在过去更好，然后，我们思考它是为了来把握未来的道路应该如何。右边这块是城郊，我们称之为郊野地，比较适合于居住，城市几乎没画，只在最右边露出一点边缘。当然这不是一张例外的绘画，在中国的山水绘画里，画城市的画就极少。因为，在这样的一种文人的强大的观念系统里头，城市根本不值得画。你可以想象，如果今天中国人还有这样的观念的话，那将是另外一个景象。所以你也可以说，从魏晋以来的中国的这种关于城市建筑的文化里，它有一个很强大的主流倾向是反城市的。中国人以前经常说，我就是为了博取功名利益，或者实现个人奋斗的一些目标，暂时到城市里来一下，之后还是要回去的。如苏东坡所说，最后还是有人要召我回去的。

那么，还有很重要的一个观念，实际上我们称之为运动，就是这些画都不是可以简单静观的。我们看一下另外一张图，是这个传说但有争议的李公麟°的《山庄图》。*这张图，你可以看到他所画的距离的变化。从起首的两张图，你不仔细看，看不出中间有蹊跷，因为这个山形结构是类似的。然后你再仔细看会发现实际上那个村落在左边那张图上，就是一个总平面的状态，很小。右边这张画是人的在其中的真实的那个尺度。我们称前面这张图是1∶2000的比例，右边那张图基本上是1∶100，就是这样的两张图放在一起，直接连接，没有任何过渡。而那个人的高度，暴露了这张图的比例，红的这个人是按真人的尺度来衔接。那么他在说的什么呢？我觉得很重要，就是我称之为叫"远观"和"近观"。为什么我们说不能用简单地用"立面"这样一种观念去描绘呢？它实际上是一种远观和近观的关系。而且这个关系是可以同时的，它不是简单地分成这是远观，这是近观。那么有趣之处在于它是同时存在的，如果它不同时存在，那个"远"的意思是造不出来的。当然这里面还有一些其他的要素，你可以看到，这些东西都不是以静态的方式存在在那里，它是连续的、连绵地一个运动的。这里面暗示着这个连绵运动的，我们称之为水的方法，和里面连续的气的描绘的方法，其实它是在暗示这件事情，就是它是运动的。它一直在运动，人在里面一直在走。

○ 1049—1106年
* 图12

We can refer to another topic, such as gardens. According to Tong Jun,° in the Chinese garden, even if there is only building, even if there are no rocks or trees, it still can become a garden. I think what he says is particularly interesting. From this point of view, this actually is a kind of method to construct a quasi-nature space between the artificial and natural. At this moment, the construction of architectural space inside becomes very important. In fact, piling up mountains to make a landscape inside gardens, perhaps only Chinese do such kinds of things in the world — making a garden in artificial ways repeatedly on a large scale. It is a belief. It is our understanding of the way of Tao that enables us to grasp a language which is coherent to a maximum degree with the laws of nature. It is only with such a kind of sense that we would do such a thing. Of course, we can also refer to another topic — how can you make a building, in which even if there is no one inside, you will still sense the operation of your consciousness in a natural way. You still have an awareness that the architecture is made in a scheme of movement. Architecture is a dynamic process. It is very difficult for us to use simple ordinary static methods to describe such aspects.

○ 1900–1983

Finally I will give you one example of my own work. What kinds of effects will the above discussed awareness cast upon an architect like me? We are talking about traditional Chinese architectural activities, such as those depicted in the painting *Watermill in the Gate** by Wei Xian of the Five Dynasties. It is a relatively rare painting, similar to *Along the River During the Qingming Festival* by Song Dynasty artist Zhang Zeduan.° We can see that in the upper left corner, a scholar official is organising affairs and a lot of people are doing manual labour forming a construction scene. In the lower corner, you can see that it is a temporary construction possibly for a celebration.

* FIGURE 13

○ 1085–1145

Before telling my own stories, I would like to speak a little about what kind of consciousness should be generated for you to think about all of these things.

I'd talk about 'Memory and Reality', because in today's architecture, it is rare to use such word as 'memory'. We use the word 'tradition' frequently, which is associated with some self-claimed symbolic elements of tradition. In fact, what we really need to discuss is the word of 'memory' which is about the feeling of being inside a situation. How do we remember? It is an important part of human nature. There is another one: 'reality', which is also rarely discussed in architecture.

Regarding 'memory', it reminds me of the Northern Song Dynasty artist Guo Xi,° who wrote *The Lofty Ambition in Forests and Streams* and painted *The Colour of Trees in Level Distance*.* One day, I walked in the suburbs of Zhejiang.° When I saw it, I almost cried. As you have seen in the painting, the spirit still exists from a thousand years ago to today. Such temperament can exist for so

○ 1023–c.1085
* FIGURE 14
○ Zhejiang Yongjia

王澍
WANG SHU TOPOGRAPHY 196

我们可以扯出另外一个话题，比如园林。童寯°先生说，中国园林里面哪怕只有建筑，哪怕没有山石树木，也可以成为园。我觉得他说的特别有意思，从这个角度看，这实际上是一种介于人工和自然之间去建构一个类自然的空间的一种方法，这时候，建筑的空间的建造在里面就非常的重要。其实在世界上，园林里面平地堆山、造园，恐怕只有中国人干这样的事情，真正大规模地反复地用人工的方法来造。因为这有一个信念，就是我们对道的参悟使我们可能掌握一套语言，这套语言可以跟自然的法则做到最大程度的相符。有这样的一种意识在里面，才会做这样的事情。当然我们也可以扯出另外一个话题，你怎么样能够做出一个建筑，就是哪怕没有人在，你会觉得整个的意识是在如自然般运动的，你仍然能够意识到这个建筑是个运动的做法，在动态的过程当中。这些我们很难用简单的习常建筑的静态描绘手法去描绘。

° 1900—1983 年

我最后讲一个我自己的工作的例子，谈一下以上的这种意识，会对比如我这样一个建筑师产生什么样的影响。我们所说的那种传统中国的建筑活动，你看五代卫贤的《闸口盘车图》,* 这是一个比较少见的图，类似于《清明上河图》。我们可以看到左上角的角落里，有个文人官吏就在那组织这件事情，很多人在劳动，形成了一个建造场景。你可以看到下面这个角上，可能是为了某个庆典，临时搭建的一个构筑物。

* 图 13

在讲自己的案例之前，我想讲一点，就是什么样的意识产生，使得你会去想这些问题。

我讲一下"回忆与现实"。因为今天的建筑学里，比如"回忆"这个词就很少讲，动不动讲"传统"，然后呢还有一些所谓传统的符号要素。其实真正可以讨论的词是"回忆"，就是我们真实地身处在一个感受之中：我们还能够记得什么，我们怎么来回忆。这是人性里面重要的一块。还有一个就是"现实"，这个也是建筑学里极少讨论的。

那么这张，我们如果说作为"回忆"的话，就是北宋郭熙，就是写《林泉高致》的郭熙，° 他的一张《树色平远图》。* 有一天，我走在浙江的乡下，° 看见的时候我差点哭了。因为你看过的这个东西真的存在，那个意思还在。一千多年前到今天，这样一个气质居然能够存在如此之久。那么，它就让我会想到"现实"。大家看一下这张图，* 编号一，很美，这是一个村庄的完整的中轴部分的景观系统。但问题是，这是我用四个村子的照片拼成的。今天稍微存在完整一点的村子都

° 1023—c.1085 年
* 图 14
° 浙江永嘉

* 图 15

建造一个与自然相似的世界
BUILD A WORLD TO RESEMBLE NATURE

long. It makes me think of 'reality'. Let's look at this picture.* It is very beautiful. This is a landscape system of the complete central axis of a village. But the problem is that it is actually a collage of photos from four villages. Today if there remains a relatively complete village, it will all be turned into a tourism hotspot and I can't take a photo of those. If you want to take a picture which has a good sense, it requires photographs of four villages collaged into such a kind of continuous scroll. And for a scroll like this, 30 years ago in Zhejiang, you can find at least 10,000 villages that remained in this state. But today, there are only about 200 villages that remain. This is what has happened in 30 years. You can see that when the culture of an area changes its beliefs, it has great influence on reality, almost unimaginable.

* FIGURE 15

Let me talk about our campus, Hangzhou Xiangshan Campus. In fact, at the beginning, I also had no strategy. I just felt that Xiangshan Hill was there, speechless, but it was heavier than anything I would design. I could not just ignore it. Its scale was close to architecture, but it also represented something more important than building. In the beginning, I made my research like a small game. I took photos of Klippe and of Lingyin Temple and collaged those photos. The scale of Klippe and Xiangshan are similar, the building is very close to the mountain, and there is a meandering pedestrian system among the stone caves. The location of the building limits the positions where you are able to take photos. So sometimes you can take photos here, sometimes cannot take photos at all, sometimes you have to walk closer, and sometimes you cannot approach it at all. Finally, I don't zoom in and zoom out of those photos. I simply put them together according to the original scenes. I think it is very interesting, because I seem to have found the secret of our traditional landscape painting. This kind of distance is like looking at a scroll. Rivers and trees create obstacles which actually transform distances and sizes.

This is my sketch.* After you have this consciousness, your design consciousness will be changed. For example, Chinese calligraphy is written from right to left and when I drew this picture, according to present habit, I drew it from left to right. After repeatedly thinking about this picture in my mind, I painted some draft drawings, thinking about atmosphere, location, orientation, scale, rhythm, materials and components, and events to be expected, etc. Then came that moment. It was after about four hours, drawing continuously from left to right. It is very important that it can't be stopped, because the whole feeling of movement, the control, and the changes during the process cannot be done separately after thinking. Although I used a pencil, it had a similar consciousness, that is, from side to side, drawing a picture in one movement. There is no need to make many further changes. Why? Because there exists a movement of spirit inside. It made the building 'alive'. This building group is comprised

* FIGURE 16

变成旅游业，不能拍了。你要想找张画，就是意思比较好的，要用四个村子的照片拼出这样一个连景轴。而这个轴，在三十年前，在浙江，至少可以找到一万个村子保持着这个状态。而今天，只剩下两百个左右，就是在三十年间发生的事情。你可以看到一个地区的文化改变了它的观念之后，它对现实产生的影响有多么的巨大，多么不可思议。

我讲一讲我们的校园，杭州的象山校园。其实我当时一开始的话，一时也没办法，只是觉得那座象山在那里，质朴无言，但比任何我将设计的东西分量都重，就是无法忽略它。它的尺度和建筑接近，又代表着比建筑更重要的事物。一开始我做了这么一个小的游戏般的研究，就是我把灵隐寺前面那个飞来峰拍照片，拼了一下。飞来峰和象山尺度差不多，建筑离山很近，山上又有一个蜿蜒起伏的石窟步道系统。因为有建筑，所以限制了你的拍摄的位置，使得你一会要在这儿拍，一会无法拍，一会可以走近，一会又走不近。而最后，我把这些照片，我不去放大缩小，我按照它的原片把它拼合在一起。然后我就觉得很有趣，因为我好像发现了我们传统画面的大幅江山图的那个秘密。这种距离就如同看长卷，河流和树木制造着障碍，实际上就是在这种大小远近之间来经营和转化的。

　　这是我画的草图。* 因为你有这个意识之后，设计的意识就会变化。比如我画这张图，中国人用毛笔字，从右向左，我按现在的习惯，从左向右画过去。这张图在脑子里反复想过之后，画一点潦草的控制图，思考气氛、位置、形势、朝向、尺度、节奏、物料与分量、可期待的事情，等等，之后就迎来那个时刻。大概用四个小时的时间，从左到右连续地画，这非常的重要，不能够停止。因为整个的运用的感觉、控制和中间的那种变化，不是去想了之后一块块摆上去的。尽管我用的是铅笔，但意识是一样的，就是从这边到那边，一气画，轻易不会再改。为什么轻易不会再改呢？有气的运动在里面，它使建筑"活着"。这一套房子画过去十几栋，大概八万平方米，这个长度从这头到那头建筑大概是三百五十米左右。光第一栋房子展开的长度就是三百米，折叠过之后是一百。这样的一组组，十几栋，一口气画过去。之前想了两三个月，想得极痛苦。

　　你们再按这个草图的原比例把它拉近看，发现这张图画在一张 A3 的纸上，当然这是我的一种自我锻炼，想跟古人

* 图16

of over 10 buildings. It is about 80,000 square metres. The length from beginning to end is about 350 metres. The outline of the first building adds up to 300 metres, and 100 metres after folding. In such groups, a dozen of the buildings were drawn all in one movement. Before make this drawing I thought about it for two or three months. It was very painful.

If you enlarge this sketch to its original scale and have a closer look at it, you will find this drawing was drawn on A3 paper. Of course, this is my own exercise. I want to compare my capacity with the ancients. They drew on scrolls, the hand scroll. They painted so small, but exercised such subtle control. I don't think this is entirely a problem of proportion. It is actually what's called 'discreet'. It is like the precision of your mind. It is reflected in the hand. We call it 'skilled hands match with the mind'. So I used a piece of A3 paper. This is the size. After that, you can see its bearing. It is the basic control of architecture. You can see that after enlarging the drawing, it only needs to be fixed by adding little details to explain the material, and the next stage is detailed drawings.

Our Xiangshan Hill, in fact, is a hill 50 metres high that can be seen everywhere in Hangzhou. But when you are in awe of it, things change. Although this is a small hill, although it's nothing in today's environment, even if it is only a piece of green space, it is like a shrine standing there. You have a different feeling. When you are aware of your whole consciousness, your relationship of control towards architecture is changed.

You can see it from the master plan.* A lot of people ask how this master plan was arranged. It is strange, because on the main road, you basically do not see the front façades of the buildings. All the façades are broken down into side views. Although the hill is not that special, it is still more important than the buildings. So the gesture of the buildings is totally different. It is important for people to 'tour' in architecture. The walking system is precisely controlled, to allow different backgrounds and different positions for viewing. When one sees the mountain again, our experiences will be merged together as a whole. This is the whole experience of the effect. At this moment, the mountain suddenly becomes a sacred object, and becomes beautiful. There will be a whole change of perception. * FIGURE 17

I simply took a picture, because a lot of people said that they did not know how to take photos of this campus. There was one professional photographer who spent two days there and said he could not take one good photo. It is a cluster of buildings which does not provide enough distance for you to take a picture of even a single façade. This is my own interpretation of how this building should actually be looked at.* This directly relates to the whole idea and handling of paintings that I just talked about. We look at it section by section. Of course, you can do it in this way because our studio controls architecture, landscape, interiors and all things. The things created by the imaginable view are like that. There are some analogies, an action of jumping and missing links. Every time when I try to reassemble the * FIGURE 18

比一比。他们画面这种卷轴、手卷，画得这么小，能做这么精微的控制，我觉得不完全是比例的问题，实际上是叫"心细如发"，是你整个的心灵的精度的问题。它反映在手上，我们叫心手相应，心手一致。所以我用一张"A3"的纸，就这么大。完了之后，它的气度，就它的建筑的基本控制，你可以看到，放大之后，基本上就再补一点点细节就可以了，交代一下材料。所以，我另外一张图就不放了，就是我对所有的材料做过的一个批注，接下来就进入方案制图阶段。

我们的象山其实是个小山，50米高，在杭州可以说比比皆是。但是当你对它产生敬畏的时候，事情就发生了变化。尽管这是一个小山，尽管它没有什么，但是在今天这个周遭的环境里面，哪怕就剩下这么一块绿地，它就像神庙一样地呆在那里。你会产生一种不一样的感觉。当你整个意识起来之后，你对建筑的控制就会发生变化。

你可以看到这个总图。* 很多人会说你的这个总图到底是怎么摆的，很奇怪。因为在主要的道路面上，基本上看不到建筑的正立面。所有的正立面都被分解成了侧面。因为这个山尽管很一般，但它比我的建筑重要。所以我整个建筑的姿态是完全不同的，重要的是人在建筑中的"游"。这个流线系统精密控制，出现不同的背景、一个一个不同的位置，当他再看到这个山的时候，他的经验才会被连贯在一起。这是对效果整个的体会。这时候，这山突然就变得有点神圣，就会变得好看。它整个会发生一个变化。

* 图17

我简单地做了一张图，因为很多人都说，这个校园，不知道该怎么拍照啊。有一个专业摄影师在这里面转了两天，说拍不到一张好照片，它不提供你任何一次足够的距离，是连一张完整的立面都没有办法拍的一组建筑群。我们可以看到，这是我自己的观法，这个建筑实际上应该怎么看。* 而这和我刚才谈到的那个绘画的整个的观念、方式是直接有关的。我们一段一段地看。这在同一栋建筑里，看这栋建筑应该怎么看。我们再来看第二段，第三段，第四段。当然能够这样做，也是因为我们的工作室控制了建筑、景观、室内所有的事情。观想出来的东西就是这样，就是有类比，中间有些跳跃，有些环节缺失。每一次我把它重新拼图都怀疑：这是我做的么？

* 图18

当然还有一种有趣在于，你看到人到底怎么和这个建筑发生关系，其实这个更重要。就是我说的真的生活。当你重建这个系统之后，人怎么和它发生关系呢？人和它在一起的

jigsaw back together, I always wonder, have I done this?

Of course there is another interesting point—you can see how the relationship between people and architecture occurs. In fact, this is more important aspect. This is what I call 'the real life'. When you rebuild a system, how do people develop a relationship with it? With what kinds of feelings towards it? A few days ago, we invited the Portuguese architect Alvaro Siza to come here to have a look. I took some photos* during the viewing. You can see the special feeling of the people passing around the inside and outside of the building. The specially designed stairs have caused some unexpected situations. The stairs are very strange. They go up by half a floor first, then go down through a ramp, and after that you walk two steps up. I call these stairs 'the mountain'. It is the stairs which are similar to the mountain, because they are not stairs for the building, they are a scene for a natural encounter. It happened here. They showed me the photos they took. Behind that image is the high distance, I call it a 'scholar tower'. Coincidentally we are working on an exhibition of studio work about the transformation from landscape painting to modern architecture. We will all enjoy that exhibition. In that particular space, you'll find special things happened. Siza and I love smoking and like to give a light to each other.* In this scene, some people told me that the courtyard is a temple, a negative form of temple. Inside the courtyard, everyday life occurs. This contrast is very interesting. Of course, this is an illusion. You can tell what is really going on after seeing the next picture.* Therefore, it has the same kind of issues that arise when we look at Chinese landscape paintings. You must check two similar images at the same time for you to understand a scene. We usually use the first one as reference. In fact, the following image is the real story in reality, and the story of what will happen next.

* FIGURE 19

* FIGURE 20

* FIGURE 21

Finally, I want to go back to the idea of distance. The few photos shown just now all have this sense.

In *Autumn Pond** by Zhao Lingrang° of the Northern Song Dynasty, the appearance of 'hidden distance'—one of the three new types of distance, is the major change in this Chinese painting. The painting uses a small scene method, rather than a big scene method, with massive volumes. It also has brushwork with a Zen flavour. If the previous works are about the establishment of large volumes and the whole idea, this kind of painting appears with a negative sense. So it makes these clues even richer.

* FIGURE 22
o 1070–1110

This is the realistic version of such a scene.* I took this photo by chance. Before I had a dinner with Siza in a restaurant at the foot of the hill, I saw this table. It was the moment of sunset. The weather was very bad. I saw this image on the table. In fact, up to now I haven't been able to figure out how to explain this picture. So in the end, I leave the picture with a sense of distance for everybody. Everybody can make associations with their own imagination. Thank you.

* FIGURE 23

感觉是什么样的呢?所以我拿几张照片出来,大家可以看一下。前两天我们请葡萄牙的建筑师西扎来转转,我拍了几张我们一起转的照片。*你可以看到人在建筑中内外穿行之中的那种感觉。这个特殊设计的楼梯就发生了它期待的事件:因为这个楼梯很怪,先上到一层半,然后朝下走一个坡,朝上再走两步。那么这段楼梯,我称之为山,类似于山一样的楼梯,因为它不是一个建筑本身楼梯的成分,你可以发现自然有一个相遇的场景,在这里发生了。他们把拍好的照片给我看。背后就是我们那个高远,我叫学者之塔。正好我们在做一个从山水画转译到现代建筑的一个作业展览,我们都看得特别高兴。在那个特定的场景中,你会发现发生特殊的事情。我们两个人都喜欢抽烟,互相点个火之类的。*在这个场景里,有些人跟我讲,这个院子,就是一个寺院,它是一个负形的寺院。就在这个院子里头,可以发生很日常的事情,这种反差很有趣。当然这是一个假象。你们看下一张照片就知道实际发生了什么事情。*所以,看我们中国的那种山水绘画的时候,它就有这个问题。你一定要类似两张照片同时看,你才能够理解它。我们一般都会用第一张的方式,将这个挂在这里,但实际上后面跟的是这张。这是在现实里面真实的故事,它是接下来发生的故事。

* 图19

* 图20

* 图21

最后我想再回到远意。包括刚才的几张照片,都会有这个感觉。

这张照理应该算是"迷远",是"新三远"法里面的。
  这是北宋赵令穰°的《秋塘图轴》。*这个景象的出现,是中国画里面重大的转变,就是用这种小景法,不是用大块大块的大景法,这就有点禅意的笔触。前面那个是关于整个大体积的建立和整个观念的建立,而这种图的出现带有消极性,所以它又使得这个线索变得更加的丰富。
  那么它的现实版就是这一张。这是我偶尔拍的一张。跟西扎在山边的餐馆吃饭之前,看到了这张桌子。*正好夕阳即将落山,天色特别不好,我就看到了桌子当中的景象。其实我到现在为止也没有想好,如何来解释这张照片的说辞,所以我把最后这张照片以一种带有远意的感觉留给大家,大家可以去联想。谢谢。

○ 1070—1110年
* 图22

* 图23

FIGURES

1

1 黄公望（元）《天池石壁图》。
Huang Gongwang (Yuan Dynasty),
The Stone Cliff of Tianchi.

插图

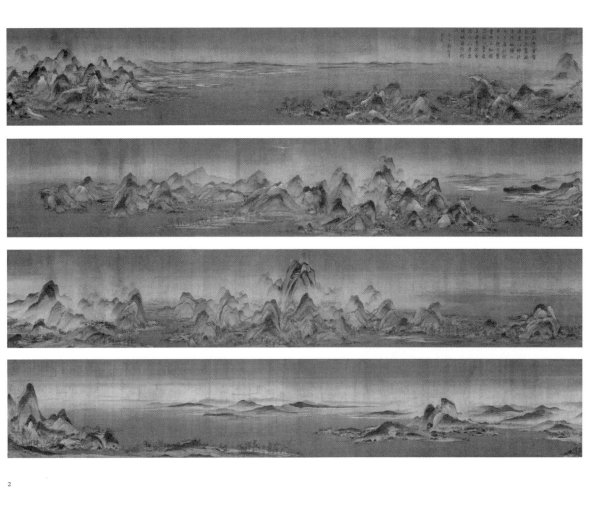

2　王希孟（北宋）《千里江山图》。
　　Wang Ximeng (Northern Song Dynasty),
　　A Thousand Miles of Rivers and Mountains.
3　夏圭（南宋）《溪山清远》。
　　Xia Gui (Southern Song Dynasty),
　　Pure Creeks and Mountains in Distance.

4

5

6

4 赵伯驹（北宋）《江山秋色图》。
  Zhao Boju (Northern Song Dynasty),
  The Autumn Scenery of Rivers and Mountains.
5 董源（五代）《溪岸图》。
  Dong Yuan (Five Dynasties),
  The Creek Bank.
6 董源（五代）《溪岸图》细节。
  Detail of *The Creek Bank.*

7　卫贤（五代）《高士图》。
　　Wei Xian (Five Dynasties),
　　Hermits.
8　刘道士（五代）《湖山清晓图》。
　　Taoist Monk Liu (Five Dynasties),
　　Pure Dawn on Lake and Mountain.
9　关仝（五代）《秋山晚翠图》。
　　Guan Tong (Five Dynasties),
　　Autumn Mountains and the Green at Dusk Grand View.

10

Left half　　　　　　　　　　　　　　Right half
Past — Future — Inside　　　　　　　Now — Outside

11

12

10　李嵩（南宋）《西湖图卷》。
　　Li Song (Southern Song Dynasty),
　　West Lake Dynastic Renaissance.
11　李成（北宋）《茂林远岫图》。
　　Li Cheng (Northern Song Dynasty),
　　Lush Forest with Distance Peaks.
12　李公麟（北宋）《山庄图》。
　　Li Gonglin (Northern Song Dynasty),
　　Lust Forest with Distance Peaks.

王澍
WANG SHU　　　　　　　　　　　　　　　　　TOPOGRAPHY

13　卫贤（五代）《闸口盘车图》。
Wei Xian (Five Dynasties),
Watermill in the Gate.

14　郭熙（北宋）《树色平远图》。
Guo Xi (Northern Song Dynasty),
The Colour of Tree in Level Distance Grand View.

15

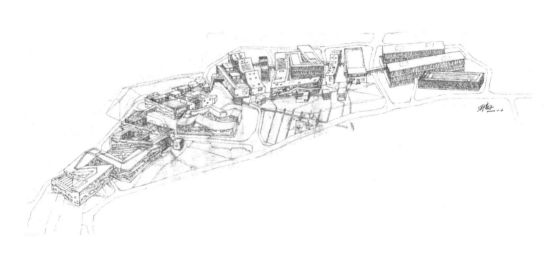

16

15 一个村庄的完整的中轴部分的景观系统。
摄：王澍
A landscape system of the complete central axis of a village.
Photo: Wang Shu

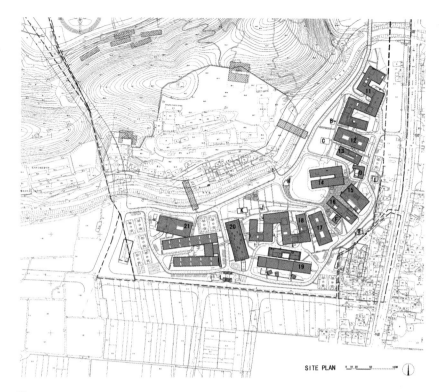

17

16 象山草图。
   画：王澍
   The sketch of Xiangshan Campus.
   Sketch: Wang Shu
17 象山总图。
   画：王澍
   The master plan of Xiangshan Campus.
   Sketch: Wang Shu

建造一个与自然相似的世界
BUILD A WORLD TO RESEMBLE NATURE

18

19

20

21

22

23

18　象山校园一隅。
　　摄：王澍
　　A part of Xiangshan Campus.
　　Photo: Wang Shu
19　葡萄牙建筑师阿尔瓦罗·西扎参观象山校园一景。
　　图片提供：王澍
　　A scene of Portuguese architect Alvaro Siza's visit to Xiangshan Campus.
　　Photo supply: Wang Shu
20　象山校园一隅。
　　图片提供：王澍
　　A part of Xiangshan Campus.
　　Photo supply: Wang Shu
21　象山校园一隅。
　　图片提供：王澍
　　Part of 'Hidden Distance', Xiangshan Campus.
　　Photo supply: Wang Shu
22　赵令穰（北宋）《秋塘图轴》。
　　Zhao Lingrang (Northern Song Dynasty), *Autumn Pond*.
23　象山校园不远处的山边餐馆里的桌子。
　　摄：王澍
　　The table in a restaurant at the foot of the hill near Xiangshan Campus.
　　Photo: Wang Shu

DISCUSSIONS AT THE SYMPOSIUM TOPOGRAPHY

会议讨论
地形学

TIME
28th May, 2011

PLACE
1st Floor, East Hall
The Main Auditorium
SEU

In addition to participants'
translation, other works are
by Li Hua, Shen Wen &
Yu Changhui.

时间
5月28日

地点
礼东二楼报告厅
东南大学

在参与者的翻译之外，
其他的工作是由李华、
沈雯和于长会完成。

THE SCENES OF
THE DISCUSSION

SCENE I

IS LANDSCAPE
ARCHITECTURE?

SCENE II

TOPOGRAPHY
HISTORY &
THE LIE OF THE LAND

SCENE III

A VIEW OF TOPOGRAPHY
FROM MY REAR WINDOW

SCENE IV

TERRAIN & XING SHENG
MISTY RIVER,
LAYERED PEAKS
BUILD A WORLD TO
RESEMBLE NATURE

讨论场景表

场景壹
《景观是建筑吗?》

场景贰
《地形学、历史及土地的谎言》

场景叁
《从我家后窗望见的地形学》

场景肆
《地形和形胜》
《从〈烟江叠嶂图〉看中国山水画》
《建造一个与自然相似的世界》

THE PERSONS OF THE DISCUSSION

Mr. Mark Cousins ················· COUSINS
Mr. David Leatherbarrow ·········· LEATHERBARROW
Mr. Shiqiao Li ··················· SHIQIAO
Mr. Junyang Wang ················ JUNYANG
Mr. Dongyang Liu ················ DONGYANG
Mr. Hui Zou ····················· HUI
Mr. John Dixon Hunt ············· HUNT
Mr. Andong Lu ·················· ANDONG
Mr. Jiang Feng ·················· JIANG
Mr. Shu Wang ··················· SHU
Mrs. Yongyi Lu ·················· YONGYI
Mrs. Wei Chen ·················· WEI

出场人员表

马克·卡森斯 ················· 卡森斯
戴维·莱瑟巴罗 ··············· 莱瑟巴罗
李士桥 ····················· 李士桥
王骏阳 ····················· 王骏阳
刘东洋 ····················· 刘东洋
邹晖 ······················ 邹晖
约翰·狄克逊·亨特 ············ 亨特
鲁安东 ····················· 鲁安东
冯江 ······················ 冯江
王澍 ······················ 王澍
卢永毅 ····················· 卢永毅
陈薇 ······················ 陈薇

场景壹
《景观是建筑吗？》
入场

卡森斯 我发现这个演讲可以帮我们弄清楚很多议题。就某一点说，如果把地形学从你的论点中抽离，你所说的很像孩子们玩的一种游戏，他们在一张纸上随意画一些点，然后用铅笔把每个点连起来。这一游戏的乐趣在于从中看到某种东西的生成：用连线来表现每个点之间联系的逻辑关系，而这些连线又导致了某些形态的产生以及某种力量的出现。这个想法又与"已在"形成了相互的支持，不论是在知识的形式之间，还是在身体介入的形式之间。另外一点是，当你刚才谈到场地和调研的时候，显然，"所予"这个概念根本靠不住。它让我想起了数据。数据本身不表明任何意义，或者说它是一种给定的"礼物"，一个容量的概念而已。你所说的地形学，恰恰是在描述那种"礼物"的各种特征。

莱瑟巴罗 是的，我所说的地形学或许和你所描述的点与点之间的连线很像。不过，我想说明的是，对于地形学的讨论，要求我们重新思考所谓"自然"与"艺术"之间泾渭分明的区分。为此，我使用了一些专门的术语以表明对这种区分的超越：如介入、参与、以及你所加上的相互支持。如果我们试着把设计看成是一种介入，那么它就是对我们生于其中的世界的一种特别的关注。在我看来，这种关注既不是理所当然的，也不是在证实既有的存在。它是一种介入的形式，试图寻找相关的表达，澄清不明之处，或使被忽视的东西焕发活力。我认为这种概念与设计的某些意义是一致的，但是在我演讲的总结里，我试图把它与近年来的实践中我们妄称的权威和控制做一个区分，因为如今城市、景观和建筑的管理问题是如此普遍，以至于我们认为他们该由我们来掌控。但我认为，在其中我们忽略了可以丰富我们的艺术以及使作品更富创意的潜力。

卡森斯 因此，我们可以说，第一种地形学行为是语言学。实际上，在某种意义上，它是在给景观命名。我记得读过一本马里姆的小说，讲的是一群乘客搭乘一艘驶向非洲西岸轮船的故事。途中，每个人都陷入了抑郁之中，船长解释说这样的情况很常见。他说，那是因为他们眼前的陆地从没有被命名。命名应该是第一个语言学形式的设计。

莱瑟巴罗 我知道，刚才提及的图景还需要继续研究，但是我觉得经过了一天的讨论，我们可以在某些时候重新回到呈现的问题上，或者我称之为一个呈现的谱系上来。我认为，约翰·迪克森·亨特昨天演讲中所说的书写，跟拍照、绘画、雕塑与构成一样，也是景观设计的一部分。我把它们称作为呈现的层级或程度。其中有些逐渐变得越来越透明，而有些则相反，更加具体化。我想把不同的艺术看成是这种呈现谱系中的一部分。我希望可以把你所说的命名与这个想法联系起来，即认为不同的艺术门类之间并不存在截然的差异，而更像是属于各种介入我们所在世界的方式的大家族中。

李士桥 戴维，谢谢你的演讲。今天会议的主题让我想起自己学习几个互相关联的英语单词的经历。英语是我的第二语言，我首次遇到单词"topic"时，它的意思是人们要去写或讨论的话题。后来，在软膏一词中，我遇到了"topical"，那是用来治疗身体表皮问题的东西。进入建筑学院之后，我学到了地形这个词，它有点像是这两种意思的结合，即一个关于表皮的话题。这个经历让我思考这样一个问题，在中国景观的创造中，无论是绘画还是园林，是否曾经使用过这些术语。不同的文化传统会不会有不同的解读土地的方式？当面对异域的土地，我们对它的解读或者说再解读是不是和以前的解读一样，只不过用的不是我们现在熟悉的术语？简而言之，中国的景观是地形学吗？

莱瑟巴罗 我相信有一些思考的方式可以达到某种程度的抽象度，而忽略你所说的那种基本性。

李士桥 忽视。

莱瑟巴罗 我想是的，如果思想达到了一定的透明度，似乎就可以超越你所描述的基本差异。但是我想说的是，有些时候，甚至在我们进行最为欢畅的学术交流时，总有些事会出其不意地发生，譬如打哈欠或者脸红之类的身体反应。有人也许会说：天哪！我多希望我们可以没有身体，那样我们就能对事物有完全清楚的理解了。但事实是我们无法脱离自己的身体，我们必定身处于其中。总有些时候我们会意识到我们身体的存在，而正是所有项目中这种残余的身体性是我想特别说明的，因为在我看来，我们的脑海中总是充满着对形式的意愿、对控制的欲望、对透明和一个有条不紊的世界的希冀。我们希望火车总是准点运行。然而，我们都知道，生活远不止此。我同样想说的是，在我们的设计中，不管是景观建筑还是建筑，我们都需要知道实践需要介入到那些身体性的环境中，他们先于高层次的呈现而存在。所以，恐怕，我的回答是亚里士多德式的。我喜欢柏拉图，但是他也需要脚踏实地。

王骏阳 为什么是六点而非五点或七点呢？

莱瑟巴罗 我想把它包含在我所说的关于地形学的内容里。我不想把时间抽象成在"不相干"的钟表上的时间。我想让时间成为我们生活的环境。我放了一张门把手的幻灯片，因为在我看来，它实际上是一种过去的印记，并通向未来。除了通过这些器具，我没有别的方式来知道我何时打开了窗户。我想把如门把手的这些元素称为地形呈现的工具。它们引导着我们的生活。通过这些呈现，我们获得了时序和空间上的方向感。我相信，如果没有它们，我们将会迷失。

刘东洋 我的这些问题其实都来自学生。不管是学建筑设计的还是学城市规划的，中国学生常常看到诸多新城像是按照效果图形象直接造起来似的。如果您到处走一走就会发现，好多新城建设对待土地的态度像是从零开始的，过去曾经存在过的东西根本不重要。也就是说，设计总在设置一个新的起点，一种新的秩序，非常人工化，就是"拍脑袋"的结果。学生因此总会问到：在这样的条件下，重思历史还有用吗？我们是不是可以完全构建一个新的故事？或许，我们的设计只是象征性地纪念一下过去？或者，我们可以从被伤害的土地那里学点东西，然后进行再造呢？概括地说，我们将如何进行这种对于地景的重新阅读呢？

莱瑟巴罗 在我看来，你的问题已经包含了它的答案：通过重新解读，根据我们现在的兴趣点去重新解读。我们感兴趣的并不是它们过去是什么，而是现在的它们与过去的它们并不是完全不同。你说你要建个房子，但是今天的房子不是五十年前的那个样子了，但也不是完全没有关系。连续性要求变化，但不是通过彻底的改变。总有些以前环境中的东西会残留下来。然而，只是这样是不够的。所以，是庆典吗？不。是忽视吗？不。是重新解读吗？是的。

邹晖 我有几点感想。我很认同你对"地形"概念的定义以及此概念如何相关于艺术、自然与建筑的通常关系、景观设计与一般定义上的设计等。如果我可以把自己对你的思想的理解延伸进东西文化的差异中，我觉得你的"地形"概念与中国的传统"山水"观有相似性。当我们在中国说到"山水"这个概念，意指与原初自然的某种互动。它要求人对自然的参与和交流，以图把自然理解为情感投射中的山山水水。我关注"山水"概念的理论操作层面，它也许可以在现代的设计观与我们通常称之为大地母亲的原初自然之间的隔离架起桥梁。在此我想把中国人对"大地母亲"的传统理解延伸到柏拉图的"chora"概念。* 当试图描述"chora"这个不可见的盆状存在时，柏拉图先将它比喻为"护士"，然后比喻为"母亲"。

* 姑且将它翻译成易于现代理解的"原初空间"或"空间的空间"。

当我们谈论设计和通过设计来理解自然时，这种比喻的类比总是指向那个超越简单主客体之分的原初世界，并唤起我们的情感回应。在类似于"chora"的想像层面上，中国的"山水"观提示我们建筑师应该带着同情与爱去展开设计，如此我们可以诗意地且可持续地栖居于自然。

莱瑟巴罗 我想对中国传统绘画中的山水有更多的了解。就像你所听到的，我首先根据传统的分类，将形式与材料对立了起来。但是我认为我们不应该将建筑师仅仅看成是形式的赋予者。我觉得我们需要认真思考这个假设。为什么我们讨论它的时候需要古典的根基？简单地说，这是前提，已存在了两千年了。但是，我仍然觉得我们需要超越这个观念。在我所理解的山之形和水之势的那个部分，山水的无形、创造性与关联是一个重要的条件，你可以称之为雾霭、氛围、某种变化或形变。环境的这种变化特征与达尔文描述的适应性很相似。在我看来，这有利于我们弄清楚，或者为我所说的介入领域的框架给出一个更为宽泛的定义，参与其中并接受有形与无形的结合。在我看来，这是一种更好的思考设计实践的方法。因此我乐于对山水有更多的了解，我认为这种思想可能也隐含在西方思想的传统中，但被其他形式的或具象化的概念遮蔽了，比如说原生态自然。这些都是我们需要超越的观念。建筑可以从景观中学到如此多的东西。昨天在我的演讲中，我将建筑当作景观来描述。那是我坚守自己论点的方式，我坚信这个世界所具有的潜力，我认为，景观建筑介入这个世界的方式能让建筑学到一些东西。

场景壹
落幕

场景贰
《地形学，历史与土地的伸展》
入场

李士桥 我很想听听您怎么看善意的谎言，或者说土地的谎言，与迪斯尼化之间的差异。这里，我不仅是指迪斯尼乐园，还指那些有点像"楚门世界"里虚拟现实的那种越来越流行的住宅区开发模式，在中国，我们随处可见这种土地伸展的模式。

亨特 问得好。我的意思是你说得对。对此，我需要做进一步的研究，当然，迪斯尼乐园也存在着谎言，迪斯尼乐园，从某种程度上，就是一个谎言。但你知道人们能够接受这种谎言。

李士桥 但迪斯尼乐园是在有意隐瞒真相。

亨特　它不是故意的。有些人去了那里，而且相信了它。

李士桥　所以说，那不是迪斯尼乐园的问题而是个体的问题……

亨特　它是一个接受的问题。我谈的是接受。很多去过迪斯尼乐园的人都认为这是一件完全可以接受的事情。

李士桥　我在谈的是城市问题；在亚洲，存在着一种日益增长的建立被认为是真实意象的趋势，人们确信不疑他们是生活在真实而不是谎言中。

亨特　是的，你说得对。我的意思是，我讲的话题有些方面还有待进一步探究。让我暂时回到"lie"这个词上。英语中，我们将有些东西称为善意的谎言——不会伤害你的谎言。它不是一个恶意的谎言，所以还好。要搞清楚你所指出的迪斯尼世界和我给的成功案例之间的差异，我还需要对善意的谎言做进一步的探究。我的意思是，总有些时候我们会撒一些善意的谎。有时候没有关系，而有时候你撒了个善意的谎，而两周后，却被戳穿了，"噢，天哪！我都做了什么！"你不得不与谎言共存。所以探究善意的谎言，不论是在生活中还是对景观建筑师来说，都是很棘手的事情。

鲁安东　我有一个小问题。在我看来，你的报告对莱瑟巴罗的报告在某种程度上是一个挑战，因为你定义了一个强调经验、想象和参与的"叙事"的概念，这和莱瑟巴罗的理解相当不同。在我看来，莱瑟巴罗排除或者试图隐藏景观的叙事。而我想问您的是，您是否认为叙事是、或者曾经是地形的一个必要或核心的概念。

亨特　这个问题有两个答案。第一个是，我受的是文学评论和历史学训练，而他不是，尽管我早已放弃了文学，不再和它有什么关系了。但是，我认为一个更好的答案是，有个法国人类学家说人就是叙述者，人们以他们自己的方式进行叙述。我们都处理自己的生活中和周围的世界里各种各样的故事。我的意思是，甚至我们生活的政治景观中都充满了故事。所以我们总会问，我真的相信电视上看到的东西吗？这背后还有什么故事？我们生活在一个叙述的世界里。因此，我会比莱瑟巴罗更看重从叙述的角度分析景观建筑学。现在轮到他说他的想法了。

莱瑟巴罗　简单地说，这只是我们理解叙事以及叙事采用形式的程度问题。我觉得叙事可以采用不同的形式。口头表述是其中的一种，但不是唯一的。或许还有其他形式，比如说空间叙事。

亨特　所有的空间叙述都需要语言表达。你不也是一直在使用语言嘛。

莱瑟巴罗　但是感知的方式多种多样，并不总是语言学层面的。

亨特　那是现象学的问题。但你必须使用语言。

莱瑟巴罗　我们都经历过这样的事，一个人说一件事，但他说话的方式指的却是指另一件事。比如说，我说：你，过来！但是我声音语调的真正意图是说：呆在那儿！这是非语言学层面的。这是另一种表达的形式。这层意思不是用语言表达出来的。

亨特　可是，你的思想中还是有语言存在。

莱瑟巴罗　感知不需要语言。

亨特　我还是讲一个简短的故事吧。一个德国朋友告诉我，有个家庭抚养了她的长子。这是一个做事有条理的德国家庭。一切都十分井然有序。这个孩子已经有一年没有说话了。两年后，他们有点儿担心了。他还是不说话。三年、四年、五年后，有一次在他们一起去餐馆就餐时，他弹着手指说，"盐"。他想要盐。于是，他母亲问为什么你突然说话了。他说，"嗯，这是一个奇怪的世界，在这样的世界面前，我不需要任何语言。"所以，因为需要盐，他就说话了。我想对于你的问题，答案是我们的角度非常不同，但是我觉得我们确实认为我们相聚在一起时，是土地、是历史背景将我们完全分开了。要是没有语言的话，我们就不会有像这样的一场会议。

卡森斯　很明显，地形学融合了历史和记忆。但是你举的例子似乎是在说，有一些地形学的介入不是为了记忆而是为了忘却。在传统意义上，与敌方和解需要达成共识并且"忘却"一些事情。因此总会有这样一些地方，一直存在着领域的争端，但是现在实际上希望做到的是，忘记争斗。

亨特　不对。我认为战争纪念碑在某种程度上是出于纪念，但对很多人来说，他们只是想忘记它。我现在想不到一个具体的例子。芬雷或许是一个。我知道他就有这个想法。他曾经谈到过碑文。他说，"在花园里我不需要碑文。要是我们有什么想法，那么就不要忘记它吧。"但是，对他来说，景观遗忘或者说不涉猎引发联想的词语。不管怎么样，他想让我们看到，我们有过提议，这不是忘记，在没有计划这个、这个或者这个的接收者看来，这是空间组织上的忘记。比如说，古英格兰人的印记，凯尔特人的标记，那些从英国南部消失的直立的石头，在某种程度上，他们象征着一种缺失，尤其是对那里的当地人来说。那些石头标志着为什么他们不想离开，甚至标志着他们不愿意讨论什么。现在我想不起一个例子了。这是一个有意思的挑战，我会发电子邮件跟你讨论讨论。

莱瑟巴罗 在一些古罗马遗址的历史上，存在着反建造的事例，为了建一堵重新定义城市范围的墙而拆除周界的墙；为了重新定义边界而遗忘之前的边界，在建造新的墙之前进行彻底的清除，几乎就像奠基与落成的反向活动，建造这座古城建筑的逆向行为。

场景贰
落幕

场景叁
《从我家后窗望见的地形学》
入场

王骏阳 谢谢。这个讲座带来了一个学术议题，并以追溯到中国现实的对于地形学意义的详尽探索而开始。那是我们与之共存的现实。我们有几分钟用于提问。

邹晖 听了你的关于某特定区域的地形景观在历史中的变迁的述说，我想提一个关于历史文本的作者的问题，即你引用的记载地区性地形地貌的传统文本类型"县志"。在历史上这些县志的作者往往是文人官员。虽然他们任职为官僚，但对当地景观的历史文化渊源相当了解。是否介绍一下当代这类地区性地形文本作者的一般情况？

刘东洋 方志的编撰人员如今仍然是学者。其中的多数人员是。要想成为某个县志或是市志办公室里的编撰人员，中文或是历史系的大学文凭还是需要的。这是我有限的一点了解，很少听说建筑师今天去写方志的。

冯江 我试着把刘东洋老师的这个讲座同前面两位教授的讲座结合起来，觉得有些困惑。先说说我对前面两位教授的讲座的一点印象。我觉得他们都讲了两个部分，第一个部分是对地形学的理解，第二个都用了一些案例，去说地形学的理解如何在设计或者实践当中去呈现。我觉得好像他们的选择刚好是相反的。因为亨特教授讲了许多知识和历史的内容，但是他在后面的例子里面更多呈现的却是一些"trace"，一些我们能看得见的"痕迹"。而莱瑟巴罗教授，我觉得他有一些细腻、敏感的、情感上的体验，前半部分很多是讲您说的"地貌"，而亨特教授更多地是讲"地形"，有趣的是，莱瑟巴罗教授在举的例子中却提到了很多次"order"，结构性的秩序。我觉得他们的理解和他们讲的例子似乎应该掉个个儿。这是一个感想。

另外，结合到您讲的中国传统上对地形的理解，我觉得理解是一方面，而呈现则是另外一方面，很多时候，会把那个理解故意摁回去。历史上，我们会觉得地形当中，譬如说，会谈风水的知识，我们认为那是天然的，这个"自然"存在着一个隐藏的结构，它是隐含在本来的那个地形当中的。但是我发现，如果在私家园林的一个小尺度里面，我们会去体现自然的方物和结构，但却不一定要它的"貌"。在一个更大尺度的计划例如苑囿、皇陵或者依山水而建的聚落里，我们会发现，把它的这个结构勾勒出来以后，往往并不去过分地呈现或者是刻画对地形学的理解，它自然而然就在那儿了，所以会看到前人把这个地形的结构又放回地下去了，似乎只在不经意间得到了自然的造化，不然，反而会觉得有太多人力穿凿，未尽其妙。我觉得这可能是几个矛盾的理解，想听听你的看法。

刘东洋 亨特教授真的更看重地景中的内在秩序而莱瑟巴罗教授更看重基地的特征和容貌吗？我真不太肯定。我没有太大把握去评价亨特教授在这一问题上的立场。倒是谈到莱瑟巴罗时，我想，他更多的该是持有阐释学态度的人吧。他会强调把解读地景当成一种动态的过程，既是对场所结构或者说"结构化过程"，也是对场所环围的一种逐渐深读。设计也一定是这种深读的一部分，如他所言，是一种对地景的刻画。而将设计再度放回环围去，或者如您所言，让"刻画出来的秩序"不那么具有刻画的痕迹感，也许就是古代文人造园时所追求的理想境界。那既需要时间也需要设计师们的不刻意的刻意。在那种状态，地形研究或是地景再造都不能只是"设计遵从自然"那么简单，而是要更进一步。

冯江 我觉得莱瑟巴罗教授提的那些例子，是很充分地表达了那种理解的，把它显现在这个设计过的地方。但是它是不是通过这个肌理的东西去呈现一个秩序，我觉得好像……

刘东洋 其实，我特别感兴趣的场景是，假如让亨特和莱瑟巴罗二位教授去阅读和设计在中国某处的一块场地，不是在某个遥远而美丽的乡村，而是那种地貌都被推平、等待开发的大都会里的基地。那里，之前的河流没了，道路就降落到地面上。他们从这样的基地条件出发会怎么做？这是我们经常遇到的现状。时不时就有学生请教说，在这样的土地上，我们该怎么做？我们连基地周围的邻居是谁、将是怎样都不知道。在这样的条件下，无论是地质构造的线索还是地表的环围都很难抓到。

场景叁
落幕

场景肆
《地形和形胜》
《从〈烟江叠嶂图〉看中国山水画》
《建造一个与自然相似的世界》
入场

王骏阳　谢谢。现在是提问和讨论时间。首先，我想问鲁安东和李士桥。在翻译"形胜"这两个字方面，你们可以给我们一些建设性的意见吗？

鲁安东　我想应该是"formal excellence"。●

李士桥　"形胜"的翻译要依情况而定。"形胜"与力量变化有关。"形胜"是可被感知的形式力量，所以其形式是静态的，但是拥有在其他条件下得以实现和设定的动态的力量。"Modulation"可能比较合适，寓于形式中的转变。这的确很难，完全是不同的语言。

邹晖　我想从历史的角度评一下王澍的这个设计作品。这是一个很令人启发的作品。我赞赏他把对宏观世界的理解与北宋的传统感知模式联系起来，以及他运用诠释的绘画与图示来体现其理解。他的设计是对某个历史时期美学观的想象性拓展。我在此应和前面林海钟先生讲的，宋代是中国绘画史上的重理性的时代。人们开始关注形体的细节并理性化对自然的感知。在那之前的历史时期，我们没有看到对细节如此关注。同时，正因为观察与梳理形体细节的理性方式，哲学家得以在更高的形而上学的层次上对传统的形而下的理论发出疑问。这在方法论上显得有些似是而非，但可以从大的哲学背景中去理解。在宋代，有强调"理"的新儒家思想的扩散，也有集佛道教于沉思的禅学的树立。这个在形体细节的理性考辨与对大幅山水的情感投射之间的矛盾张力很有趣。对这个关系的理解也许可以帮助我们评论王澍的这个作品并反思当代中国的城市主义。

鲁安东　我对王澍最后放映的那张图片特别动心。同时我也对王澍对这个建筑项目的整个解说过程感兴趣。开始时，你给我们看了传统绘画并介绍了一些远的东西和一些近的东西。此时，你提醒我们山作为一种暗示之物一直都在、而建筑则作为一种线索提示着我们那是山，这体现了一种对山特殊的尊敬感。而后，我感觉到在建筑自身内有许许多多美妙的场所，在对建筑的游历过程中●山也或多或少被遗忘了。然而最后那张图片忽然令我们意识到那一不在场之物才是重要的、不在基地里的东西才是重要的，它提醒我们已经迷失在建筑中、或者被建筑捕获。我意识到这个建筑并非关于建筑，而是关于如何脱离现实。

莱瑟巴罗　我想问一个关于材料的问题。我还不确定问谁这个问题。我希望这个问题对三个演讲人都有意义，从他们身上我学到了很多。但由于是通过翻译，我显然只理解了报告的一部分。有个吸引我的地方是那个画家诗人对着石头沉思了三到五个小时的画面。我想以这样的方式提问：是不是有一种距离，不仅仅存在于绘画或园林的空间组织结构中，还存在于物质的具象表现中？石头会不会比树叶或水更远？我倾向于认为石头是远的，它们有一种特定的距离，让诗人几个小时沉浸于其中。我认为有另外一种对景观的介入，它是变化的能力，它就像散文中的物质关系，等等。我想知道人们说得那么美妙的、有关"远"的东西，是否与我所认为的介入的框架有关，它使我们用不同的方式、通过不同的距离来融入这个世界。它也有可能产生于材料之中。我的问题是关于最后一个项目，其中给我触动最深的是材料的多样性。我推测，我们是不是可以这样说，石材比木材更远，更需要人们去深察，并且更与想象力有关。那么，可能有一种关于材料组织的逻辑，它与作为不同距离模式的空间结构有关。因此，关于园林或建筑中的身体性的问题，与人们说的空间、或远、或画家详细描绘的那种远，紧密相联。在材料中，是否存在着一种距离的暗示？

我能再谈一个我的观点吗？我一直在想的是材料的加工方式，而不仅是材料本身。如果我理解了今天对材料的解释的话，那么，由于石材可以被抛光，也可以被加工得粗糙，所以石材能以不同的方式产生某种氛围，并且使他们显得远近不同。因此，材料即是材料加工完成的方式。

王澍　确实有这个问题。整个建筑应该说是考虑比较远的感觉。这一时期我的建筑有一个特点，比较黑，也就是说从这一时期开始，我没有做过纯白的建筑，都比较黑。完了之后我称之为它又有点脏，我还说过现代建筑不长毛，而我的建筑还长毛。它是这样一种，好像是一种动物般的，好像会呼吸的这样一种房子。它跟整个现在的背景对比，整个感觉就很远，这是第一层。那么第二层，因为我们这个校园里主要要用的砖石材料全都是旧建筑的回收料，很多材料都已经有五十年、一百年，甚至两百年的历史，它是一种回收的材

● 注：在会后与陈薇老师的讨论中，我认为更好的译法是"topographical excellence"。

● 为了介绍这一建筑

料的建筑。当时很多人反对，说新建筑怎么能用旧材料。因为中国人现在都不喜欢旧的，都喜欢新的。那么我们说这是一个很划算的生意，时间就是生命，我们建好之后，我们校园其实立刻就有了五十年的历史，这是一个多划算的事情。所以它只是从时间的一个角度来说，带有远近。至于说第三个问题：是关于石头的远不远和木头近不近的问题。我觉得它是一个相对的问题。当石头以某种方式用的时候，它会感觉是远的，但是如果你用寻常的方法用的话，其实它也近。比如，我在一个房子里用混凝土造了一个类似于石头一样的铁架，塞在一个房子之中，连混凝土都会感觉是远的，并不是说它就会变近。木头我觉得是这样，总体上你会感觉它很温暖，很亲切，好像是近的。但是实际上现在中国的建筑很少有这样用木头来做的，所以人们会接触到它，它就好像是一个遥远的事情，这怎么可能去实现呢？所以我们整个的校园很多人都说它是一个乌托邦。但是我觉得它的意思是，在中国现在的条件下，乌托邦是可以实现的，这很重要。

刘东洋　王老师，我想问问你的二期里，那个村落跟它有什么关系？它们是共生关系吗，而且也没有设围墙……

王澍　因为我觉得这种状态，它们可以共存。

刘东洋　将来会是什么状态？

王澍　现在忍这么久之后，学校还是决定把它拆了，但是现在希望用一个新建筑给它补上。

刘东洋　我看到村民的房子是各种各样的……

王澍　有七八栋。我是不愿意拆它，因为我觉得那是一段真实的历史，我们不会用美丑来判断它。因为这段历史发生过，这很重要，我们不要把所有的历史都给干掉，然后编造一个新的出来。

王骏阳　我想把王澍的这个项目与今天早晨莱瑟巴罗教授对形式主义的批判联系起来看看。莱瑟巴罗教授认为，地形学既不是形式，也不是单纯的地的翻译。为此他举了两个例子，一个是埃森曼的，还有一个是西扎的建筑。他批评了埃森曼的卫克斯纳艺术中心。在我看来，那是个非常强的形式，非常强烈的形式。而西扎的那个建筑，用一个时髦的话来说，可能是一个"弱形式"。我觉得这是一个很有意义的差别。我在想这样的差别是否也存在于王澍崇尚的中国山水画中的理念和他的象山校区。前者在我看来属于一种"弱形式"，而王澍的象山小区则充满了强烈的形式，甚至有点拜物教的感觉。我很想听听王澍自己是怎么看这个问题的。当然了，我并不预期王澍会同意我的观点，因为他的报告已经告诉我们，他对象山校园及其建筑的设计是非常满意的，似乎不觉得有什么可质疑的。

卢永毅　王澍老师以对这么多古代山水画的阅读，来引导我们理解你的作品。我关注到这里面你是要建立一些概念上的关联性，其中有几个好像是你特别感兴趣的：一个是总体上的把握，这似乎容易理解；另一个是你多次讲的、这些山水画的抽象性；还有一个也是你强调的，说这些山水画艺术是有一种精确性的。我想知道，这种抽象性和精确性是指什么？如果这两点也是你有意识地要在象山校园这个作品的设计实践里有所呼应的话，那究竟是以什么来体现的呢？

卡森斯　我想就建筑师如何使用"形式"这个术语谈一点看法。十年来，我们对于什么是形式争论不休。但是我发现，建筑师之间讨论形式时，通常指的是一个作品中他们喜欢的部分。那个部分被理所当然地称为作品的形式。事实上我认为，如果他们直接说是什么使我感到愉悦或使你感到愉悦，问题会简单得多，而且我们可以就此更多地在理论层面上讨论形式的问题。我们都可以对形式意味着什么发表自己的看法，但这些讨论中似乎存在着这种奇怪的现象。

莱瑟巴罗　我也喜欢各种各样的东西。今天我们学到了很多。回想一下我们之前对于地形学意义的讨论，的确存在着一种将心智与身体相分离的观念。因而在某种程度上形式是不可或缺的。然而即便如此，形式仍只是问题的一个方面，身体总是存在的，而身体具有实体性。我唯一的问题是，希望在对形式的讨论中，能将问题的另一个方面带回到对话中。物质性确实出现在这些项目和园林之中。对于我来说，目前我还没弄懂但希望能尽快弄懂的是，材料及其处理方式与空间直觉之间的关联，而这种关联在园林设计和建筑场景的构筑中发挥着作用。我认为，这个话题要好好思考，因为我相信空间可以变近或变远，不只是通过绘画中描述的方式，而是通过那些既存的物质性：薄雾，水，抛光的石头等等。我觉得我对这些建筑的认知并不局限在这里的材料本身，而在于和人们的行为相关的材料，或是用来眺望的阳台，或是用来烹饪的房间，或是用来读书的书房。就距离而言，材料很重要，并且这二者共同构筑了目前为止我所能描述出的这些场景。这就是我看这个项目时的直觉。

邹晖　巧合的是，正是在宋代人们开始绘画"千里江山"。这是在历史上首次出现对大幅山水充满信心的感知把握与表象。由此导致后来的中国

5月28日
28TH MAY, 2011

园林中出现的"小中见大"的设计原则。

王澍　我都害怕我把一开始的问题给忘了。一开始的时候，王骏阳提到两个建筑师，我觉得挺有意思，一个是艾森曼，一个是西扎。说埃森曼做的建筑形式很强，但我记得他恰恰写文章研究弱的形式。这是挺有意思的一个反差。那么西扎的东西呢，表面上不动声色，会有一点点小小的戏剧性，但是他的恰恰是特别有尊严的事情，他的房子永远能够感觉到是有尊严。因为这个世界上能把房子做得感觉有尊严的建筑师其实不多，所以说，它其实内在很强大。我觉得王骏阳可能稍微理解错了点我的意思。你觉得我骄傲的是，我说它会好像是我做的吗？我的意思其实是说，我确实有时候解释不太清楚。我讲的这种东西在思考当中很精确，但中间是带有跳跃性的。我很难用很逻辑的办法把它说得很明白，所以最后产生的结果是不可以想象的，比如象山这个项目，我的业主没有见过我最后的方案，他们没有见过效果图，也没有见过模型。我所有的控制全是在我的意识中，在最标准的建筑师的平面草图中，所以当它建成的时候，连我自己都会惊讶，我们当然都可以享受这种感觉。

刘东洋　您今天下午讲座里所呈现的观点相当成熟。我想问，在象山这个项目的哪个阶段，您就有了今天下午讲座的这个境界？

王澍　最早是从 2001 年开始的。

刘东洋　您今天所讨论的问题在那个时候就已经差不多……

王澍　我最早开始思考这个问题是从 1983 年开始的。刚刚提到关于形的强弱的问题，一方面我同意王骏阳的意见，连我自己都觉得可能形态上还是很强烈的。所以，我特别地给我自己界定，我自己判断我现在大概是活在 17 世纪晚期，这是我整个意识大概能达到的高度，这是我对自己的评价。我还没敢说，我已经到了南宋、到了北宋，我还真不敢说，因为明清是一个带有文艺复兴色彩的时期，整个一代人重新整理过去的文献和图像，是再思考的那样一个状态，但是确实和北宋、南宋还是蛮远的一个距离。当然反过来关于形的强弱，有个问题就在于我更在意的是形和形之间的变，而不在于一个形的强弱，这才是主要的。整个大和小，就是从大到小的变化，你把它切分了，像拍电影一样，切割分成了很多集之后，那么怎么样从一个环节到另外一个环节，是用一种不连续的叙事来进行讨论，所谓抽象就是这么产生的。它并不是一个连续的叙事，它有点像卡尔维诺的意识流那种关系，它带有那种抽象性、叙事性在里面，所以它从一个形到另一个形以后，从一个空间到另一个空间，从一个现场到一个现场，它带有很强烈的突变性，但同时还带有一种难以表达的相似性。

刘东洋　王老师，关于您讨论的抽象的写实，那在很大程度上体现了文人绘画的优势。他们可以用想象、用对游观之后的整体体验去超越时空固定地点的约束，他们把远处的景物同样画得非常仔细。这是绘画。您在建筑创作中又如何做到类似的时空叠加，能够呈现"既在此地、又在别处"的感觉呢，会不会觉得这个绘画的意境和方式是不可译的，有没有这个时刻？

王澍　当然有。比如说，因为有一个远的距离，大的空间使你在远距离观看的时候，整个形式就已经被控制，但是你走到跟前的时候，你根本看不到那个大的控制形。这才是我做东西算是有点讲究的地方，走在跟前时，那个控制形就不见了，这时候是一个现场，现场的形，而不会说远处看的是这个房子，到跟前看还是这个房子。我把建筑和建筑之间的间距控制到规范允许的最小范围，所以你根本没办法看，基本上是这样的一种手段。建筑和建筑之间摆放的方式也采用的是这种手段。但是除了这条线索之外，还有另外一条线索，我称之为叫讨论，就是"dialogue"，就是说这个房子是带有自己内部的一个讨论。因为它离山的距离有点远，而那山比较平，所以我很难建立起这个房子和山的直接的关系，这时候建筑就开始它自己的讨论。

王骏阳　那是不是陈老师说的第二种，势比较弱用形来补？

王澍　我的好像不是这个关系。我说的是另外一种。我觉得，陈老师的说法有点给建筑师设的一个理解，是关于不在场和在场的讨论，就是那个山尽管不起眼，也不高，但在这一组里面一切都是次要的，就它最重要。所以所有的反应，看上去的反应，总有它在那个地方起的作用，这才是一个主要的考虑，有点像博尔赫斯的说法，在《交叉小径的花园》里面，唯一没有突出的词是时间，他通篇谈的都没有时间的内容，通篇谈的都是空间，最后像谜底一样揭开说，原来他是讨论时间的问题。

刘东洋　王老师，前面的那条河后来有没有处理过？这条河当中还有一个岛？我说的当时它是在那儿，还是你为了……

王澍　这个实际上是一个错误。一开始看原始总图的时候，我以为那有一个池塘，于是我就产生了这个观想。现在去看是一个工厂的围墙，很常见，但是我最后还是把它挖成了池塘，因为那是我探寻的起讫。

王骏阳　我想问一下陈老师，刚刚我说王澍的象山有一个很强的形，这是不是形胜，用这个强的形来弥补周围有点平庸感的山？当然即便是形胜，也有做得好和做得不好的，但那可以是一个思考的方向。

陈薇　我觉得我们在讨论王澍老师作品的时候，有两个误区。

第一，就是明清的园林和唐宋的园林有一很大分野。如果按照唐宋当时园林正常发展的趋势，应该更加纯粹。比如，日本枯山水园林*其实是受到唐代园林意向和宋代画派技法的影响。而中国因为有一个元朝夹在中间，发展走向就出现了变化，中国园林在唐宋后没有走上抽象纯粹的道路，有一个历史原因，即当时南宋后的江南文人被元朝宫员强迫到宫廷里去作画，而南人在元朝身份低下，所以他们心中充满郁闷。因此，在作画的时候往往要把纷繁复杂的情绪表达出来，一幅画难以解脱，还要加上诗等等，多重的表达才能把他心中的情结疏解出来，这个转变是非常重要的，对园林也有很大影响，就是要用很多要素来表达意趣，建筑也在园中增多，当时园林叫"园亭"，我觉得这个分野非常重要。而日本枯山水园林和建筑能走上今天的道路，一方面是受到中国唐宋重要影响，另一方面中国传播过去的禅宗提倡的特有的此岸性，也与日本自身文化相合拍，在日本显得尤为突出并深入人心，即将"空无"的"形而上"概念转换为主体内心自我实现的一种自律、一种"形而下"的心理体验和一种思维方式。这就如莱瑟巴罗教授给我们介绍的，像槇文彦，他会把一种极端的情绪，用纯粹的材料简洁地表达出来。回过头来说，当我们把唐宋的园林境界和明清的园林要素捆在一起说的时候，会产生矛盾。我一直觉得，如果园林的根本作用是给精神一种疏缓的话，唐宋时期是最高境界。明清只是在手法上给我们一些启示。当我们两者兼顾，都要学习，做在一个作品中的时候，这是非常困难的。你要么学习它的技巧，就是用空间的方法来做；你要么就是取大势，弱化你的建筑。如果你想两个都要，都放在那里，它肯定会产生矛盾。这是我的一个理解。

第二，我觉得建筑在一种环境中的强势和弱势，是和它的功能分不开的。王澍老师的象山校园是功能性非常强的一个建筑。如果西扎在里面走一圈，他是可以感受到随着时间的进程周遭空间的变化，但如果我是一名学生，要很快进入教室，没有这个游赏需求和心境的话，我可能会感受不到，我可能感受的只是它的一个局部，所以我觉得这会带来矛盾：一个是境界的东西，一个是手法的东西；一个是游赏的心情，一个是功能的需求，可能会难解难分。从我对园林在当代发展研究角度来讲的话，我更希望走上日本曾经走过的道路。因为现代人生活节奏快，常感时间紧迫，在没有经过很长的距离和时间去体会的情况下，建筑师或艺术家若已经把那种理解过的东西层叠在一起，给你一个爆发力或者给你一个想象，我觉得这可能更加接近现代生活的需求。

场景肆
落幕

● karesansui

SCENE I

IS LANDSCAPE ARCHITECTURE

ENTER

COUSINS I find the talk helpful in terms of sorting out a number of issues. I am thinking that at one point, in your argument without topography, what you were saying is rather like the game that children play when they make a series of random dots on a page and then use a pencil to link each dot. The pleasure they get is seeing something grow where the logical existence of every dot having a connection is materialised in the line, which leads to the emergence of certain shapes, and certain intensities. This idea cooperates with 'what's there', either between the forms of knowledge, or between the forms of some kind of physical intervention. The other point is that when you were talking about sites and investigations, it is extraordinary how corrupted the idea of the 'given' becomes. It makes me think of data. It's entirely bereft of signification, or that it's a gift and is the dimension of capacity. What you are calling topography is precisely recognising the kinds of lineaments of that gift.

LEATHERBARROW Yes, what I am saying can be compared to drawing connections between the dots, as you've described. I tried to explain that part of my argument about topography requires that we consider again the sharp distinction between so-called 'nature' and so-called 'art'. I introduced a few terms to point beyond this distinction: involvement, participation, and you added cooperation. If we try to see design as involvement, it would appear as a particular kind of concern for what exists around us in the world. The kind of concern I have in mind does not simply take for granted and affirm what it is here. It is a form of involvement that seeks articulation, clarification, or making vivid what has been overlooked. I think this conception is congenial to some senses of design, but in my concluding comments I tried to distinguish it from that sense of authority and control that we have arrogated to ourselves in recent practice, because the matter of administrating cities, landscapes and buildings is now so common that we take it to be our proper remit. But I think in so doing we neglect capacities that would enrich our art and make the work even more creative.

COUSINS In the service of that, we might say that, the first kind of topographical act is linguistic. It is actually, naming, in some sense, the landscape. I remember reading a novel by Prosper Mérimée which has passengers being taken on a ship down the west coast of Africa. Everyone falls into a depression, and the captain explains that this is very common. It is because, he says, the land they are looking at has never been named. Naming would be a first, linguistic form of design.

LEATHERBARROW I know the image that was just introduced requires quite a bit of work, but I thought that through the course of the day, if seems sensible, we could try to occasionally return to this matter of articulation, or what I call a spectrum of articulations. I think writing, as John• showed yesterday, is as much part of landscape working as are picturing, painting, carving and forming. I want to call each of those practices *levels* or *degrees* of articulation. Some are increasingly more transparent, and others more opaque, more fully embodied and concrete. I want to see the arts as part of such a spectrum of articulation. Your sense of naming, I hope I could relate to this idea, a notion that doesn't call for a sharp distinction between the arts, but sees them more like siblings in a family of involvements with the world that we have been given.

SHIQIAO Thank you David for your presentation. The title of this session somehow reminds me of my experience of learning a few related English words. As someone who learned English as a second language, I first encountered the word 'topic' to mean something to write about. Later on, I encountered the word 'topical' as in topical cream, for a surface of the body where there seems to be a problem. After entering the school of architecture, I learnt the word topography, somewhat a combination of the two meanings, a topical topic, so to speak. All these make me wonder if these terms ever existed when the Chinese landscape, either as a painting or as a garden, was created. Do different cultural traditions read land differently? When one encounters a foreign land, do we read it, or do we re-read it as it has probably been read before, just not in the terms that we know? In short, is Chinese landscape topography?

LEATHERBARROW I believe that there are styles of thought that achieve a degree of abstraction that neglects that foundation.

SHIQIAO The negligence.

• John Dixon Hunt

LEATHERBARROW I believe so, kinds of thought that operate with such transparency that reference to what you describe as the ground is seemingly unnecessary. But what I've tried to say is that sometimes even in our most convivial, collegial and communicative moments, something else, something unthought obtrudes itself, something like a yawn or blush, some consequence of embodiment. Some might say: Dear God! I wish we could be without our bodies, then we'd understand things perfectly. The truth is we are not disembodied. The truth is that we are situated. Every now and then, we are reminded of that and it is that remnant of embodiment in all projects that I am making an apology for, because I think the will to form, the desire for control, the hope for transparent, well-administered world is insufficient. We'd like the trains to run on time. But there is more to life than that. We know that. I'm also trying to say that in design, in both landscape architecture and architecture, we need to acknowledge practices that require involvement in those embodied conditions that antecede the higher articulations. So my answer is Aristotelian, I'm afraid. I like Plato, but he needs to keep his feet on the ground.

JUNYANG Why does time explain six points?

LEATHERBARROW I meant it to be subsumed into what I said about topography. I don't want to abstract time as something 'over there' on the clock. I want time to be the milieu of our lives. I showed the slide of the door handle because for me, it is really an inscription of the past and an opening into a future. I have no other way of knowing when I am opening the window, than through these instruments. I want to call elements like the door handle instruments of topographical articulation. They orientate our lives. We are orientated, chronologically and spatially, through these articulations. Without them, I believe, we are lost.

DONGYANG My question is actually from students. Most students in China, both in architectural design and urban planning, constantly face new cities built directly from renderings. If you travelled around, you would see many new towns were planned almost from scratch as if nothing there previously mattered much. In a sense the design creates a new start and order of its own, very artificial and arbitrary. Thus students constantly ask, what is the usefulness to relearn the history of a place? What are we going to do with this situation? Shall we remake a story anew? Shall we commemorate the past only in a symbolic way? Or shall we learn again from the deformed land and then create something else? After all, how should we proceed with this rereading of landscape?

LEATHERBARROW Your question contains its own answer: by rereading, rereading from the vantage of our current interests. Our interests are not what they once were, but they are not wholly different. So you say that you are going to build the house, but the house today is not what it was 50 years ago. But it is not unrelated. Continuity requires change, by means of not thorough change. There is always some remnant of the antecedent condition. And yet, it's inadequate. So commemoration? No. Neglect? No. Rereading? Yes.

HUI I would like to comment on some of your points. I appreciate how you defined topography and how that concept could be related to art, the taking-for-granted relationship between nature and architecture, or landscape architecture and design in general. If I can extend my understanding of your thoughts into the cultural difference between East and West, I feel your definition of topography holds a certain similarity to the traditional concept of Chinese landscape, so-called 'shan shui', literally 'mountains and waters'. When we talk about the term 'shan shui' in China, it means an involvement in a certain way with nature, raw nature. It requires human participation and engagement into nature in order to understand nature as emotionally bounded mountains and waters. I like the theoretical level of operation of this concept, which helps us bridge the gap between design in the modern sense and nature which we habitually call 'Mother Earth'. I also like to refer this Chinese understanding of 'Mother Earth' to Plato's concept 'chora'.* When describing the invisible receptacle of chora, he depended on metaphors—first 'nurse', and then 'mother.' When we talk about design and what we understand about nature through design, such a metaphoric analogy always hints at a primordial world which goes beyond mere subject and object, and calls for an emotional response from us. The Chinese concept of shan shui, on a similar imaginative scale of chora, suggests we implement design with compassion and love so that we can dwell in nature poetically and sustainably.

LEATHERBARROW I want to learn about the mountains and water depicted in classical Chi-

• probably, the 'primordial space' or 'space of space' in the modern sense.

nese paintings. As you heard, I began by polarising form and material, according to the classical topos. But I believe we need to get beyond the idea of the architect as the form giver. I think we need to think through that assumption. Why do we want the classical foundation for this? Let's just say it is a 2000 year old premise. But I still think we need to get beyond it. In part of what I understand to be the profile of mountain and generative potentiality of the water, the formlessness, the creativity, the conjunction of mountain and water, is an important condition, what you call the mist, or atmosphere, or some kind of change or metamorphosis. The metamorphic character of that milieu is analogous to the adaptations Charles Darwin described. That for me would be a very helpful insight to help us clarify, or give greater definition to the framework that I'm saying is the domain of our involvement, answering the invitation and accepting the gift of this conjunction of profile and formlessness. That to me is a better way of thinking about design practice. So I'm happy to learn more about it and I think it is probably implicit in the tradition of Western thought, but it has been covered over by other conceptions of form or embodiment, such as raw nature. These are the conceptions we need to pass beyond. Architecture has so much to learn from landscape. Yesterday, I gave a lecture in which I described the building as if it were a landscape. That's how committed I am to this line of argument, to the potentiality of the world that I think architecture can learn from the way that landscape architecture is involved in the world that we've been given.

SCENE I
EXEUNT

SCENE II
TOPOGRAPHY, HISTORY & THE LIE OF THE LAND

ENTER

SHIQIAO I'm interested in your position to further articulate the differences between a good lie, or the lie of the land and Disneyfication. Here I'm not just referring to Disneyland but to the increasingly dominant residential development formats that are a bit like the *Truman Show* set, and that we find occupying vast stretches of land in China.

HUNT It's a very good question. I mean you're quite right. But I need to, and I really haven't done the exploration that I need to do, but there is certainly a lie that exists in Disneyland which is somehow a lie. You know one can accept it.

SHIQIAO
You know that Disneyland lies intentionally.

HUNT Ah, it doesn't lie intentionally. Some people go there and believe it.

SHIQIAO But it's not the problem of Disneyland but the individual...

HUNT Well, it's a problem of reception, and I talked about reception. Many people who go to Disneyland think it is a perfectly acceptable event.

SHIQIAO I'm talking about urbanism; in Asia, there is an increasing trend to build images that are thought to be authentic, and people are absolutely convinced that they live in the truth rather than in a lie.

HUNT Yes, you're right. I mean that there is a subarea of my subject which has to be explored much more. Let me stick to the word 'lie' for a moment. In English we talk about something called 'white lie' — a lie won't harm you. It's not a black lie, and it'll be all right. I think I need to explore a lot more white lies to pull out this territory you've rightly pointed to between the Disney world if you like, and examples which I gave, which I think, are successful projects. I mean we all use white lies sometimes. And sometimes it doesn't matter and sometimes you know you use a white lie and about two weeks later it comes back to you, oh, my god, what have I done! And you have to live with it. So the white lie is very tricky thing to explore even in life as well as for landscape architects.

ANDONG I have a small question. Your presentation seems to me a small challenge to David's, as you've defined a concept of narrative which stresses experience, imagination and participation in a quite different way from David's. It seems to me that David excluded, or tried to hide away, the narrative of landscape. And I'd ask you whether you think narrative is, or was, an essential or hardcore concept to topography.

HUNT I have two answers to this question. One is that I was trained as a literary critic and historian. He was not, even though I had given up literature and put it behind me. However I think the better answer is that some French anthropologist talks about human being as homonarrators, people narrate in their own way. We are all dealing with stories in our own life, in the world around us, I mean even the political landscape in which we live is full of stories. So we actually have to say, do I really believe what I've just seen on television, what's behind the story? So we live in a world of narratives. And I will give much more weight to an analysis of landscape architecture in terms of narrative than I think David would. Now it's up to him to say what he wants.

LEATHERBARROW Very briefly, I will say it's just a matter of degree how we understand narration and what forms it takes. My proposal is that it takes different forms. And verbal articulation is one among them, but not the only form. Maybe there is something called spatial narrative.

HUNT All spatial narratives need words. You are using words all the time.

LEATHERBARROW But sense is conferred in different ways which are not always linguistic.

HUNT That's the phenomenological flaw. I think you have to use words.

LEATHERBARROW There is direct experience that one can say one thing and mean something else in the way it is said. How does that happen? I say: You, come here! But what my tone of voice really means is: Stay there! It's non-linguistic. It's another form of expression. The sense is non-verbal.

HUNT But you are then assuming that there is a word in your mind.

LEATHERBARROW Words aren't required for sense.

HUNT Let me tell you a very quick story. A German friend told me about her family bringing up her elder son. The family is a very organised German family. Everything was very perfectly organised in the right order and this child didn't speak for one year. After two years they were a bit worried. He didn't speak. After three years, four years, five years, and when they were in a restaurant together, he clicked his finger and said 'salt'. He wanted some salt. And the mother said, 'Why have you suddenly spoken?' He said, 'Well, it's such an odd world. I didn't need any language before this.' So it is the demand for salt that requires words. I mean to answer your question; we come from very different sides. But we do assume a hint of land, a background that divides us ultimately when we get together. We couldn't have a conference like this without words.

COUSINS Obviously topography is saturated with history and memory. But what would you say, about examples, where there is some topographical intervention not to remember but to forget. Traditionally making peace with enemies entails making agreements and 'forgetting' something. And so, where there would previously be a contest over territory, now actually what one wants to organise is the forgetting of the conflict.

HUNT No. I'm thinking of a war memorial which is in a sense to memorialize, but also for many people they just want to forget it. I can't think of one particular example. I'm thinking of Ian Hamilton Finlay. I know he plays with the idea. He was talking about inscriptions. He says, 'I don't need an inscription in the garden. If the idea is in the mind, let's not forget it.' But the landscape forgets or does not include the triggered word for him. He wants us somehow not to see where we have been, or have a suggestion and this is not forgetting but it's forgetting in the organisation of the space for the receiver who will not have plotted this, this, or this. I mean one can think of ancient English markers, Celtic markers, those standing stones go away through southern England, for example, which represent in a sense, especially for local people, an absence there. They mark something where they don't want to go and what they don't want to talk about. Now at the moment I can't think of an example today. That's a nice challenge, and I'll email you.

LEATHERBARROW In the histories of several ancient Roman sites there were instances of unbuilding, destroying the perimeter wall in order to rebuild a wall that would redefined the city; forgetting the former definition in order to redefine the boundary, a clearing before building the new foundation wall, almost like un-founding, un-inaugurating, reversing the founding act of this ancient city building.

SCENE II
EXEUNT

SCENE III
A VIEW OF TOPOGRAPHY FROM MY REAR WINDOW

ENTER

JUNYANG Thank you. The lecture brings forward an academic issue, and begins with elaborations of topography in its meaning referring to the Chinese reality. Yeah, that is the reality we are living with. We have a few minutes for question.

HUI Following your narrative of the change of a specific site in history, I have a technical question about the authorship of the historical texts, namely the gazetteers which you quoted as a traditional documentation of a regional topography. We understand from history that those writers of gazetteers were usually scholarly officials. Though serving as bureaucrats, they were very knowledgeable in the genealogy of their local landscapes. What about the authorship of the modern documentations of a regional topography?

DONGYANG They are academics too. Most of them are. At least a university degree in Chinese or History is probably required for the current compilers of gazetteers either at the local county level or municipal level. That's the part I know, no architects involved.

JIANG When I try to link Dr. Liu Dongyang's lecture with what Professor Dixon Hunt and David Leatherbarrow talked about before, I feel a little confused. In my impression, both professors' speeches conclude two parts, the first part on the comprehension of topography, and the second on the expression of the comprehension in the design through some cases. But, to my opinion, it looks that they should choose each other's cases. Professor Dixon Hunt introduced mainly about knowledge and history on what you defined as *dixing*, however, he showed many cases to express visible traces. Professor Leatherbarrow had many sensitive and emotional experiences on what you defined as *dimao*, but he mentioned order many times in his cases. That's why I got confused.

You also discuss the comprehension of topography in traditional China, I think another important issue we cannot ignore is the expression of the thinking of *dixing*, usually those in charge, or the designer, chose to hide their discovery of topography in our tradition. In the long history of China, people think there is a hidden structure inside its nature. Naturally, it lies where it lies. It's legible but invisible. This is a part of geomancy knowledge. In a small scale site like a private garden, ancient intellectuals prefer to represent the natural property and its order rather than its appearance. In large scale projects like royal gardens or settlements that lie between mountains and waters, we can see that the literati might read and sketch the structure of the geography first, and then put it back into the earth after the layout of the landscape has been made. So, they maintained or created a sense of natural landscape. Although manmade elements were very common in traditional Chinese landscapes, the over-expression of topography was never the best choice.

DONGYANG Does Professor Hunt emphasise more the innate order of a site whereas Professor Leatherbarrow focuses more on the features and appearances of a site? I am not sure. I am not confident enough to comment on Professor Hunt's point of view. But turning to David, I would tend to see him as a hermeneutist who emphasises the reading of landscape as a dynamic process, which involves a gradual understanding of both the 'structure' or 'structuring' of a place and its atmospheric aura. Designing will have to be a part of a probing process, as he said, an articulation of landscape. To insert design back into nature, or as you have just said, to make the articulated order less articulated, might be an ideal pursued by ancient Chinese literati in their garden making. It requires both time and the effortless efforts of designers. In a way, it is true that topographical studying or remaking requires one step further than merely 'design-following-nature'.

JIANG The cases mentioned by Professor Leatherbarrow express the understanding adequately, by articulating it at the right positions. But, does design really express a structure through texture? I do not think so…

DONGYANG What really attracts me is the idea or situation to have both Professor Hunt and Leatherbarrow to read and design a site in current China, not in some remote beautiful village, but on levelled land for development in some metropolitan region. Where previous rivers are removed, roads are arbitrarily laid on top of the earth. What will they do, with this kind of conditionality? Because that is the situation we constantly come across. Every now and then, a student would approach me, and ask what he or she shall do with this kind of man-made site and with your neighbouring projects. No one knows exactly. In this case, neither innate order nor atmospheric features are graspable.

SCENE III
EXEUNT

SCENE IV
TERRAIN & XING SHENG
MISTY RIVER, LAYERED PEAKS
BUILD A WORLD TO RESEMBLE NATURE

ENTER

JUNYANG Thank you very much. It is time for questions and discussion. First of all, I would like to ask Lu Andong and Li Shiqiao, would you give us some helpful suggestions for the English translation of 'xing sheng'?

ANDONG I think it should be 'formal excellence'.•

SHIQIAO The translation depends. It is about a change in force. It is force in the form that is perceived, so the form is static, but carried through dynamic force that is realised and set in other conditions. Modulation could be a better word, modulation embedded in the form. It is really difficult. It is a different language.

HUI I have some comments on Wang Shu's project from a historical perspective. This is an inspiring work. I appreciate how he relates the understanding of the macro world to our traditional perception originated from the Northern Song Dynasty and how he uses interpretive paintings and drawings to demonstrate his understanding. His design is an imaginative development of the aesthetics of a historical period. I agree with Mr. Lin that the Song Dynasty was an age of rationality in Chinese painting history. People began to pay attention to physical details and rationalise the perception of nature. We do not see that level of detail in Chinese paintings in previous history. At the same time, because of this rational approach of observing and categorising physical details, the philosophers were able to challenge the traditional materialist theories with more advanced metaphysical thinking. Methodologically this seems paradoxical, but it can be understood in a broad philosophical context. In the Song Dynasty, there was the dissemination of Neo-Confucianism with its focus on the concept of 'li', which we usually translate as 'reason'; meanwhile, there was also the religious movement of Chan,• a mixture of Buddhism and Daoism focusing on meditation. What is interesting is the paradoxical tension between the reasoning of physical details and the emotion towards massive landscapes. This understanding might help us critique Wang's project and reflect on Chinese urbanism.

ANDONG I'm quite fascinated by the final image that Wang Shu has shown. I'm also quite happy with the overall presentation of your° project. In the beginning, you showed traditional paintings and you showed something distant and something close. At the time, you remind us that the mountain is always virtually there as a hint and the building exists as a kind of trace and reminds us that it is a hill, the very sense of respect towards the hill. And then I have a feeling that there are a lot of fascinating places within the building itself, and the hill is more or less lost in the architectural trip.* But the final image suddenly lets us realise that the absent things are more important, something not on the site is more important, and reminds us that we are getting lost or getting trapped by architecture. I realise that it is not about architecture, but something about getting out of reality.

LEATHERBARROW I wanted to ask a question about materials. I'm not sure to whom I'm directing this question. I hope it might make sense to each of the three lecturers, from whom I've learn quite a bit. But obviously in translation, I understood the talks only partially. What is fascinating for me in part is the image of the painter poet contemplating a stone for three to five hours. I want to pose the question in this way: Is there a sense of distance not only in the spatial organisation of the painting, or gardening, but also in its physical embodiment? Are stones more remote than foliage, or water? I tend to think that the stones are remote and they have a certain kind of distance which structures the poet's involvement for several hours. I believe there is a different kind of involvement with the landscape, it's capacity for change, it's material relation in prose, etc. I wonder if what had been said so beautifully about distance relates to what I tend to see as the framework of involvement, that we are involved in the world in different ways, through different distances. It might also occur in materials. My question is about the last project. One thing that struck me was the range of materials. I have a hunch that

---

• Following discussion with Professor Chen Wei after the Symposium, I think the better translation could be 'topographical excellence'.

• or Zen
o your *building* project

* for describing the building

5月28日
28TH MAY, 2011

if one can say:stone is more remote than timber, more distant, inquiring more, living more with imagination, etc. then there might be some logic to the configuration of materials that relates to the structuring of space as a pattern of distances. So the question about the physical body of the garden or the building has related to all that's been said about space or distance or the kinds of distances that the painters elaborated so well. Is there a distance implied in materials?

Can I add one more comment? In these comments I've been thinking about the ways materials are finished, not just the material as such. Because stones can be polished, and can be roughened, they take the atmosphere in different ways and make the stones more or less distant, if I understood the explanation today. So material is also the way it is finished.

SHU Yes, it is. It can be said that we have taken the sense of distance into the whole building. One feature of my building during this period was black. From this time onward, I haven't built any white buildings. All of them are more or less black. After completion, they even look a bit 'dirty' in my terms. I have said that a modern building wouldn't get mouldy, but my building gets mouldy. It is kind of an animal-like house that can breathe. In contrast to the whole setting now, it gives the feeling of remoteness, which is the first and fundamental reason of using the material. The second is that all the bricks and stones used in our campus are recycled materials of the old buildings, which had 50 year, or 100 year, even 200 year histories. It is a building made from all kinds of recycled materials. When it was under construction, many people opposed this idea. They could not understand why a new building would use old materials. Chinese people like new rather than old. However we insisted that this was a good deal, as time is life. As soon as the buildings are completed, they will have a 50 year history. So at the level of time, the materials also have a sense of distance. As to the third question: is stone more distant than timber? I think it is a matter of relation. When stone is used in certain way, it gives a sense of distance. However, if you use it in a normal way, it could be close, as close as timber. In general, timber has a feeling of warmth and kindness, and perhaps closeness. But in reality, very few buildings in today's China are made of timber. So when people encounter this material in buildings, they feel rather distant or far away. That is why many people say that our campus is like a utopia. To me, the significance of it is that in the current condition of China, utopia could come into being, which is of great importance.

DONGYANG Professor Wang, Sorry to interrupt you. What were your thoughts to the existing village located on your campus, abutting the second-phase buildings? Because there isn't any fence, are they supposed to be co-existing?

SHU Yes, I think they can coexist.

DONGYANG Will this be so in the future?

SHU After tolerating the village buildings for quite a while, the university got them removed, and has now decided to make a new building on the site instead.

DONGYANG
I see a variety in those village houses…

SHU There are seven or eight left. I don't want to clear them up, as I think they are embedded with a real history. We can't judge them by beauty or ugliness. They are very important as part of the history which happened there indeed. We shouldn't delete all the histories once present and then make up a new one.

JUNYANG I'd like to relate David's criticism of formalism this morning to Wang Shu's Xiangshan project. In David's criticism of formalism, topography is neither form nor land. Saying so, he referred to the Wexner Centre by Peter Eisenman. Of course, form is important for architecture. It is something we cannot do without. But my understanding of David's criticism is that Peter Eisenman's form is simply too strong. It's a strong form. By contrast, you can say the house he showed in the slides by Alvaro Siza is sort of a weak form well integrated into the landscape. Now I wonder if there is any similar disparity between the Chinese ideal that Wang Shu is much fond of, and his own project of the Xiangshan campus. The former, as I see it, is on the week side, while Wang Shu's projects often hold a strong form, you might say on the verge of fetishism. I would like to have Wang Shu's own ideas about it, but I guess it is very hard to hear any self-critical view from Mr. Wang since his presentation of the project was full of confidence and satisfaction, no doubt at all.

YONGYI I can see that Professor Wang Shu tries to get us to understand the Xiangshan project through a profound reading of the ancient Shan

Shui paintings. It seems to me that you intend to establish certain coherence between a number of concepts, in some of which you seem to hold particular interest. One is the control of the whole, which seems easier to understand. The other two are the abstraction of Shan Shui painting, and the precision of its art. I would like to know what you exactly mean by such a kind of abstraction and precision. If you are conscious to respond to them in the Xiangshan Project, how do you realise them?

COUSINS I want to make a point about how architects use the term 'form'. Now we could debate for ten years what form is. But I think when one listens to discussions amongst architects about form, what they often mean by form is that aspect of the project that they really like. It's certainly called its form. I think we can actually often be very much simpler, if they just said what gave me pleasure and gave you pleasure, and reserve the question of form in a sense for more theoretical discussion. We can all argue about what the form means. But it has this sort of strangeness.

LEATHERBARROW I like things too. We learned a lot today. If I could recall the meanings of topography that we unfolded earlier, one conception separated the mind from the body. So in a sense it might be that form is necessary. Yet, even if it is necessary it is not sufficient. The body is also there. That's physicality. My only question in anticipation of a question about form is to bring the other side back into the dialogue. The materiality in these projects and those that lead in the gardens really happen. For me, the thing that I haven't grasped yet, and hope to understand in time, is the correlation between materials and their finishes and the spatial intuition that is at work in the design of gardens or the configuration of architectural settings. For me, this is a wonderful topic to try to think through, because I believe spaces can be made more distant or closer, not only through the means described in the paintings but through the physicality that's there: the mist, the water, the polish of the stone. I am conscious of the material of these buildings, the material with respect to what people do, whether it's a terrace for gazing, whether it's a room for cooking or whether it's a study for reading. Materials matter as do distances, and both make up the configurations so far as I can tell. This is my intuition about this project.

HUI Coincidentally, it is exactly during the Song Dynasty that people began to paint '1000 miles of mountains and waters'. It is the first period in history to grasp that expansive scale of perception and representation in confidence. As a result, in Chinese gardens there later emerged the design motto, so-called 'seeing bigness through smallness'.

SHU I'm afraid that I have forgotten the question raised by Wang Junyang. The two architects he mentioned, Eisenman and Siza, are very interesting. He said that Eisenman's form is too strong. But I remember that Eisenman wrote an article about his research on 'weak' form. It's an interesting contradiction. By contrast, Siza's buildings seem humble in appearance, but are actually quite dramatic. Siza's houses always give you a feeling of dignity. That is to say Siza's buildings are very strong in essence because very few architects can make a house that feels dignified. His houses always feel located with dignity. Siza's form in essence is very strong. I think Wang Junyang perhaps misunderstood what I said. You may think that when I said that I could not believe it was my work, I was proud of myself. But what I mean is that sometimes I could not explain it in a very clear way. It is precise in my thinking, but there are some jumps. I can't elaborate it very logically, so what it presents is kind of beyond what I imagined. Take the Xiangshan Project for example. My clients didn't see the final scheme, neither any renderings nor models. The whole project was under control in my mind, in the very architectural drawing of plans. So when the buildings were completed, they even astonished myself. We can certainly enjoy this kind of feeling.

DONGYANG The ideas delivered in your lecture this afternoon seem very well-thought out. I am curious about at which moment in your Xiangshan Project you reached such a level of coherence?

SHU It began in 2001.

DONGYANG What you talked about today was almost done then…

SHU I began to think about the whole issue in 1983. About form discussed just now, on the one hand, I agree with Wang Junyang. I also think that the shape of the Xiangshan Project is probably very strong. So I think I am probably a person who is living in the late 17th century, which is the level my mind could reach.

This is a self-judgment. I do not dare to say that I have reached the level of which the Southern or Northern Dynasty achieved. Because the Ming and Qing Dynasties are periods with certain cultural-revival characters, the whole generation at that time recollected the documents and images of the past, and was in a state of re-thinking and reflection, but still remained far from the Southern or Northern Song Dynasty. On the other hand, what I am really concerned with is not the strength or weakness of the forms, but the transformation from one to another, from bigger space to smaller space. It is something like shooting a movie, you cut the whole into a series of episodes. What really matters is how to link one episode to another, how to narrate them in a kind of discontinuous way. Then abstraction is generated. It is not a continuous narrative. It's a bit like Italo Calvino's stream of consciousness, abstract and narrative. So from one form to another, from one space to another, from one setting to another, it has strong mutability and also a kind of similarity that is beyond any words.

DONGYAN Professor Wang, when you talked about the abstracted realism of Chinese literati paintings, for instance, they would in a highly imaginative manner and co-presenting what they have experienced to resist the fixation of time and space, painting any scenic features at a far distance with thorough details. This is what painters used to do. But in architectural design, how could you make one's experience of a place simultaneously coexist with experiences of other places? Or would such a manner and mode from painting become un-translatable into design? Is there such a difficulty?

SHU Yes, I have. For example, because there is a long distance, the whole form of a big space is controlled when you look from a great distance. But when you get close to it, you can't see the controlled big form at all. It is the special part that I construct. When you get close to it, the controlled form is gone. What you can see is a setting, the form of the setting. And you won't say the house that you see in distance is exactly the same as the one in front of you. I narrow down the space between the buildings to the allowed minimum, so you could not see the whole form. It's the way in which the buildings are organised. Apart from this, the other way of organising the complex is what I call dialogue. That is to say the building itself has an internal dialogue. Because it's a bit far from the mountain, and the mountain is a bit flat, it's difficult for me to establish a direct relationship between the building and the mountain. So the building would start its own dialogues between different parts.

JUNYANG Is that what Professor Chen Wei called the second category, namely, when you come across with a situation consisting of some sort of Shi which is weak, you need to strengthen it with a strong form?

SHU No, I don't think so. What I said is different from what she said. I think what Professor Chen Wei said is something like setting up an understanding for an architect. It's about the absence and the presence. That means although the mountain isn't eye-catching and high, everything else is less important than it. It is the most important element in the whole arrangement. So it always takes effect upon every response. Everything has to respond to it, which is the major consideration. It is something like what Jorge Luis Borges does in the *Garden of Forking Paths*,• in which the only word that he does not emphasise is time. What he talks about is not time but space through the whole novel. But in the end, it is revealed that time is the real matter of his discussion.

DONGYANG Professor Wang, what about the creek in between Xiangshan and the first-phase buildings? Did your design re-create part of the river scape? For instance, the small island?

SHU In fact it's a mistake. In the very beginning when I looked at the master plan, I thought there was a pond, so I had that kind of idea. But it's actually an enclosure wall of a factory, which is quite common. Nevertheless I dug a pond in that, because it's the beginning and also the end of my exploration.

JUNYANG I would address a question to Prof. Chen Wei: is it possible to use xingsheng to understand Wang Shu's Xiangshan project, which I in an earlier statement described as strong. Is the point to strengthen the weak topography with a somewhat strong form? Of course, you need to be very careful in order not to do too much, but could it be in any way something worth trying?

WEI Apparently, we unconsciously fell into two traps when discussing Mr. Wang Shu's work.

First, the landscape gardens built in the Ming and Qing Dynasties are very different from the

• 1941

ones built in the Tang and Song Dynasties. That means the landscape gardens built in the Tang and Song Dynasties should have been purer, if they had gone along with the natural development. For example, the Japanese dry landscape garden* actually takes the effect of the landscape gardens built in the Tang Dynasty and of the landscape paintings that prevailed in the Song Dynasty. Unfortunately, this trend was interrupted by the Yuan Dynasty that occurred in-between them. China's landscape gardens made a turn from the pure and abstract path it used to take following the Tang and Song dynasties. There is a historical reason in this. After the Southern Song Dynasty, the painters coming from the southern region of the country were forced to draw paintings in the Royal Palace. Southerners' low status in the Yuan Dynasty made these painters laden with gloomy feelings. In consequence, they tended to deliver complex feelings when working on a painting. Sometimes, a poem would be added, as a painting may not be sufficient to release the feeling. Then, people wondered how heavy a way of expression had to be to release the anger and wrath. This makes a very important change, a change that also affected the way of building a landscape garden. That means one has to use more elements to vent what he wants to air in public. Consequently, more buildings were built in landscape gardens. At that time, a landscape garden was also called a pavilion garden, which implies an important difference. The development of Japanese dry landscape gardens and buildings, on the one hand, took in the influence of the gardens and buildings built in the Tang and Song Dynasties, and on the other hand, the philosophy of localisation advocated by Zen Buddhism that spread to Japan from China. This influence integrated with Japanese culture, and has been absorbed deeply by the Japanese, who turned the philosophy of 'emptiness' and 'metaphysics' into a psychological experience as well as a way of thinking featured with self-discipline and physicality to realise his or her self. As Professor David Leatherbarrow showed yesterday, Maki Fujimoto could express an extreme emotion through pure material. Coming back to what I said at the beginning. When we bind the philosophy behind the landscape gardens of the Tang and Song dynasties to the elements used in Ming and Qing gardens, we see certain divergences/contradictions. I always believe that if the fundamental function of a landscape garden is to ease one's mind, Tang and Song gardens represent the best model for this, while Ming and Qing gardens only lend us some techniques. It would be very difficult to find a balance between the two, and learn from both, while we work on one project. You either learn the techniques — organising space by buildings, or you have to take care of the whole propensity and weaken the effect of buildings. If you want both, in having both of them there, you will certainly see a conflict. This is my understanding.

Second, a building in an environment, strong or weak, is associated with its function. Mr. Wang's Xiangshan Campus is a complex with strong functionality. If Alvaro Siza was to make a tour around the campus, he would feel the spatial change over time. However, as a student, he or she may not be able to feel that, as he or she wants to get into the classroom quickly rather than wandering around. So what he or she could feel is a small part of the campus rather than a whole. In this context, I can see the conflict between the two: one refers to a mentality, and the other to a technique; one at the campus in an appreciating mood, and the other merely there for functional needs. This kind of dilemma is not easily solved. In terms of the development of contemporary landscape gardens, I hope we can follow more closely the path taken by the Japanese. The fast pace of modern life does not allow people to take time to digest things at a leisurely pace. If an architect or artist gathers good things together to give people something that evokes their imagination, I believe this could be closer to the needs of modern life.

SCENE IV
EXEUNT

• karesansui

MENTAL SPACE

心理空间

PROJECTION

投射

马克·卡森斯
Mark Cousins

I want to talk this morning about the relationship between architecture and what I'm calling projection. I will also be talking about a term which is not its opposite, but might seem to be, which is introjection. So you can put it in your mind the two terms go together: projection and introjection. The only other thing I should say by way of introduction, some of the paper is concerned with Sigmund Freud and with the theory of the unconscious. I will adopt his usage and you will hear me talk about the subject and the object. In a sense, the object is anything that's not the subject. The subject is the source of subjectivity or the locus of subjectivity. So an object is not just things like a table. It can be things like your mother.

Now as you are aware especially in Europe and America, one way of understanding architecture, is as part of a general field of projection, at all levels. Projection is the basic mechanism of what architecture does as a practice. To put it very crudely, it conceives internally the idea of a building, which is represented. It draws it. Drawing can be a first or intermediate stage of projection and then it is put or as we would say built into the world. The term in English projection comes from the Latin, meaning you put forward or actually you throw forward or throw outside. So a fundamental category but often a fundamental metaphor in English is this idea of projection. It's often thought to be closely linked or indeed in some sense identical with more kind of technical issues. So in the case of per-

今天上午，我想谈谈建筑与投射°之间的关系。我还会再谈一个可能听上去像投射的反义词，但事实上并非如此的概念：内摄。°所以，你可以在心里把"投射"和"内摄"这两个词放在一起。在此，我需要说明的另一件事是，这篇论文的一部分与西格蒙德·佛洛伊德以及无意识理论相关。我将采取他对相关概念的用法，你将会听到我谈论主体和客体。从某种意义上说，客体是任何不是主体的东西。主体是主观性的来源或主观性之所在。因此，客体不仅仅是像桌子那样的物体，它也可以指你的母亲。

　　你可能会注意到，尤其在欧洲和美国，有一种理解建筑的方式是，把建筑在所有层面上都理解为投射领域中的一部分。投射是建筑实践的基本机能。粗略地说，它内在蕴含了关于一个建筑的想法，并被表现了出来。它描绘着那个想法。绘图可以是投射的最初或中间阶段，然后建筑在世界上被建造出来。在英语里，"投射"这个术语来自于拉丁文，意思是推出或向外扔出去。投射的概念是英语中一个基本范畴，但更是一个常见的重要隐喻。而且，它通常被更多地认为是或等同于技术性问题。因此，就透视而言，我们处理的实际上是投影几何。在某种程度上，透视画法是核心问题。

○ projection

○ introjection

投射
PROJECTION

spective, we're dealing with what you might call or what is called projective geometry. In some sense perspective, perspectival representation is a central issue.

Perhaps the most elaborate and consistent account of architecture as projection is the text perhaps by Robin Evans called 'The Projective Cast'. I'm not here going to talk in any detail about Robin's book but it can stand as a major statement of this. Of course central to that idea, of architecture of projection, is his making central to architecture the practice of drawing. So as it were, drawing becomes the means of transforming something internal to something partially external and then finally purely external. So projection involves certain techniques. It's not only a concept. Many practices which think they are using projection often use architectural metaphors to show that. Such things as an economic plan, the way it's used in as having a kind of projective character, is probably taken from architecture. Now in all this, the question remains — what is this internal space that does the projecting. Let's call that here the subject, or the subject of architecture.

Now I'm going to leave this question to one side for a moment, because I want to prepare the ground for asking the question, the following question. It wouldn't be the first time in the history of thought that someone manages to crystallise the idea of projection in the way that Robin Evans did, at precisely the historical moment when that system is about to change. One of my questions will be: has that traditional idea of projection given way to an emerging practice within architecture which we might characterise as being that of introjection? Are we leaving the field of projection and entering the field, or the kind of epoch, the age of introjection? I'm going to take what in English is called the scenic route to get there.

First of all, by posing the question: what is the internal space? I found that the discussion yesterday was extremely interesting and valuable. But in a sense I also found that it was what might be called symptomatic. It's not that everyone has exactly the same ideas. I'm sure many of them would've disagreed amongst themselves. But let's notice that everybody thought that the issue of topography was crucial and central. It's always interesting when people from different positions nonetheless seem to accept that the issue is important. It seems to me that if you were asked yesterday 'what does topography mean?' It was that topography was the past in the present. It was interesting in that many issues which a hundred years ago would've been expressed in the term tradition, are now in an age which doesn't like referring to tradition, some of its issues are displaced. I don't use the term in a critical sense but simply displaced onto the issue of topography. So topography somehow perhaps figured yesterday as the past in the present. You might imagine someone thinks 'well we've lost these traditions, etc., but there's topography'. That would explain why many people's relation to topography was full of anxiety, as though in a way topography, perhaps especially

关于把建筑视为一种投射的观念最详尽与最连贯的表述可能是罗宾·埃文斯的《投射之模》○一书。在这里，我不会谈埃文斯这本书的任何细节，但它可以作为这个问题的一个主要论点。当然这个观念——投射建筑学——的核心是，他把制图实践作为建筑学的核心。像从前一样，制图是一种把内在的东西先转化成外在的一部分，进而最终将其完全外化的手段。因此，投射牵涉到一些特定的技巧。它不仅仅是个概念。许多认为他们在使用投射的实践人员，常常会用建筑隐喻来进行表达。比如经济规划，其所使用的方法有些投射的特性，可能主要是从建筑学中借鉴来的。问题是什么是进行投射的内部空间。在这里，让我们把它叫做主体，或者建筑的主体。

○ Projective Cast

现在，让我们把这个问题暂时放在一边，因为我想为下一个问题做些准备。在思想史上，罗宾·埃文斯并不是第一个精准地把握住一个系统即将变化的历史时刻并提炼出关于投射的理念的人。我的一个问题是：传统观念上的投射是否已经让位给了一种正在建筑学中兴起的实践，一种我们或许会把它赋予内摄特征的实践？我们是否正在离开投射的领域而进入了内摄的领域或时代？我将会通过英语中"场景路径"○一词的来回答这个问题。

○ scenic route

首先，我要提出的问题是：什么是内部空间？我发现昨天的讨论非常有意思也很有价值。在某种程度上，我也发现它很有代表性。每个人的观点都不尽相同，我敢肯定这些观点之间也存在着相互的对立。然而，值得注意的是每个人都认为地形学*的问题是关键和核心。当来自不同立场的人都认为某个议题很重要时，这总是很有趣的。对我来说，如果你昨天被问到"地形学的意义何在？"每个人会以不同的方式表达同一个观念：地形寓过去于现在。有意思的是，一百年前，许多问题都不会以传统的名义来进行表述，而现在在一个不愿意参照传统的时代，这样一些问题却以传统的面目出现了。我使用"传统"这个词，并没有任何批判性的意思，而只是想说明它的问题被转移到了对地形学的讨论中。因此，在昨天的讨论中，在某种程度上寓过去于现在被看成是地形学的重要特征。你大概能想象有人认为"我们已经失去了那些传统等等，但是还有地形。"这大概可以解释为什么很多人和地形的关系是充满焦虑的。尽管，在某种程度上，地形尤其是在中国的地形是某种通向失去的道路。

* 在这一段落，演讲者主要谈到了关于地形学讨论在中国流行，以及中国人在实践中对于地形的态度和处理方法——一方面认为地形含有过去的信息，拥有或只是谈论地形就能够与传统相联系，但在实践中却完全不顾原有地形，抛弃原有地形。这与真正的地形学的理念是背道而驰的。原文中谈到地形和地形学时使用的都是"topography"一词。

投射
PROJECTION

dramatically in China, is something which is en route to being lost.

One presentation yesterday morning studied an area and the knowledge of the area, which was held by the local inhabitants. The reconfiguration of that area is simply the provision of a physical platform for development. What I think concerns him is not the transformation of topography but in a real sense the abolition of topography. One thing about a platform is that it can be anywhere. Its description shows that it's one of those elements which belong to anywhere. If you are to interpret a lot of contemporary planning practices as merely the provision of platforms for a developer, you can see that in some sense it would be inevitable that you were contrasting topography to contemporary development. I suspect that's a view which is representative. I suspect the many architectural teachers and students in China take pleasure in the discussion of topography, pleasure looking at these pictures, in talking about them, the pleasure comes as it were in contrast to the state in which we find ourselves, which we all find ourselves in, but the Chinese find themselves in even more. One's asking: what is the unprecedented growth of building and urban development? How does it affect the architect? I don't mean in their designs, I mean affect them as a human's subject. We saw yesterday an interesting and clear conception of this contrast between development on the one hand and topography on the other. And a concern with loss.

My question is: what might Freud's account add to that? When Freud comes to define the unconscious, he gives it three different meanings. I'm not going to have time to elaborate them. The first is what he calls the descriptive unconscious. By that he simply means those things we do and say in which we're not aware of. We speak grammatically but we are not conscious. The third is what Freud calls the economic, that is to say, the quantitative. I say quantitative yet not numerical analysis of the instincts as they convert into psychic material, into beginnings of mental activity. But the one on which I want to concentrate on is what he calls the topographical sense of the unconscious.

Now the topographical unconscious for Freud is essentially an analysis of the relations between the subject and all the relevant objects in the subject's existence, in terms of how the subject does or does not get pleasure from those objects. Essentially this is what Freud called the pleasure principle and putting aside complications, his account is that we seek objects of many different kinds, sometimes humans, sometimes tables, in order to get a kind of a quantum of pleasure or, which to him is saying the same thing, to discharge any kinds of instinctual tension. It is a gradual mapping, within psychoanalytical practice, as if the patient were a landscape, and you are trying to chart them. You are trying to make a map of the patient's passions, ultimately what we would call the relations to objects, on the assumption that you will begin to find certain distinctive qualities in those relations.

昨天上午的一个演讲，研究了一个地区，而关于这个地区的知识是掌握在当地居民那里的。

那个地区的重构的只不过是为开发准备一个平台。我认为他所关心的不是地形的转变，而是对地形的抛弃。这样的平台可以建在任何地方，这就是问题。其描述显示这个平台是那些可以属于任何地方的元素中的一个。假如你将很多当代规划实践解释为只是给发展商提供的这类平台铺路，你可以看到，在某种程度上，你不可避免地将地形与当代的开发相对立。我怀疑这是一种具有代表性的观点。我猜许多中国的建筑学教师和学生很愉悦地讨论着地形学，他们看着那些照片，谈论着地形，获得了很大的快乐。这种愉悦和我们所发现的自己所处的境地形成对照，我们所有人都处于这种境地，但对中国人来说更是如此。有人或许会问："什么是建筑与城市前所未有的增长？它是如何影响建筑师的？" 我的意思不是指在他们的设计中产生影响，而是对他们作为一个人类主体的影响。我们昨天看到一个有趣且清晰的概念，一方面大力开发和另一方面热衷讨论地形学之间形成的反差，对于所失去的东西的关切。

我的问题是，弗洛伊德的理论会对这个问题有什么帮助？当弗洛伊德定义无意识时，他给予它三种不同的含义。我没有时间去详细阐述它们。第一种被他称作"描述性的无意识"。所谓"描述性无意识"，就是指那些我们所说、所做却不自觉的事情，就像我们不觉得我们说话时在使用语法一样。第二种是弗洛伊德称之为"经济的无意识"，也就是"量化的无意识"。我说的量化的不是指对本能的数据分析，而是其被转化成为心理素材，成为心理活动的开始。但是我主要想说的是他称之为"地形学的无意识"的这一种。

"地形学的无意识"对弗洛伊德来说，本质上是对主体及在主体存在中的所有相关客体之间的关系的一种分析，基于主体是如何从那些客体中获得愉悦或无法获得愉悦。本质上，这就是弗洛伊德的愉悦原理。撇开那些复杂的东西，他的看法是我们寻求很多不同类型的客体，有时候是人，有时候是桌子，为了能够获取得某种程度的愉悦，或者说是释放任何类型的本能的紧张，对他来说这两者是同一件事。这是一种在心理分析实践中逐渐形成的拼图，就好像病人是一种景观似的，而你正试图绘制完成这幅拼图。你试图绘制的是一幅关于病人的激情的地图。最终，我们会把它称为与客体的关系，以期你将开始在这些关系中发现某种独一无二的品质。

Freud is well aware that he has a reason for using the term topography, for we might say that at the level of the infant, you could say the first significant experience of the world is as a landscape; and usually that landscape is called the body of the mother, if you think about the relations of an infant, even really before the infant comes to recognise that he or she has a body. Because the infant has to somehow learn that he has a body by which I mean a certain coherence what sometimes in English is called a body image, or what the French analyst Lacan would call the mirror stage, the recognition by the infant of his body. I'm not sure the mirror is very helpful, because the mirror is rather active and we normally call this active mirror a mother. A mother has a function of mirroring back to the child in a way which will promote development. The first body as a landscape is the mother. We say the first and the most fundamental grasp of topography comes in those valleys and hills which the infant explores in the body of the mother.

I recognise that lots of people find this almost ridiculous and exaggerated but if it's true it's rather important and it's rather important to architects. So you're welcome to kind of dismiss it, but you should hear the argument out. There was a famous architectural critic writer in the 1940s—50s, Adrian Stokes, to whom this was an essential proposition. For him, architecture was a strong analogy to the maternal body. And he knew perfectly well that infants also have fantasies of destroying the mother. So we could say we all have unconscious fantasies of destroying architecture. So, architecture, for him, was not so much a conception of design. Rather he thought of all architecture as what he called a form of repair. All architecture, to him, was an act of repair, an act of restoration, even though you don't know what you're restoring. Even though it may take the form of a new building, it still may be conceived of as a projection in the service of restoration. Perhaps it's a helpful way to locate the discussion on topography, in a way which respects the ambivalence of the architect, perhaps particularly in China at the moment, where the architect confronts a situation full of opportunities, but also full of anxieties. I don't have a cure for this. It is the painful condition which architects and which students have to suffer. Anyone seriously engaged in practice will always find themselves pulled apart by ambivalence. We should beware of people who have very easy solutions. Actually the heart of the practice is the experience of being torn apart.

The other thing that Freud may contribute to the discussion of topography is this idea that topography functions as the past. It's necessary in Freud's account to see why this is linked to pleasure. How often in China does the chairman of a planning committee say: Let's just think for a moment. Do you really think this development will be a real pleasure for people? I think like bureaucracy anywhere, this discussion is rather rare, even if it should be the central discussion. Planning actually should be the organisation of pleasure. Now

弗洛伊德非常清楚他使用地形学这个概念的原因。从婴儿的层面上，你可以说，在这个世界第一个意义非凡的经验就是一种景观；并且，通常来说，那个景观被称为母亲的身体，假如你思考一下一个婴儿所有的关系，尤其是在婴儿真正认识到他或她有一个身体之前。因为婴儿需要以某种方式知道他有一个身体，我的意思是指某种整体的一致性，这种一致性在英语里有时被称为身体意象，° 而法国心理分析学家拉康会称之为"镜像阶段"，即婴儿对他的身体的认知。拉康学者解释了"镜像"的观念，我不确定这个镜像是否很有帮助，因为这个镜像是活的，我们通常把这个活的镜像叫做"母亲"。母亲有一个功能就是作为孩子的镜像反射来促进他／她的成长。第一个作为景观的形体就是母亲。我们说最初和最基本的对于地形学的理解来自于婴儿在母亲身上探索那些山谷和山丘。

° 体像

我知道到很多人认为这几乎不可理喻并且很夸张，但是假如这是真的，它就相当重要，尤其是对建筑师而言。因此，欢迎你驳斥它，但是你应该听一听相关的看法。在20世纪40年代到50年代，有一位著名的建筑评论家兼作家叫阿德里安·斯托克斯。我刚才所说的对他来说是一个根本主张。对他来说，建筑与母亲的身体非常相似。而且他很清楚地知道婴儿也同样幻想着毁掉他们的母亲。所以，我们可以说，我们都下意识地想要毁掉建筑。因此，建筑对于他来说，并不是设计。相反，他认为所有的建筑是一种修复的形式。所有建筑，对于他来说，都是一种修整、复原的举动，即便你不知道你修复的是什么。尽管可能是以新建一栋建筑的形式出现，这仍可以被理解为是一种修整。也许，一种有益的讨论地形学的方式，是尊重建筑师所面对的这种矛盾，尤其是在现在的中国，建筑师正面临一种充满机遇同样也充满焦虑的状况。我对此没有解决方案。这是一种建筑师和学建筑的学生们必须得忍受的痛苦状况。任何严肃的投身实践的人总会发现他们被周围环境所抽离。我们应当小心那些有很容易的解决方法的人，因为实践的核心就是一种被撕裂的经验。

弗洛伊德对于地形学探讨的另一贡献是他关于地形学作为过去的功能的看法。因为在弗洛伊德的想法中，探究为什么这种看法与愉悦相联系是必需的。在中国，规划部门的领导说"让我们再想一下"这种情况并不常见。你们真的认为这个地产项目会给人们带来真正的愉悦吗？我想就像任何地方的机关一样，这种讨论是相当少的，即使它应该是讨论的

of course we don't permit this to be said or thought often, because it would threaten the disciplinary worlds and the divisions of labour which we have established. The fact that it is not going to happen doesn't mean that it isn't true. And behind the anxiety of topography functioning as the past, you can detect it is the way the Chinese nation considers the unity of Chinese people, a question of not only topography, but also a question of return. China is one of those cultures which have added to literature on the theme of returning home.

So I thought I might say something about topography, the past and the return. As they might be discussed in Europe, maybe there are Chinese parallels. This is the question of topography, when you feel that you are in exile from it. Obviously the exile I'm immediately talking about is an exile in space. But I will come on to say that we are in a sense already experiencing another kind of exile which is an exile in time. Strangely enough, the idea of what we might call the pain of wanting to return is usually to return where you were born, and then it becomes possibly the country.

But it emerges first of all within military history in Europe. It stems from a particular point in the 17th century, when warfare in Europe was fought by mercenary soldiers, soldiers who didn't belong to one side or the other, but who were paid. The mercenaries who had a high reputation were the Swiss. But generals acknowledged that there was a problem in using the Swiss. Sometimes something would happen, like the distant sound of a cow bell and that was it. They lay down their arms and started to walk home. This became classified as a disease, a disease whose name in English, we still use—nostalgia, from the Greek meaning: the pain—*algia*, to return—*nostos*. It was thought to be very serious and there was no cure for it except to send the soldier home. I think the 17th century was quite progressive. The illness is brought on by hearing or seeing a sign of home. We can generalise this as a systematic experience of exiles. In a sense we fear that we are going to be exiled from home not by physical separation from home, but by the transformation, almost the erasure of home. It is as if we are being projected into an entirely alien condition. I don't think there is any solution to this problem. But it's important to be able to identify it. Certainly it reflects many conversations I've had in China. This is a way of trying to summarise them.

What we have to of course bear, in this ambivalence, is what we normally call reality and games. This is one of those terms about which I'm going to pause for a few moments. I'm not here interested, in fact I'm never interested in any what you might call philosophical accounts of reality. I don't really find in life much need for the term reality. I'm very pleased I've never been there. It sounds absolutely horrible. Philosophers say that: reality is in one way or another the sum total of all it is, whatever that means. And from human sciences, you get a version of that, its reality is the sum total of what everyone thinks it is. Freud was not concerned with a reality

核心。规划应该是关于愉悦的一种组织。当然，现在我们不允许经常谈论和思考这个问题，因为这会威胁到已经规范化了的世界以及我们已经建立的劳动分工。事实是，还没发生的事情并不代表就不是真的。在地形学作为对于过去的焦虑的背后，你能察觉到这是一种思考中国民族统一性的方式，不仅仅是一个地形学上的问题，而且是一种回归的问题。在各种文化中，回家的主题是文学里经常出现的主题，中国文化正是其中之一。

所以，我想谈谈地形学，有关过去与回归。就像它们在欧洲被谈论的那样，也许还有与中国类似的情形。地形学问题的提出，是当你感觉到你从地形中被驱逐出去的时候。关于放逐，显然，我马上要谈论的是在空间中的放逐。但是我要说的是，在某种程度上，我们已经在经历着另一种放逐，时间上的放逐。有意思的是，我们或许可以把这称为想要回归的痛苦，而回归通常是回到你出生的地方，也很可能是你的国家。

这种痛苦最早出现于欧洲的战争史中。它始于17世纪的某个特殊时间点，当时的战争是由雇佣兵来打的，士兵们不专属于交战的一方或另一方，只属于付钱的那一方。在这些雇佣兵中，享有盛誉的是瑞士雇佣兵。但是，将军们知道使用瑞士人有个问题时有发生。比如远处传来牛铃声，那就完蛋了，他们会放下武器开始往家走。这被定性为一种疾病，在英语中，我们今天还在使用的词——"nostalgia"，° 来源于希腊语的痛苦和回归。° 这在当时被认为是一种非常严重的疾病，除了送士兵回家没有别的办法医治。由此，我认为17世纪是相当进步的。这种病症由听见或看见家乡的符号引起。我们可以将它归结为一种关于放逐的系统化的经验。在某种程度上，我们都害怕从家中被放逐，这不是物理上的离开家，而是一种转变，几乎是抹掉了家，仿佛我们被投入到一个全然陌生的环境中一样。我认为对这个问题没有任何解决办法，但重要的是能够找到问题之所在。当然，我在中国进行的很多对话都反映了这个问题。这是试着去总结它们的一种方式。

在这种环境下，我们不得不忍受的是我们通常称之为现实和游戏规则的东西。这是其中一个我需要暂停一下来说一说的概念。事实上，我永远不会对任何人们认为的哲学上关于现实的概念感兴趣。我认为生活中其实并不那么需要有现实的概念。我很高兴我从来没有处于其中。这听上去很糟糕。

○ 思乡、怀旧的
○ algia & nostos

as a certain type of existence or being. For him reality is one thing only. It's when your desires bump into an obstacle. As long as a desire meets no obstacle, we are completely in the realm of fantasy and desire, even if we are outside. Reality for Freud is whatever functions as obstacle. I should say this is not some complex philosophical issue.

I learnt this actually while going shopping. I was in a clothes shop in London called Harvey Nichols, which tends to be rather expensive. And while I was kind of waiting patiently, or not so patiently, I was watching people pay with credit cards for the clothes they bought. I noticed the way which you can distinguish there was a difference between how rich people paid and middle-class people paid. It's very unfair but rich people pay very elegantly. That card goes straight through, no problem, a very elegant gesture, just straight through. Middle-class people, you know people who work in universities and sort of things, they pay in a very ugly, disjointed way. Their card would go 'eh-hum' because as it were they are aware that this is money they are talking about. This is next month's salary. And this anxiety gets translated. The gesture moves between yes or no, should I buy it or should I not? My point is that for the rich, price is just something you sign for. You don't really pay! It's not an obstacle. So what's an obstacle for one person isn't an obstacle for another, so, the rich occupy a larger domain of pure fantasy and desire, because they do not bump into any obstacles. The poor face only obstacles. Even the next meal is an obstacle.

From this, Freud had the idea of reality testing. And this returns to a kind of architectural issue. In architecture, we might call this reality testing, the acceptance of constraints. The architecture accepts the need to work within constraints. Whether the constraints are issues about the finance, or about planning regulations, that has to enter into the design. So as it were, the design is not purely a kind of elaboration of a fantasy. It has already, at least in thought, met constraints. All architecture, is at some level a compromise. And of course the moment you start reading architectural history, it will seem to be that the history of architecture is nothing but constraints. What's important to us here is it means that external phenomena are taken to be part of or worked through by what is projected.

It's at this point that I want to use and to turn to the question of introjection, rather than projection, as perhaps the name of a new kind of destiny for architecture. I'm just trying to understand what's going on. And I'm going to try to understand it by the rise of certain aspects of digital architecture. I want to try to keep my argument as simple as possible. One way of talking about constraints is that traditionally they involve the architects coming to know what the constraints are. It involves the architects having to find out, or listen when they're told. We don't have any problems about this. This is a traditional kind of picture.

在某种程度上，哲学家们说：现实是某种对于所有事的总称，不管那意味着什么。从人类科学的观点来看你会得到这样的看法，现实就是被每个人认为就是那种情况的总和。弗洛伊德不关心作为具体存在或此在的现实。对他来说，现实不过是一件事而已。当你的欲望受到阻碍时，它就会出现，当欲望没有遇到任何障碍时，我们是完全处于幻想和欲望的王国中的，即使我们好像是在这王国之外。现实对弗洛伊德来说其所起的作用就是障碍。我应该说这不是什么复杂的哲学问题。

事实上，我是在购物的时候明白这点的。我们在伦敦一家叫做哈维·尼克尔斯的服装店，那儿一般来说相当的贵。当我耐心等待的时候，其实有那么点不耐烦，我在观察人们用信用卡为他们买的服装付钱。我注意到你可以分辨出有钱人和中产阶级付钱时的区别。这很不公平，但是富人付钱的方式十分优雅。信用卡一次刷过，没有任何问题，以非常优雅的姿势，流畅地刷过。而中产阶级，你知道那些在大学和类似地方工作的人，他们以一种难看的、哆哆嗦嗦的方式付款。他们的卡总会出些问题，因为他们意识到他们付的是钱，这是下个月的工资。那种焦虑从他们的动作中体现出来。那种姿态在是与否之间彷徨，犹豫我该买还是不该买？我的意思是，对于富人，价格只不过是一个签名而已。你不是真的在付钱！所以这里没有障碍。对于一个人是障碍的东西对另一个人来说完全不是障碍。因此，富人占有更大的对于纯粹的幻想和渴望的自主权，因为不会有任何障碍冒出来。穷人面对的却只有障碍。甚至下一顿饭都是障碍。

由此，弗洛伊德有一个关于现实考验的想法。这就又回到建筑学的问题上。在建筑学中，我们也许会把这称为现实考验，对于限制的接受。建筑学接受在限制中进行工作的需要。不论这种限制是资金，还是规划法规，都需要代入到设计当中。因此，设计从来不是纯粹对于幻像的一种精细化劳作后的成果。它至少已经在思想上受了限制。所有具有建筑学属性的，在某种程度上都是一种妥协。当然，当你开始阅读建筑史时，看起来建筑学的历史都是关于限制的历史。对我们来说，这里重要的是它意味着外部的现象参与到投射中，并通过被投射的观念起作用。

正是在这一点上，我想转到内摄而非投射的问题，而内摄也许是建筑学新命运的名称。我试图去理解到底发生了什么，并将通过讨论数码建筑中的某些方面来试图理解它。我尽可能使我的论点保持简洁。传统上，一种谈论限制的方法

The question now, however, is how our relation to what used to be called the external world is being transformed. We can follow that change by just thinking about the term which I raised yesterday as a kind of token, the word—data. What on earth is data? Now at a naïve level, you might think, well it's the same as knowledge, facts, used to be. But, of course, it's not. It's now something which appears to be about external reality. But it is fed into what you might call the subject of design. So at the very least, the data becomes part of the kind of formulation of a design. I want to point out that this is very different. It's one thing to, in a sense, be preparing a design and adjusting to constraints, which we think are external, but which we have to compromise with. It's another to try to incorporate something about those constraints as if they can lead to the design.

I say introjection because it would mean taking stuff from out there and feeding it into the very process which is designing. It's clear that for some people, it leads to the kind of fantasy that with the right software and a little editorial tweaking, you could have a kind of self-designing design process and that somehow you create a kind of look, almost that you would have automatic design. In some sense, some of these ambitions are coded in the term you are all used to by now, which is the parametric. Although I must say I ask the AA each year to nominate the word, which is the word they've heard most of that year and still don't understand. The parametric has been the winner for some years now. Now I want to look seriously at some of those terms, because I think they are extremely instructive and symptomatic.

In a sense within the parametric, or kind of automatic modes, kinds of design, I realise it's more complicated than that. You have a form of what I'm calling introjection. We don't have to take it purely at this conceptual level. You can turn to certain aspects of practice. And here I'm going to quote from the work of a Ph.D. student of mine, who is part Chinese and who worked in Chinese offices. Her argument was that in Chinese offices, it was now possible and obviously kind of established at all stages of design. There was a kind of way of meshing together the software on plans and the software from renderings. So these orders of kind of renderings and plans are so crucially separated in the history of architectural practice. You have only to recall Corbusier saying 'above all we hate renderings'. You know, as it were you have a practical form in which the rendering, instead of representing a design, can be part of the design. Now I have no way to tell you exactly how to correct that. But it conforms to my general idea that a process as it were is occurring which is called introjection. We need maybe to discuss what the consequences of that are. But it's very different from the traditional form of projection.

就是,在某种程度上让建筑师参与进来知道限制有哪些。这包括到建筑师去弄清楚或倾听他被告知和阅读的信息。我们对此没有任何问题,这是一种传统的处理方式。

然而,现在的问题是,我们与外部世界之间的关系正在改变,或者很可能已经改变了。我们可以追踪这种改变,通过思考我提到过的一个概念,单词"数据",把其作为一种表征。"数据"到底是什么?在一个最幼稚的层面上,你或许会想,这就和知识、事实一样。它曾经是这样,但是现在当然不是了。我们现在认为,它是关于外部现实的东西,但它被融入设计主体之中。因此,最起码,数据变成了设计的一种配方。我想要指出这两种方式非常不同。一种是,在某种程度上,数据是为设计做准备,并且对它进行调整以适应我们以为是来自外部的、但不得不与之妥协的限制。还有另一种方式,即试图整合这些限制中的一些元素,就像它们能够导向设计似的。

我说内摄,是因为这意味着从外面吸取一些东西,并且把其注入到设计进行的过程中。对一些人来说,很明显这会使他们产生一种幻想,即使用正确的软件再加上一点编辑上的调整,你就可以拥有一种自行设计的设计过程;你便可以创造一种外观,几乎就有自动化的设计了。在某种程度上,这类野心中的某些部分体现在一个词上,即你们现在都已经很熟悉的"参数化"。然而,我要说的是,在"AA"每年我都会让学生们提名一个他们这些年听得最多但却不理解的词。"参数化"总是这几年的赢家。现在我希望认真地看待这些情况,因为我认为它们非常有指导性和代表性。

在某种意义上,在参数化的、或是所谓的自动模式的那种设计范畴内,我意识到这件事并没有那么简单。这里有一种我称之为内摄的形式。我们不必只对其做纯粹概念上的理解,可以看看实践的某些方面。在这儿,我将会引用我的一个博士生的研究,她有一部分中国血统并且曾经在中国工作。她的论点是,在中国的建筑事务所里,现在可能或者说在某种程度上很明显,内摄已经融入到了设计的各个阶段。他们可以把做平面图的软件和渲染的软件整合在一起。于是,在建筑实践史上,渲染图和平面图的关系发生了改变。这让我们联想起柯布西耶说的"我们尤其痛恨渲染图"。可是,这里有一种实践形式,渲染图不是用来表现设计,而是作为设计的一部分。现在我没有办法确切地告诉你如何纠正这种现状。但这证实了我总体的想法,即一直存在着一个内摄的过程。我们或许需要讨论随之带来的种种后果,因为它和传统形式的投射有着很大的区别。

THE FIGURE AS HOME

象与家

李士桥
Li Shiqiao

TRANSLATORS
Zhang Tianjie
Li Shiqiao

翻译
张天洁
李士桥

This essay draws a parallel between the figure and home; they are both understood as locations of final emotional and intellectual retreat, as places where their inherent forms, stemming from that of the womb, outweigh the strength of logic in instrumental reason. Extending from this parallel, the figure assumes its ontological status against the plane of truth, and the home constructs a realm in contrast with the sphere of the public. This parallel is crucial to ground a discussion of a central feature of the writing system in China, which strategises the figure as language. Once this linguistic strategy is set in place, the entire relationship between meaning and form begins to shift in a distinct direction; it is in this direction that we discover consequences of constructing homes, which provides us with a revealing understanding of Chinese architecture and cities. In this sense, the Chinese writing system presents an 'empire of figures' whose territories go far beyond those traditionally associated with the emblematic and the Gestalt—categories of the figural in the context of alphabet-based languages in the Western tradition; it constructs both meaning and space according to a figural logic. The initial parallel of figure and home brings us to a discussion of cultural constructions and anxieties of 'leaving home' and 'returning home' respectively, leading to very different conceptions of home on the one hand and public space on the other.

本文指出象与家存有相似之处；它们均被理解为情感和思想的最终隐退地点，在这些地方，它们的先天形式似乎反映了子宫的原形，其影响力超过工具理性的逻辑。从这一相似之处延伸开去，象的本体性与真理层面形成了对立，而家的领域与公共空间也形成了对立。这一类比对于展开基于中国文字系统中"以象为文"这一策略的讨论至关重要。这一策略建立导致了整个意义与形式关系的发展方向；沿着这个方向，我们发现建造家的结果，给我们提供了关于中国建筑和城市的一种揭示性的理解。中国书写体系呈现出一个"象的帝国"，其殖民地远远超出了传统的象征理解和格式塔完型心理学的范围，即西方拼音文字中可以出现的形象类别；它根据象的逻辑，既形成了意义也建构了空间。象与家的相似之处引发我们探讨在中国文化中"离家出走"与"回家"的建构和焦虑，导致不同的家与公共空间的概念。

THE UNHOMELY

Sigmund Freud's 1919 essay, 'The Uncanny', is perhaps a particularly useful point of departure, because it placed in psychology a fundamental principle of what may be described as 'the Western social-spatial form': the city is only possible when one leaves home. An early version of this proposition can perhaps be found in Aristotle's *Politics*, in which he stressed that home is where emotional and hierarchical authority dwells, while the city—housing, so to speak, a political life°—is where equality and freedom becomes possible. The *polis*—the assemblage of spaces that physically manifest the notion of a political and public life—became the embryonic form for the Western city; the language of urbanism—natural rights to free open space and community-based life—in the West still reach all the way back to that of the *polis*. Neither the imperial imaginations of ancient Rome, nor the missionary ambitions of the Christian Church have deviated from the urban goals of the *polis*; the public realm, together with its urban-based institutions such as squares, city halls, museums, cultural centres, libraries, universities, remained the most important site to construct a public life. This is an extraordinary conception, particularly when we consider that our natural loyalty tends to gravitate toward the family, and towards institutions modelled on the family as a prototype. If we postulate that the urge to build a public life runs against a biological intuition to strengthen and protect the family, then what force has been at work to effect such a rebellion of nature? This question perhaps highlights the importance of Freud's work, not that it pointed to a universal truth, but that it captured an intellectual construct that seems to be so puzzling to the Chinese reader and yet so reassuring to the Western reader. The force that worked to incite an act of leaving home in the Western cultural context is constructed by Freud in the notion of the uncanny, the feeling that there is something familiar yet strange about home, both as a real space and as a signifier for the womb and for thought. The desire of the public realm is cultivated so intensely that it seems to have produced a feeling of repulsion for the familiar, a strangeness in the well-known. Freud explained that the uncanny is nothing new or foreign, but something familiar and old; what creates the feeling of the uncanny is rooted in a process of repression which transforms the familiar into the uncanny.

○ *bios politikos*

REPRESSION OF THE FIGURE

Repression works in relation to both home and the figure. Home in Freud's formulation was never just about domestic spaces that all of us grew up in and remember with some details. Home, here, is a mental condition; it is a place where undifferentiated biological life weaves its intellectual justifications with various degrees of elaboration. The act of thinking bears a striking resemblance to the con-

陌生的家

西格蒙德·弗洛伊德在 1919 年所著的《暗恐》一文,或许是特别有用的出发点,因为它把一项基本原则置于心理状态之中,一种"西方的社会／空间形式":城市的形成只能建立在个人离家出走的基础上。这一论点的更早版本或许可以在亚里士多德所著的《政治学》中觅得,在书中他强调家是情感和等级权威的居所,而可以容纳政治生活°的城市是平等和自由成为可能的地方。城邦从物质上显现政治和公共生活观念的空间组合,并成为西方城市的初期形式;西方都市理论中对自由开放空间和社区生活权利的追求,其语言仍然来自希腊的城邦。无论是古罗马的帝国想象,还是基督教会的传教雄心,都未曾偏离古希腊城邦的城市目标;公共领域建立在城市机构之上,例如广场、市政厅、博物馆、文化中心、图书馆、大学等;它们是构筑公共生活最重要的场所。这是一种不平常的观念,尤其是当我们考虑我们天生对家庭,以及以家庭为原型机构的忠诚。如果我们提出建立公共生活的强烈欲望违背了巩固和保护家庭的生理直觉,那么是什么力量在导致这种对本性的反叛?这一疑问体现了弗洛伊德著述的重要性,并非它指向万能真理,而是它抓住了一种思想建构;这个建构往往令中国读者感到困惑,却消除了西方读者的疑虑。在西方文化背景中,激起离家出走行为的力量被弗洛伊德建构为暗恐,即是感觉家既熟悉却又陌生;这里家既是如同子宫似的空间,又是一种具有同样安慰感的思维。对公共领域的强烈渴望似乎导致了一种对所有熟悉东西的排斥,一种在所有众所周知里的陌生。弗洛伊德解释到,暗恐并非新的或外来的感受,而是某种基于熟悉的和陈旧之中的感受;暗恐的产生也许来源于对熟悉的抑制,及其将熟悉转变为暗恐的过程。

° *bios politikos*

对象的抑制

抑制在家和象的领域中同样存在。弗洛伊德所系统阐释的家绝非只是我们每个人长大并记住某些细节的家庭生活空间。在这里,家是一种精神状态;它是生命在混合状态中以不同的详细程度解释其特定状态的地方。思维这一行为与家的特征有着惊人的相似之处;这种相似之处表现为思维的象,即思维沉淀为一种最终的和完整的形式,能用来锚固支离破碎

ditions of home; here, the core of this resemblance is found in the figure, a sedimentation in final and complete form which can be used to anchor fragmented thoughts. Often, it is the final and complete form of the figure that demands conformity from empirically derived fragments of knowledge. This is where the suppression of the figure emerges; in the history of Western philosophy, the figural form has been systematically suppressed in favour of empirically derived fragments of knowledge, unless both could be accomplished. One of the purest expressions of the suppression of the figure can be found at the time of the emergence of early science, in the works of Francis Bacon. In his *Advancement of Learning*,° Bacon argued that the hieroglyphic appeared before the alphabetic as a distinct stage of the development of the human intellect, a stage when thoughts were only present in figural forms. The goal of knowledge is to purge the hieroglyphic from its domain. Bacon identified three kinds of hieroglyphic thinking, three vanities of thought: the fantastic, the contentious, and the delicate, exemplifying different ways of the form of thinking substituting the substance of thought.

    As a follower of Bacon, Robert Hook contributed crucially to the rise of science and the practice of architecture. In his *Micrographia*,° a collection of surprising images of very small things and insects drawn from his observations through the microscope he invented, Hooke exploited the visual sensations of the microscope. In this book, Hooke paraded the triumphs of science and its suppression of the figure. The first image of the book was a deliberate provocation and an aesthetic act; it is a drawing of three common things: the tip of a needle, a printed period mark, and the edge of a razor. Hooke showed them to be different from what we normally thought: the sharp needle point is seen to have a '*broad, blunt,* and very *irregular* end', the printed period mark looks like a 'great splatch of *London* dirt', and the edge of a razor appears to be 'almost like a plow'd field'. The figural thoughts would repeat comfortably statements such as 'as sharp as a needle's point' or 'as sharp as a razor's edge', but there is, in Hooke's mind, no truth in them. These are, essentially, figures of the mind. If we see *Micrographia* as an aesthetic adventure, much of 20th century modern art demonstrates a similar aesthetic effect of 'suppressing the figure' by progressively removing physical and intellectual conceptions such as those of landscape° and the body,° often resulting in nothingness.°

    The figure for the intellect could be seen to be what home is for society. Here, the problematisation of the figure—a process, like a scientific report, of dissolving figures of thought—is equally intense. Western epistemology is antithetical to the figure; it consists of acts of defamiliarisation in the form of epistemological criticality. The figure, however, is never completely expelled from the intellect; its attraction lies in its extraordinary capacity to put thoughts in final and complete forms.

° 1605

° 1665

° Monet, Seurat, Cézanne
° Picasso, Matisse, Bacon
° Rothko, Creed

的想法。这种最终完整的形式往往要求源自经验的知识碎片来服从其完整的形式。这是对象的抑制的产生之处；在西方哲学史中，象一直被系统性的压制而源自经验的知识碎片往往受到重视，除非两者均能实现。抑制图像的最纯粹的表达之一可以从早期科学出现时弗朗西斯·培根的著作中找到。在他的《学术的进展》中，°培根指出象形文字的出现先于字母，是人类智力发展的一个阶段，在此阶段中思想只能以象来呈现。知识的目标是要将象从其领域里清除。培根辨别了三种象形思维，三种虚荣的思想：幻想型、争辩型和精美型，展示了以思维形式取代思维内容的不同方式。

○ 1605 年

作为培根的追随者，罗伯特·虎克对科学兴起和建筑实践的贡献至关重要。他所著的《显微术》°是通过他发明的显微镜所观察的一些极小物体和昆虫的意外细节的图集，充分利用了显微镜的视觉冲击力。他炫耀了科学的胜利和它对象的抑制。书中的第一幅图像既是一种挑衅也具有美学意义；它是三件普通物体的图画：针尖、印刷的句点、剃须刀刃。虎克所展示的与我们通常认为的并不相同：锋利的针尖看起来有着"宽的、钝的、极不规则的端头"，印刷的句点看起来像"伦敦污垢的大斑点"，剃须刀刃显得"几乎如同犁过的田地"。象形思维会容易重复如"像针尖一样尖"或"如剃须刀刃一般锋利"等陈述，但在虎克看来，这些并非事实。这从本质上讲是象形思维。如果我们把《显微术》看做一次美学的冒险，那么 20 世纪的现代艺术大多示范了一种类似的"象抑制"美学效果，通过逐渐去除如风景°和身体°等物质和知识概念，最终导致虚无。°

○ 1665 年

○ 莫奈、修拉、塞尚
○ 毕加索、马蒂斯、培根
○ 罗思科、克里德

象与思维的关系可以被视为如同家与社会的关系。这里，象的问题化如同家的问题化一样强烈；这里，科学报告溶解了象形思维。西方的认识论同象形成对立；它的认识论的中心内容是其批判性，也就是从熟悉中寻找陌生的行为。尽管如此，象似乎从未在思想中清除；其吸引力在于令思维呈现出最终完整形式的非凡能力。

THE EMPIRE OF FIGURES

At the start of his comparative study of Greek and Chinese thought *Detour and Access*,° François Jullien posed questions like this: ° 2004

> What if the world were not an object of representation, and figurative meaning did not tend to represent something—symbolically? What if generalisations were not the goal of thought, or speech tended not to define° but to modify itself—to reflect the circumstances? In short, what if consciousness did not strive to reproduce the real in order to ground it in transcendence?° And what if the purpose of speaking about the world, to make it intelligible, were not to arrive at Truth?

° to build a universality of essences

° of being or of God

This is an important understanding for us to imagine a second conception of the figure, which takes place entirely within its own territories: an empire of figures. This understanding is derived from the 'character-based' writing system in China in which a graphic unit° is used to indicate meanings. Although this graphic signifying system makes compromises with phonetic symbols, its intention is essentially different from that of phonetic systems of writing, in which the writing system is fundamentally a recording of the sounds of language. The morphemic principle of the writing system dominated the creation of the Chinese language, dictating to some extent grammatical and phonetic principles of the language. The character-based writing system using graphic units produces a large number of final and complete forms as signifiers; the *Kangxi Dictionary*° numbers them to be about 49,000, but the current usage by a literate person in Chinese would typically range from 6,000 to 7,000 of them. Compared to less than thirty abstract symbols in phonetic scripts, these characters in Chinese lay out a drastically different strategy of meaning; these characters, separate from the spoken language and demanding deep adaptation of the spoken language, are the entire realm that produces ethical, philosophical and aesthetic meanings. To commit them to memory and to reproduce them perfectly demands a system of cultivation that possesses unique features: much of which are apparently features of the writing system itself, particularly when compared with the phonetic writing systems. The primary task of memorising requires long periods of drill and repetition; the value of this practice could not exist without first establishing a system of values that legitimises it. The cultivation of positive values in drill and repetition is a cultivation of endurance, one of the central virtues of Confucianism and Daoism. Western perceptions of this practice of repetition and endurance have always been critically formulated, both within their own moral values and in the response to features of Asian cultures. Philosophically, this set of final and complete forms functioning as characters

° morpheme

° 1716

象形帝国

在《迂回与进入》°一书对希腊和中国思维的比较研究中，弗朗索瓦·于连提出了这样的问题：

° 2004年

"假如世界不是表征的对象，象形表现的含义亦不趋向表现什么——象征性的，会怎样呢？假如归纳并非思考的目的，或言语并不倾向于定义°而是自我修改——以反映情况，会怎样呢？简言之，假如知觉不再为了打好°超验的基础而努力复制现实，会怎样呢？或假如谈论世界或者将世界呈现得可以理解的目的，并非要达到真理，又会怎样呢？"

° 为建立本质的普适性

° 人的或上帝的

这是一种重要的理解，让我们思考象的第二个概念，它完全发生于自身的领域内：即一个象形帝国。这一理解源于中国象形书写体系，用图像单元°来指示含义。尽管这种符号表示体系在语音符号方面做出了妥协，其目的从本质上不同于语音书写体系，即书写体系基于发音记录。书写体系的语形素原则统领了汉语的创造，在一定程度上支配了汉语的语法和发音原则。象形书写体系利用图像单元生产出大量能指的最终完整形式；《康熙字典》°总计有49,000个，但目前受过教育的中国文通常使用6,000至7,000个字。同拼音文字不到30个的抽象字符相比，这些汉字建立了它们特定的语义策略；在语言和文字脱节的情况下，汉字不但深刻影响了中文语言的发音特征，同时也是产生道德、哲学和美学意义的全部表达途径。熟练地掌握汉字需要长期培养，需要独特的培养系统；培养系统的特征往往是汉字记忆和书写特征的直接和间接的表达，它们在与拼音文字对比下格外明显。记忆汉字的首要任务是长时间的操练和重复；这个价值观的建立必须基于更高层的价值观的建立。操练和重复正面的价值的培养是忍耐力培养的一部分，这是儒家和道家的重要美德之一。无论是在他们自身的价值观体系中，还是在对亚洲文化特征的回应，西方文化对这种重复和忍耐的实践一直是持批判性态度的。从哲学上讲，这套作为文字被完美掌握的最终和完整形式，可能确定了它们自身的认知范围和结构，将对知识的追求转向寻求有类似特征的最终和完整的形式。

° 语形素

° 1716年

to be perfectly mastered may indeed prescribe their own epistemological limits and structures, directing pursuits of knowledge in search for final and complete forms of comparable strengths.

It is perhaps in the field of aesthetics that these features of the character-based writing system work in most powerful ways. The empire of figures, in its demand for inherent endurance in the process of mastering them, cultivates the ultimate aesthetic act in calligraphy. Calligraphy represents the height of mastery of the Chinese characters; its morphemic principle is essentialised by purging all grammatical and phonetic symbols, achieving aesthetic satisfaction in both intellectual and visual dimensions. The ability to attain good calligraphy is valued as a trait of cultivation and status. In the Chinese urban environment, calligraphy in all its styles plays a crucial role not only in signifying places but in aestheticising architecture and the city.

In traditional contexts, contrary to the Western urge to harbour critical thoughts, thoughts that were considered to be legitimately formulated in the Chinese intellectual context tend to be far more systematic, displaying features of final and complete forms. It is as if in the process of obtaining empirical knowledge, the empire of figures imposes its inherent formal logic, forcing the completion and finalisation of thoughts through other means. There is nothing outside the empire of figures. In this sense, the empire of figures moves completely in the opposite direction to that proposed by Francis Bacon; instead of a purification of facts, the empire of figures establishes a purification of forms. All fragments and violence are seen to be unformed conditions in the process of becoming final and complete.

It is in this sense that the empire of figures is also the home of thoughts located in the endless procession of figures. Traditionally speaking, philosophy and aesthetics never left home; there was nothing outside the home. The power of the figure in securing its own homeliness lies in the way in which it constructs—through analogy rather than mimesis—an entire realm which approximates external references and replaces them altogether.

THE NOT-HOME

Under this condition, what is outside home can only be perceived philosophically as an unnamable 'not-home', a seemingly unique condition in the Chinese cultural context that has tremendous consequences in the production of environment in China. A novel such as Cao Xueqing's *Dream of the Red Chamber*° takes place entirely ° 1791
within the spatial compound of a socially privileged family, suggesting the primary importance of home. During the May Fourth Movement in 1919, China's first effective introduction to Western cultural concepts, many intellectuals began to experiment with different scenarios of home in the Chinese culture. Ba Jin's novel *Jia*,° a story ° *Family*, 1933
of the different fates of members of an upper class family caught up

或许可以认为，书写体系这些特征在美学领域里表现特别突出。象形帝国在要求精通其特征所需的忍耐性的同时，把书法视为其最终的审美行为。书法代表着精通汉字的最高成就；其语形素原则通过排除所有的语法和语音符号而得以精炼，获得心智和视觉双重维度的审美满足。书法的成就无疑被高度尊视为修养和社会地位的特征。在中国的城市环境中，各种风格的书法扮演了重要角色，不仅指明地点，同时也美化建筑和城市。

在传统语境下，同西方培育批判性思维的强烈欲望相反，中国知识语境中被合法接受的思维更加趋于系统，呈现出最终和完整形式的特征。似乎在获取经验知识的过程中，象形帝国强加了其固有的形式逻辑，迫使思维经由其他方式来完成和最终确定。在象形帝国之外似乎什么也没有。在这种意义上，象形帝国朝着同培根提议的经验主义知识完全相反的方向移动；象形帝国并没有净化事实，而是净化了形式。在追求最终和完整的过程中，一切碎片和暴力均被视为未成形的条件。

象形帝国正是在这种意义上成为思维的家，家中充满了象的无止行列。从传统意义来讲，哲学和美学从未离开家；家之外什么也没有。象确保家的能力来源于比喻能力而非模仿能力，以及通过这个能力所建构的整个领域，复制了并完全替换了外在的参照。

非家

在这种状况下，家之外的空间只能从哲学上被理解为难以形容的"非家"，这似乎是中国文化语境下一种独特状况，对中国环境的生产有着巨大的影响。曹雪芹的《红楼梦》° 中所发生的事件全部在一个享有社会特权的家庭空间之内，充分表现了家的至关重要。1919 年"五四运动"期间，在中国引入西方文化概念的初期，许多知识分子开始试验家在中国文化中的各种场景。巴金的小说《家》讲述了 20 世纪 20 年代"五四运动"巨变中一个上层阶级家庭成员的不同命运，探讨了中国这一根深蒂固的社会形式所面临的危机。其主角们展现了留恋家庭里和离家出走的不同倾向，体现了这个时期中国社会变革的状态。同时，亨利克·易卜生的戏剧《玩偶之家》° 在"五四运动"下的中国掀起了特殊的风暴；一名中产阶级家庭的女性° 选择离开她的家和丈夫，这激起了

° 1791 年

° 于 1879 年在欧洲首演
° 挪拉

in the upheavals of the May Fourth Movement in the 1920s, explores the crisis of this deep-rooted social form in China. Its main protagonists show tendencies of both staying at home and leaving home, highlighting a transitional condition of the Chinese society. In the meantime, the play *A Doll's House* by Henrik Ibsen° caused a particular storm in the May Fourth Movement China; the notion that a women° in a bourgeois family would choose to leave her home and husband stirred deep emotions within Chinese youth seeking changes to their society. The family as it was imagined in traditional China was without doubt one of the most critical battlegrounds for young minds inspired by revolution. Behind the revolutionary fever, Lu Xun, perhaps one of the most radical reformers in 20th century China, posed a sobering question: what happens to Nora after she leaves home? Lu Xun was speaking to an audience of female students at the Beijing Women's Higher Normal School on 26 December 1923, keenly aware of the non-existence of any sort of space outside the home for Nora in the Chinese cultural context. He speculated that after Nora leaves home, she would either fall or return home. Lu Xun's note of caution speaks of a deep perception of a social form in which home is never made uncanny or unhomely; instead, this social form cultivates the force of perpetual return. The space outside home,° at the same time, was formulated and practiced as the not-home. Nothing ethically, philosophically and aesthetically legitimate takes place in this not-home. It is perhaps not surprising that Lu Xun became an ardent advocate for the abandonment of the character-based writing system in China.

° premiered in 1879 in Europe

° Nora

° public space

This traditional conception of urban space maintains its presence despite dramatic changes which have taken place in China since the 20th century; together with economic forces, the notion of home and not-home work to produce spaces of inclusion and exclusion in most imaginative ways. The institutional home-making through *danwei* and the spatial home-making through walls remain essential urban features as China re-imagines its history and its roles in a contemporary global community. It opens up an extraordinarily promising field of enquiry.

LACONIC CRITICALITY AT HOME

It is perhaps a misconception that, in China, a different opinion is either not expressed, or it is expressed in the form of Western-style critical debates. Of course all these do happen, but here we would like to draw our attention to a more indigenous form of dissent, something which might be described as laconic criticality. The primary critical strategy here is not to articulate an analytical framework that places both the canon and dissent into 'objects of criticism', but to act within the canon. The Chinese artist Qiu Zhijie perhaps exemplified this laconic criticality through his performance piece *A One-thousand-time Copy of Lantingxu*. Between 1990 and 1997, Qiu filmed himself copying perhaps the best known and most

正寻求改变社会的中国年轻人的深层情感。传统中国所想象的家，在受革命激发的年轻人看来，毫无疑问是最关键的战场之一。鲁迅可以说是中国 20 世纪最彻底的改革者之一，他在这种革命热情的背后，提出了一个冷静的问题：挪拉离家以后发生了什么？这是 1923 年 12 月 26 日鲁迅在北京女子高等师范学校演讲，向女学生听众们提出的问题，敏锐地意识到在中国文化语境中并不存在对挪拉而言的家之外的空间。他推断挪拉离开家后，或是堕落，或是回家。鲁迅的告诫显示出对家这一特定社会形式的深刻觉察，在中国的这个特定社会形式中，家从未变得暗恐或陌生；相反，这种社会形式培养了永久回家的力量。同时，家之外的空间。在思想和实践中成为一种非家。在这种非家的空间中，所有事件缺乏任何在道德上、哲学上和美学上的合法性。这些条件也许自然促使鲁迅在这个文化变革年代中成为废除汉字的先锋。 ○ 公共空间

  20 世纪以来尽管中国发生了巨变，但这种城市空间的传统理念仍然存在；结合经济力量，家和非家的观点以丰富想象力建构了各种内部和外部空间。中国社会中的单位不但在机构的意义上，而且在内部空间的建造过程中体现了对"家"的重新想象，成为当今中国重构历史和国际社会地位过程中的重要城市特征。这开启了一处非常有前景的研究领域。

家中的无言批评性

或许是我们的误解，在中国或是不能表达不同观点，或是必须通过西方批评性的辩论模式来表达不同观点。当然这些都可以发生，但这里我们希望将注意力放在分歧的一种更本土的形式，或许可以描述为无言批评性。此处的首要批评策略不是清晰表述一种分析框架，将规则和分歧均当作"批评的对象"，而是在规则之中的批评活动。或许中国艺术家邱志杰通过其表演作品《重复书写一千遍兰亭序》示范了这种无言批评性。从 1990 年到 1997 年，邱拍摄了自己在同一张纸上临摹王羲之的著名临本作品《兰亭序》。一千遍。每次临 ○ 353 年
摹在前次的基础上稍稍偏移，以至于最初几次临摹后书法看起来有些抽象；五十次后，整张稿纸变成黑色，进一步的临摹完全是在一张黑稿纸之上。这种对王羲之书法范本的超常虔诚、临摹过程中的劳力，加上黑纸上无法辨认所体现的反讽的挑衅意义，戏剧性地显示了汉字书写体系和书法对心智的巨大控制力。徐冰的作品《天书》。同样也是基于这种无 ○ 1987—1991 年
言批评性；它有效地制作了"无家可归"的一刻，困扰了许

copied example of calligraphy, Wang Xizhi's *Lantingxu*,° one thousand times on the same piece of paper. Each copy was slightly offset from the previous one, so that the copies looked slightly abstract after a few times; after fifty times, the entire sheet of paper became black and further copies were completed over the piece of black paper. The act of extraordinary devotion to Wang Xizhi's calligraphic example and the strenuous labour in the execution, with a provocative sense of irony in the illegibility of the blackness, highlighted dramatically the powerful hold of the character-based writing system and calligraphy on the intellect. Xu Bing's work, *Book from the Sky*,° also worked within this laconic criticality; it effectively created a moment of 'homelessness' which disturbed many Chinese politicians and intellectuals. Xu Bing's work consists of 4,000 unrecognisable Chinese-looking characters, beautifully crafted in traditional 15th century Ming Dynasty imitation of 10th century Song Dynasty font and printing techniques. They were made into printed scrolls, and a set of four bound volumes which resemble traditional canonical texts in literature, medicine and law; the printed scrolls and volumes made from these fake characters have been displayed in a range of formats in museums throughout the world. The impact of Xu Bing's work on those who can read the Chinese square words was exceptionally intense. The familiarity of the form and the profound meaninglessness of the 'text' brought a moment of disbelief; it touched a core issue of Chinese culture's rootedness in 'image as text'. By maintaining the form and suspending meaning at the same time, one is made to experience a moment of seeing the empire of figures as its ghost.

° 353

° 1987–1991

This laconic criticality at home demonstrates a possible formulation of an indigenous criticality; it does not begin with critical definitions, but with critical modulations: a process of unfolding situations, a critical propensity that engages and modifies as it proceeds. With a certain degree of contrivance, Freud's psychologisation of home, as we find in his essay 'The Uncanny', can be recreated in the Chinese cultural context, but it would be ineffective to engage with the primary forces of Chinese culture. However, as we shift our attention to its indigenous criticality, what appears to be crucially important is the possibility of sustaining this laconic criticality through an establishment of its legitimacy. If the empire of figures does not conceive an 'outside', can this laconic criticality guarantee its own legitimate status, and ultimately fairness and humaneness at home? What are the possible formulations of the 'outside', the not-home and not-nature that are now subject to careless destruction?

多中国政治家和知识分子。徐冰的作品由 4000 个无法识别的类似汉字字符组成，用 15 世纪明代的仿宋体和印刷术精心制作。它们被印成卷轴，和一套四册的精装书，如同传统的文学、医学和法律典籍；由这些伪字符组成的卷轴和卷册以不同格式在世界各地博物馆展出。对于能够读懂汉字的观众而言，徐冰的作品有着异常强烈的影响力。形式的熟悉和"文本"无意义带来不可置信的一刻；它触及了中国文化根植于"形象为文字"的核心问题。通过保留形式和删除意义，徐冰使观者看到了象形帝国的鬼影。

　　这种家中的无言批评性显示了本土批评性的可能；它并非从批评性定义开始，而是始于批评性调整和渐进：一种展现情况的过程，一种在进行中的批评性参与和调整。弗洛伊德在"暗恐"一文中对家的心理分析也许能强迫性地在中国文化语境中重新创造，但它对中国文化的主要力量似乎并不是很有效。然而，当我们将注意力转移到本土批评性上，似乎至关重要的是建立无言批评性合法地位的可能性。如果象形帝国不构想"外界"，那么这种无言批评性能否在家中保证自身的合法地位，最终保证家中的公平和人道？我们如何重新建构目前正遭受草率破坏的"外界"，外界中的非家、非自然的地位？

A NARRATIVE STRUCTURE OF CROSS-CULTURAL ARCHITECTURE

交叉文化建筑的情节结构

邹晖
Hui Zou

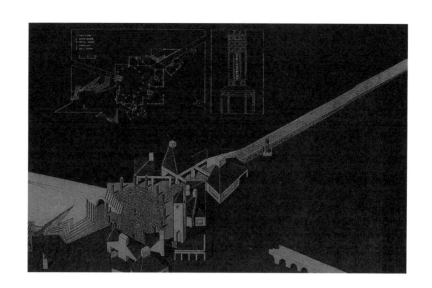

邹晖
HUI ZOU

MENTAL SPACE

Economic globalisation based on the technical worldview flattens cultural differences and propels meaningless buildings and homogeneous urban landscapes. Architecture is losing its ability to create intriguing ecstatic cultural encounters as witnessed in the 18th century. Comparative studies provide an effectual means for interpreting East-West cultural exchanges of which historical interpretation acts as a strategy for cross-cultural design development. The theoretical endeavour of the cross-cultural approach will enlighten architectural practice in preserving the poetic tradition during the process of modernisation and urbanisation.

Methodological issues of historical interpretation need to be analysed. The method and objective of historical research is related to the philosophical understanding of objectivity and subjectivity. Historical objectivity is the starting point for historical research and is the historical knowledge that is desired. Our search for historical objectivity is mixed with the expectation for the historian's subjectivity. Historical subjectivity is not only related to the historian's individuality but also human subjectivity in general, and thus integrates the author and audience through reflection. The philosopher Paul Ricoeur advances the concept of 'the historian's craft,' arguing that a historian can be a craftsman, defining historical objectivity and creating historical knowledge through searching for historical traces.* Historical reality results from investigation by the historian • 1

当今经济国际化的趋势以及建立在科学技术发展上的世界观正在消除传统文化的差别，导致建筑形式的无意义与城市景观的雷同，建筑正在失去作为文化碰撞载体的能力。历史的比较研究为我们提供了诠释中西建筑文化交流的手段。对这些建筑碰撞的历史意义的解读，为当今国际化背景下交叉文化的设计途径提供策略性的参考。这对于如何在中国的现代化与城市化运动中保持和发展本土诗意的文化传统与人居环境具有重要的实践指导作用。

对历史的解读存在一个方法论的问题。历史研究的方法与目的与我们对客体性与主体性的哲学理解有关。历史的客体性是历史研究的出发点，即我们所期盼的历史知识。对历史客体性的期盼应和着我们对历史研究者的某种主体性的期盼，这种历史主体性不仅仅局限于研究者个体，也涵盖一定的人类主体性，它把作者与读者通过反思联系起来。法国现象学诠释学家保罗·利科提出"历史学家的手工艺"概念，认为历史研究者是界定历史客体性的工匠，他通过小心追寻历史记录的痕迹来建立历史知识。* 历史的事实是通过研究者 • 1

whose responsibility is not to retrieve history as fact but rather to establish a tracing sequence towards a historical objectivity in which no absolute idea can dominate historical understanding. Any historical continuity is based on the historian's judgment regarding the value of historical events, but the criteria of judgment needs the support of theories. Ricoeur emphasises the importance of imagination in constructing historical subjectivity, which never follows a scientific model, but rather is a rare talent that on the one hand brings history close to us and on the other hand, maintains a sense of distance from history, the remote depth of time in our mind. History can be enlivened by the encounter between the present and past both of which can engage in a dialogue based on humanity. Although history reflects the historian's subjectivity, the historian's craft educates his own subjectivity. This is a process open to others and has nothing to do with the historian's autonomous self. Historical subjectivity includes not only the historian's subjectivity but also historical people's subjectivity and the audience's subjectivity, which is presented through philosophical reflection on history and is intended to transcend individuality towards general humanity. The thinker is called by history to interpret himself historically so as to produce historical meaning. The phenomenological reflective history and the historian's objective history hold different foci but interact with each other. The reflective thinker searches in history for desired meanings and isn't afraid of the incompleteness and singularity unfolded by history.

When philosophical reflection on history takes place, what is intended is not a complete and cohesive history but rather how to interweave fragments of human works. In this sense, history is composed of discontinuous representations each of which needs a completely new encounter with different representations. Such a reflective interpretation matches the historicity advanced by phenomenological existentialism and requires the historian to engage in those excellent historical works in which the cosmic world can be rebuilt based on human existence and thinking. Comparative history should present the historical events of Chinese and Western encounters, unfolding historical meanings by tracing those ecstatic cultural events rather than set up an a priori objective framework for controlling historical meanings. In the process of tracing historical events, there will emerge ambiguity and indecisiveness, but the real values of human life rather than unmistakable objective structures can be revealed through rebuilding and rediscovering history. As Ricoeur states, the historical object is the very human subject.

Philosophical reflection on history affects our understanding of cultural autonomy, cross-cultural experience and cultural encounters. In today's globalisation of technology and economics, rationality penetrates into all aspects of society and pushes towards the universalisation of human living, including the standardisation of buildings and the propagation of consumerist culture. Dwindling

的考察而被建立，历史学家的任务不是为了恢复如其所是的历史，而是建构一个回顾的次序来导向历史的客体性，没有任何绝对的观念能够统领历史的总体。历史的连续性是建立在历史学家对历史事件重要性的判断之上的，而判断的标准需要理论的支持。利科强调历史研究中的想象力正是历史主体性的体现，它从来不是科学的模式，而是一种少有的才能，它一方面将历史的过去带近我们，另一方面维系着历史的距离感、一种我们所能意识到的遥远的时间深度。历史因为今天与过去的碰撞而生动起来，今天与过去在人性的基础上产生对话。虽然历史反映着历史学家的主体性，但历史学家的手工艺反过来教育着他自身的主体性，这是一个向他人打开的进程，它与自我中心无关。历史的主体性不仅包含着历史学家的主体性，也包含着历史中的人的主体性以及读者的主体性。读者的主体性很大程度上经由哲学对历史的反思来达成。对历史的思考实际上是为了超越个体性而达到普世的人性，思考者被历史呼唤着并对自己作出历史性的诠释，由此产生的就是历史的意义。现象学家的反思的历史与历史学家的客观历史着重点不同但又相互作用，反思者在历史中搜寻着他所欲寻求的意义，他不回避历史所展开的非完整性与奇特性。

当用哲学的反思方式来反思历史时，所关注的不是紧密完整的历史运动，而是如何将片断的人与人的作品交织起来。从这个角度出发，历史是由一系列的非连续的表象组成，每个表象都需求全新的碰撞。这种反思的历史解读尤其符合现象学的存在主义的历史观，即历史学家致力于研究那些优秀的历史作品，在这些作品中，宇宙世界围绕着人的存在与思想而被重构。现象学的历史比较观要求对中西文化碰撞的历史事件产生关注，通过对这些精彩事件的不断追踪来展现历史意义，而不是先预置一个客观的历史框架来决定历史意义。在对事件的追踪中，会涌现模糊性与犹豫不决，但在对历史的重构与再发现中，人类的真实价值而非错误的客观结构得以呈现出来。这正如利科所言，历史的客体恰恰是人的主体。

对历史的哲学反思影响着我们对文化的自主性、交叉文化及文化碰撞的理解。在科技与经济日益全球化的今天，理性渗透进技术的各个层面，随之而来的是生活方式的全球一体化，包括建筑的标准化以及无处不在的消费文化。文化差异性的消失降低了文化交流的戏剧性，造就着一种低俗的世界文明，使地区文化所赖以生存的道德且神秘的主体解体。

cultural differences reduces the theatricality of cultural exchange and leads to a mediocre world civilisation, which deconstructs the ethical and mythical core of regional culture. Worldwide similarity and the non-identity of consumerist culture represent the lowest level of creativity and unavoidably results in the absolute hedonism of an easy life. Facing the disappearance of cultural complexity, Ricoeur advances his concept of cultural encounter — that if a culture wants to survive, it must respect its own origin and meanwhile seek for creativity in art, literature, philosophy and spirituality so that a culture can maintain ecstatic encounters with others and consistently redefine the cultural encounter.* The cultural encounter highlights the theatrical value of communication through which a native culture can maintain its singularity and meanwhile open up infinite imagination for other cultures. Human truth lies exactly in this lively process of creative encounters. • 2

The comparative study of cultural encounters should not be limited to morphological analogy or restricted by conceptual exchange based on formalism, but rather seek the treasures of encounters in architectural history. By locating itself in cultural reality, comparative research carries out philosophical reflection on architectural encounters, from which enlightenment for theatrical cultural exchange can be found. This endeavour opens up the interactive relationship between architecture and philosophy as well as the poetical potentiality of architectural practice implied in cultural differences.

In the 1940s, the Argentine writer Jorge Luis Borges wrote a fictional novel entitled 'The Garden of Forking Paths'.* Borges' book begins with a historical record of the First World War regarding a British attack against German-occupied France that was delayed for a few days. The delay was related to a Chinese national, Yu Tsun,° a former English professor at a German school in Qingdao, eastern China. Tsun worked as a spy for Germany during his stay in Britain. In this story, he was chased by a detective while attempting to report on a secret artillery location. In a phone book, Tsun found the name Albert, who was a stranger living in the village of Ashgrove. Arriving at this village, Tsun was told that if he took the road to the left and turned again to his left at every crossroad, he would find Albert's house. Walking down the solitary road under the full moon, he recalled that continuously turning to the left was a common procedure for discovering the central point of certain labyrinths. Tsun grew up in his father's symmetrical garden. His grandfather, the former governor of the Yunnan Province, renounced worldly power and dedicated himself to writing a novel and constructing a labyrinth for thirteen years until he was murdered by a stranger, leaving the novel incoherent with only chaotic manuscripts left behind and the labyrinth lost. As Tsun continued to hurry along the winding country road while reminiscing on his grandfather's lost labyrinth, he imagined an infinite labyrinth that would encompass the past as well as the future while involving the cosmos. • 3

o namely, You Zun in pinyin

消费文化在全球的相似性及无身份性代表着创造性文化的最低层次，不可避免地导致安逸生活中的绝对虚无主义。面对文化丰富性的消失，利科提出他的文化碰撞观，认为一种文化若要生存下去，必须忠实于自己的原初，同时必须在艺术、文学、哲学和精神层面追求创造性，如此，该文化才能维持与其他文化的精彩碰撞并赋予文化碰撞以意义。○文化的碰撞强调一种戏剧性的交流关系，通过这种交流，本土文化维系着自己的独特性，并留给其他文化无限的想象。人类的真理正是寓于这种生动的创造性的文化碰撞的进程中。● 2

　　从文化碰撞出发的比较观，将不关注静态的建筑形式的差异与相似或者建立在形式比较之上的简单概念比较，而是在历史长河中搜寻着依靠建筑为载体的文化碰撞的火花。它伫立于当今的文化现状，哲学地反思历史中的建筑碰撞，力图从这些建筑碰撞的事件中寻找文化交流的戏剧性启发，从而打开建筑与哲学的互动比较关系以及中西方的文化差异对当代建筑实践所蕴含的诗意力量。

　　1940年代，阿根廷作家乔治·路易斯·博尔赫斯发表了一篇题为《交叉路径的花园》的短篇小说。○小说的情节开始于第一次世界大战时，英军计划袭击德军占领的法国，但该计划因某种原因而推迟。推迟的原因据说与一位中国人有关，他名叫友尊，○是青岛一所德国人开办的学校的英文教授。友尊实际上是一名为德国工作的间谍。在英国一次逗留期间，他试图将关于英军炮兵阵地的情报送回德国，但与此同时，他正在被一位英国侦探追捕。从一本公用电话簿中，尊找到一个陌生人的名字叫阿尔伯特，此人住在一个叫灰林的小镇。尊设法到达了这个小镇，当地人告诉他如果沿着左边这条路往前走，在每个交叉口都向左转，他就可以抵达阿尔伯特的住所。孤独地走在圆月下的乡村小道上，尊记得有一种迷宫，如果在其中持续左转的话，就可以抵达迷宫的中心。他从小在父亲的对称布局的花园中长大。尊的爷爷是云南省的前省长，他放弃世俗的权力，潜心十三年写一部小说和造一所迷宫，最终被一位陌生人所谋杀，留下支离破碎的小说手稿，建造的迷宫也不见踪影。当尊在英国蜿蜒的乡村小道上急行时，他回想着爷爷失去的迷宫，不由得想象着一座能把过去、未来及宇宙都包容在里面的无限的迷宫。● 3　○ 音译

　　事有凑巧，这个叫阿尔伯特的人曾经是生活在天津的一名西方传教士，后来成为英国的一位汉学家。当友尊不期而来时，阿尔伯特领着他穿过前花园的曲折小径，一边走一边

By coincidence, Albert served as a missionary in the Chinese city of Tianjing and later used this experience to become a sinologist. As he greeted Tsun, they walked through the zigzagging path of the front garden and entered the library while talking about Tsun's grandfather. According to Albert's research, Tsun's grandfather isolated himself in the Pavilion of Limpid Solitude, located at the centre of an intricate garden, to work on his book and the labyrinth. Albert then showed the guest a miniature ivory labyrinth, which, he thought, was the lost labyrinth made by Tsun's grandfather. Tsun was also shown a letter in which his grandfather described his project as 'a garden of forking paths.' In Albert's view, the 'garden of forking paths' symbolised an invisible labyrinth of infinite time in which all possible outcomes occurred. As Tsun had planned, he murdered Albert and was later arrested. The murder was covered in a local newspaper and thus, those in Germany received a secret message from Tsun that the new British artillery park was located in the city called Albert.

Borges used the labyrinth as the primary structure of the plot, with multiple labyrinths at both the physical and metaphysical levels. The physical labyrinths include:

1. Geographical labyrinth
2. Village of Ashgrove
3. Albert's house
4. Albert's library
5. Grandfather's miniature labyrinth
6. Grandfather's intricate garden
7. Grandfather's Pavilion of Limpid Solitude
8. Father's symmetrical garden

The metaphysical labyrinths include:

1. Historical book of WWI
2. Phone book
3. Grandfather's letter
4. Albert's collection of books
5. Grandfather's novel
6. Grandfather's classic books
7. The novel *Dream of the Red Chamber*
8. Newspaper report of Tsun's death

谈论着尊的爷爷的身世，然后一起走进了书房。根据阿尔伯特的研究，尊的爷爷把自己关在一座叫清幽阁的房子中写书与造迷宫，这所房子位于一个错综复杂的古典园林的中心。阿尔伯特给客人看了一个袖珍的象牙迷宫，认为这就是尊的爷爷丢失的迷宫。他又给尊看了一封信，在信中尊的爷爷将自己的作品题为"交叉路径的花园"。就阿尔伯特看来，这个题目象征着无限时间的不可见迷宫，在其中什么可能性都可能发生。接下来，如友尊所计划，他谋杀了阿尔伯特，后被逮捕。此谋杀事件被当地的报纸报道出来。在德国的友尊的同事通过这条新闻解读出那个英军的炮兵阵地就坐落在名叫阿尔伯特的城市。

博尔赫斯把迷宫作为整个故事的主结构。迷宫的概念存在于形而上与形而下的多个层面上。形而下的迷宫包括：

　　1：地理的迷宫
　　2：名叫灰林的小镇
　　3：阿尔伯特的住宅
　　4：阿尔伯特的书房
　　5：爷爷的袖珍迷宫
　　6：爷爷的错综复杂的园林
　　7：爷爷的清幽阁
　　8：父亲的对称花园

与形而下的迷宫相对应，形而上的迷宫包括：

　　1：第一次世界大战的历史书
　　2：电话簿
　　3：爷爷的信
　　4：阿尔伯特的藏书
　　5：爷爷的小说
　　6：爷爷的古典书籍
　　7：小说《红楼梦》
　　8：关于友尊被逮捕及处决的报纸报道

The first physical labyrinth was on the global geographical scale. Tsun came from Qingdao, a German colonial city on the eastern coast of China. His grandfather was the former governor of Yunnan, a remote province in southwestern China. Albert, the sinologist, was a former missionary in Tianjing, which was further north of Qingdao. The site of the encounter between the two strangers was in Britain, a Western empire that was a great distance from China. The second physical labyrinth was the village of Ashgrove, where a zigzagging country road led Tsun towards the centre of the village—Albert's house. The third was the front garden of Albert's house. The fourth was the household library where Eastern and Western books were stored and the encounter between the two strangers took place. The fifth was Tsun's grandfather's miniature labyrinth which was stored in the library. This labyrinth led to the sixth one, Tsun's grandfather's garden in remote China, and the seventh one, the Pavilion of Limpid Solitude within that garden. The eighth physical labyrinth, the least described by Borges, was the exotic symmetrical garden of Tsun's deceased father, which was the foreshadowing of Tsun's death. Paralleled with the physical labyrinths, metaphysical labyrinths were presented as books and other types of writings. The first metaphysical labyrinth was the historical book that recorded the spy incident. The second was the phone book where Tsun found the name of Albert. The third was Tsun's grandfather's letter. When Albert tried to show this letter to Tsun, he was murdered and the detective dashed through the front garden to arrest Tsun. The fourth metaphysical labyrinth was Albert's collection of Eastern and Western books. The fifth was the chaotic novel written by Tsun's grandfather. The sixth were the classic books, which Tsun's grandfather, a scholar official, tirelessly interpreted. The seventh metaphysical labyrinth was the famous novel *Dream of the Red Chamber* of the Qing Dynasty, whose amorous narrative took place in the Garden of Grand View. The eighth was the newspaper report of Albert's murder and the ultimate punishment of Tsun.

If judged independently, each physical and metaphysical labyrinth had the potential to be an authentic historical fragment. Within each pairing of physical and metaphysical labyrinths, the metaphoric connection was convincing. The third and fourth pairs hinted that the cultural encounter itself was a labyrinth. The metaphoric connections in the sixth and seventh pairs demonstrate Borges' deep knowledge of Chinese culture. The most hidden connection was in the eighth pair where Tsun's deceased father's symmetrical garden, which was associated with Tsun's act of murder through the sense of death. When these labyrinth fragments, the 'illusory images' described by Borges, were interwoven into a narrative, the fictional novel was constructed. Because each fragment was precisely located in a specific historical and cultural context, the fiction presented itself as a reality.

第一个形而上的迷宫发生在全球地理的层面上。友尊来自青岛,青岛曾经是德国的殖民城市。阿尔伯特以前是天津的一名传教士,后成为汉学家。两人偶然碰面的地点在与中国相距遥远的英国。第二个形而下的迷宫是名叫灰林的小镇,蜿蜒的乡间小路将友尊带到镇的中心——阿尔伯特的住宅。第三个形而下的迷宫是阿尔伯特住宅的前花园。第四个形而下的迷宫是阿尔伯特住宅的书房,里面收藏了不少的中西书籍,而两人的交谈正是发生在那里。第五个形而下的迷宫是尊的爷爷的袖珍迷宫,它恰恰被存放在阿尔伯特的书房内。这导向第六个形而下的迷宫,即爷爷在遥远的中国云南省的古典园林,以及第七个形而下的迷宫,即作为爷爷的花园中心的清幽阁。第八个形而下的迷宫是文中最少提及的,即友尊的逝去的父亲的花园,它暗示着尊自己的死亡。与形而下的迷宫相平行,形而上的迷宫被呈现为书籍或其他的写作文本。第一个形而上的迷宫是记载了这个间谍事件的一本历史书。第二个形而上的迷宫是电话簿,友尊从中找到阿尔伯特的名字。第三个形而上的迷宫是尊的爷爷的亲笔信,当阿尔伯特把信给尊看时,尊把阿尔伯特谋杀,而这时,追捕的侦探正冲进前花园。第四个形而上的迷宫是阿尔伯特书房中的大量中西书籍。第五个形而上的迷宫是尊的爷爷留下的支离破碎的小说。第六个形而上的迷宫是尊的爷爷收藏的古典书籍,作为一个传统文人,他不懈地诠释着这些书籍。第七个形而上的迷宫是文中提到的清代著名小说《红楼梦》,书中的爱情故事发生在大观园。第八个形而上的迷宫是报道阿尔伯特被杀以及友尊被处决的新闻报纸。

如果单独地对每一个形而上和形而下的迷宫加以判断,每个似乎都可以成为一个真实的历史片断。虽然博尔赫斯没有来过中国,但这些精心设置的历史片断没有显示任何对中国历史与文化的误读。在每对形而上与形而下的迷宫之间,隐喻的联系让情节变得可信。第三与第四对迷宫暗示着文化碰撞自身就是一个迷宫。第六与第七对迷宫显示了博尔赫斯对中国文化的深入了解。最隐蔽的是第八对迷宫,尊的逝去父亲的对称花园通过贯穿全文的死亡的线索与尊自己的死亡联系起来。当这些迷宫的片断,即博尔赫斯所说的"错觉的意象",被编织进故事情节中时,一部虚构的小说就诞生了。因为每一个迷宫碎片是如此仔细地被置于特定的历史与文化背景中,虚构的情节展现出现实的意义。

If as stated in the fifth pair of labyrinths that 'the book and the labyrinth were one and the same,'• the fictional text should be understood as a labyrinth. Thus, the fiction and reality became one in the primary labyrinth which Borges called the 'labyrinth of labyrinths' or the 'infinite labyrinth' where one path of time could lead to others and they sometimes converged, just as in the story where Tsun sat beneath the English trees and reminisced on the lost labyrinth of his grandfather. The infinite labyrinth defines a cross-cultural order where physicality remains permanent, which is most likely why Borges did not differentiate the physical details between English gardens and Chinese gardens, but rather emphasised their commonality through the symbol of the labyrinth.

• 4

Borges' seventh metaphysical labyrinth was the Chinese classic novel *Dream of the Red Chamber*, whose narrative took place in the fictional Garden of Grand View.° During the 1750s, when the author Cao Xueqin wrote his novel at the foot of the Fragrant Hill in the northwestern suburb of Beijing, emperor Qianlong was expanding his Garden of Round Brightness° nearby in the town of Haidian. This garden expansion included mainly the Garden of Eternal Spring° and the Western garden designed by the Jesuits. When Qianlong's garden expansion was done, he ordered the court painters to paint a panoramic view of the entire garden complex and named this painting as the Grand View,° coincidently the same name as the fictional garden in the novel *Dream of the Red Chamber*. The old part of the Garden of Round Brightness was composed of the Forty Scenes each of which was given by the emperor a poetic name and a poem and was also recorded through a painting created by the court painters. Each named scene was an encounter between the emperor's historical mind and the specific garden view. These forty scenes were organised into a numeric sequence and the meandering movement of the sequence appeared mystic and bewildering. By connecting each scenic point numerically throughout the garden with a straight line, a diagram of the zigzagging path, like a cosmic labyrinth in the sky, comes to light.• In this complicated zigzagging diagram, surprisingly, there is neither an overlap of the lines nor a centre. The hidden cosmic order defined the emperor's poetic journey towards infinity.

○ *Daguan Yuan*

○ *Yuanming Yuan*

○ *Changchun Yuan*

○ *Daguan*

• 5

In contrast to the winding paths of the Chinese garden, the Jesuit garden was planned out on a T-square with two major axes. The Western garden view was composed of multiple exotic scenes, such as the labyrinth, multistoried buildings, mechanical fountains, geometrical pools, a geometrical hill and an open-air theatre. In a set of twenty copperplates drawn by the court painter Yi Lantai in 1786, all the garden scenes were depicted through the Western technique of central perspective. On the northern end of the first axis, there was a Western labyrinth; on the eastern end of the second axis stood an illusionary stage set of an open-air theatre, which was composed of illusionary perspective painted walls. The layout of these per-

按照博尔赫斯在论述第五对迷宫时所说"书与迷宫是同一的",虚构的文本自身应当被理解成一个迷宫,在其中,虚构与现实成为一体。他把这个原初的迷宫称为"迷宫的迷宫"或者"无限的迷宫",在其中,某条时间的路径能够导向其他的时间路径,并相互交织。比如,当友尊坐在英国乡间的树下时,他回忆起爷爷丢失的迷宫。无限的迷宫界定着交叉文化的秩序,此秩序的形态是恒定的,这也许解释了为什么博尔赫斯没有进一步区分英国园林与中国园林,而是通过迷宫的象征形象强调两者的共性。

● 4

博尔赫斯的第七个形而上的迷宫是清代的古典小说《红楼梦》。书中的情节发生在虚构的大观园。1750年代,当曹雪芹在香山脚下写这本书时,乾隆皇帝正在离此不远的海淀镇扩建他的圆明园,此次扩建主要包括长春园与耶稣士设计的西洋楼。巧合的是,当圆明园扩建完成时,乾隆让宫廷画师画了一幅巨大的圆明园鸟瞰图,命其名曰《大观》,碰巧与《红楼梦》中的大观园同名。在老的圆明园部分,有著名的四十景,每一个景都配有题名、诗和画,每一个景都是意与象的交流。这四十景按照一定的先后次序组织起来,曲折的序列让人感到神秘且迷茫。如果从第一景到第四十景依次用直线联系起来,所得到的路径图案曲折迷离,像一个天象的迷宫。令人惊讶的是,此图案中有无数的转折却没有任何交叉线和中心,显出无序中的有序。这个隐藏的宇宙秩序确定了朝向无限性的诗意旅程。

● 5

与圆明园的曲折路径相对立,耶稣士设计的西洋楼园平面由两条轴线相交成几何的"T"形。西洋楼园的奇观包括迷宫、西洋楼、喷泉、几何的水池、线法山、露天剧场等。在由宫廷画师伊兰泰于1786年所绘的西洋楼二十幅铜版画中,所有的构图都采用了西方的中心透视手法。在较短的南北轴线的北端是一座西式迷宫。在长的东西轴线的东端是一个露天的戏剧舞台,舞台上的背景由具有错觉感的线法墙构成,墙体的平面布局遵从纵深透视的原理,类似帕拉第奥在意大利维琴察的奥林匹克剧场中的舞台背景设计原理。西洋楼的透视景与圆明园的诗意四十景形成鲜明的对比,也成为相互补充的视觉体验。如同意大利文艺复兴园林中的"秘密花园"一样,西洋楼成为乾隆皇帝独自欣赏的秘密花园。园中的迷宫与错觉透视舞台将他的想象引向无限。

○ 即透视墙

● 6

○ giardino segreto

根据雍正的《圆明园记》,"圆明"这一概念的含义意味深长,它意指一种深远的道德。他对"圆明"语义的诠释

spective walls followed the principle of remote perspective depth like Palladio's Olympic Theatre in Vicenza.* The perspective views of the Jesuit garden and the poetical views of the Forty Scenes formulate a sharp contrast as well as a complementary relationship of visual experience. Like the 'secret garden'° in an Italian Renaissance garden, the Jesuit garden became Qianlong's secret garden. The Western labyrinth and the illusionary theatre led his imagination into infinity.

• 6

○ giardino segreto

According to the emperor Yongzheng's garden record, the concept of 'Round Brightness' indicated a 'deep and distant' virtue, which was difficult to understand. His approach to interpreting the meaning of the garden involved referring to ancient books, meanwhile identifying the 'virtue of Round Brightness' through his garden experience. The virtue of Round Brightness reflects the Daoist sage Laozi's concept of 'deep and distant ethics',° which can reach the ultimate harmony by 'acting opposite of the regular way,' or moves into the Tang writer Liu Zongyuan's in depth spiritual concept of 'abstruse as such'° by following the meandering path in a garden labyrinth.* In the Jesuit garden, such an abstruse depth of the mind was realised through perspective depth. It is this paradoxical movement of detouring that intertwines theatrically with the Western perspective view and the traditional Chinese discursive mind.

○ xuande

○ aoru

• 7

The Garden of Round Brightness demonstrated how the paths of varied historical times and cultures forked and converged within a garden reality where the named garden scenes acted as metaphoric images. The surrealist André Breton called the assemblage of metaphoric images a 'poetic analogy' which acted as the bridge between two objects of thought situated on different planes.* Dalibor Vesely applied the theory of poetic analogy to the art of collage in which two distant realities meet by chance. He analysed how the architectural collage of environmental metaphors can be open to a series of readings and reveal the situational character of dwelling in an overlapping context.* Interpreting the mythology of Daedalus' labyrinth, Alberto Pérez-Gómez expounds that Vitruvius took the rational order as the meaning and beauty of architecture and that Western architectural theories tend to understand the labyrinth as a type of order, namely the Ariadne's thread, but in fact, the meaning of the labyrinth is more related to theatrical rituals like the dancing ground designed by Daedalus for Ariadne.* He further argues that architectural reasoning can be understood as the ability to interpret a place, and the design idea can only originate from an in-depth dialogue with the history and culture of that place. The true objective of architecture is to find ways to express and answer the consistent needs of humanity. The reason that we can understand art and architecture throughout history is because human beings have learnt how to translate over time. Although all human beings do not share a common cosmic view, our historicity is pre-given as a gift, which defines our ability of retrieving those cultural traditions that acted

• 8

• 9

• 10

途径是反复阅读古典经书，同时又通过园林的亲身体验达到与"圆明之德"的认同。这个圆明之德反映着老子的"玄德"思想，即通过非常规的行为通达极致的和谐，如同穿过园林迷宫的迂回曲折之径，进入柳宗元的"奥如"境界。不过在西洋楼园，这种境界是经由透视的深远来实现。正是通过似是而非的迷宫般的迂回，西方的透视景与中国传统的发散思维戏剧性地交织起来。 • 7

圆明园体现了不同的历史时间与文化在园林的现实中时而分叉时而汇聚，每一个园景都是隐喻的形象。法国超现实主义诗人安德烈·布莱敦把隐语形象的汇聚称作"诗意的类比"，它把不同层面的思想载体联系起来。英国建筑现象学家达利博尔·维斯利把诗意类比的理论与拼贴艺术联系起来，在拼贴艺术中两个遥远的现实可以偶然地相遇。他认为由环境隐语片断构成的建筑拼贴可以打开一系列的解读进程，并在一种交叠的文脉中揭示居在的场所性格。1980年代，加拿大建筑史学家阿尔贝托·佩内兹·戈麦兹对作为戴达琉斯神话的迷宫及其现象学的建筑含义进行了研究。根据维特鲁威所写，建筑的意义与美取决于理性的秩序。但是，佩内兹·戈麦兹认为维特鲁威未能将仪式看作宇宙观场所的意义，并指出西方建筑理论往往把迷宫理解成秩序，即阿尔德尼红线，但事实上迷宫的意义更多地与仪式相关联，比如戴达琉斯为阿尔德尼设计的舞蹈场地。佩内兹·戈麦兹进一步指明建筑的理性应该是诠释特定场景并提出与之相适配的恰当答案的能力，而恰当的设计理念只能发生在对历史文化深入理解的对话中。建筑的真正目标是要把握住对持久人性问题给予表达与回答的方式。我们之所以能理解历史的艺术与建筑，是因为我们已学会在时间的流淌中进行翻译。虽然人类缺乏一个共享的宇宙观，但我们的历史性°是一种被给予，历史性是我们对那些奠定作品基础的特定文化传统加以重新找回的能力。

• 8

• 9

• 10

° historicity

• 11

诗意的类比在中国传统的比兴文学手法中有很好的体现。《易经》的六十四卦的"比"卦是由"坤下坎上"构成，按照八卦的宇宙学含义，意指"地下水上"，刚好构成"山水"的意味。"比"卦是吉卦，强调人与人关系中的亲比原则，即以道德诚信为基础的比较关系。在《文心雕龙》的创作论中，"比"与"兴"被归类为相近的写作手法。作者刘勰°认为"比显而兴隐"，前者通过打比方来说明事理，后者通过托物起兴来寄托情感，但两者都能起到小中见大的作用。

• 12

• 13
° 5世纪

as the foundation of meaningful works.•

The poetic analogy is well demonstrated by the popular rhetoric method called 'comparing and evoking'° in Chinese traditional literature. In the *Yijing*,° the *bi*° cosmic diagram is composed of the structure of 'lower earth and higher water,' which indicates the concept of mountains and water.° The 'comparing' diagram is auspicious, emphasising the harmonious co-presence between human beings based on ethics.• In the book *Wenxin diaolong*,° the rhetoric method of 'comparing'° and the rhetoric method of 'evoking'° are combined into one category.• According to the author Liu Xie,° the 'comparing' method acts obviously and the 'evoking' method acts obscurely. The former is to explain through analogy; the latter is to evoke poetical emotion through allegory. Both are to reveal big principles from small phenomena. He pointed out that in the *Shijing*° and Qu Yuan's poem of *Lisao*,° the method of 'comparing and evoking' always demonstrated the tradition of ironic allusion for revealing the truth, but in the works of later generations this principle was overlooked by over emphasising rhetoric composition of comparison. In the structure of comparing and evoking, the paired sides look very different but can reach ultimate honesty through interaction so as to make rhetoric effects as vivid as a flowing river. The garden theorist of the Ming Dynasty Ji Cheng° described the highest level of garden craft as: 'the garden is made by humans, but it seems to be opened by heaven.'•

The poetical analogy is very active in the theories of Chinese traditional literature, gardens and painting but is hard to realise in today's Chinese urbanism and architecture, which advocates a technical worldview of realism. The homogenous urban landscapes that make up the reality of modernisation fail to create the fictional context for ecstatic cultural encounters. Nevertheless, at the beginning of this new urbanism in the 1980s, Borges' works were seriously studied by the younger architectural faculty and their students in the inland city Chongqing, located in southwestern China. One member from this group, Tang Hua, won a design award in Japan in 1986 with a fictional project, a neighbourhood cultural plaza. A decade later, he designed another cultural plaza project that was built in Shenzhen, a coastal city in southern China. In the 1986 project, Tang Hua used the broken bridge as a metaphor, implying his design as a 'metaphysical bridge'.* He constructed an architectural fiction of a labyrinth, which was composed of mnemonic fragments such as the clock tower in an old town, the stories circulating in a pebbled plaza under moonlight, and a long stone bridge leading towards a distant place where you feel as if you have been before.• The centre of this labyrinth is occupied by a solitary man, the dreamer. The clock tower looks exotic in the Chinese tradition, but its image in silence, as in De Chirico's paintings, brings about the sense of time. Both the bridge and the clock tower create a path towards the mystical world. The composition of the design drawing

• 11
○ bixing
○ *Yijing:* Book of Change
○ *bi:* literally, 'compare'
○ mountains & water: shanshui
• 12
○ *Wenxin diaolong:* The Literary Mind & the Carving of Dragons
○ comparing: bi
○ evoking: xing
• 13
○ Liu Xie—fifth century
○ *Shijing*: Book of Odes
○ *Lisao*: The Lament

○ 17th century
• 14

* FIGURE I

• 15

他指出在《诗经》与《离骚》中比兴手法都表现出了讽喻的传统，他批评某些写作为了形式上的"比体云构"而放弃了比兴揭示高尚情操的法则。在比兴的比较结构中，比较双方"物虽胡越，合则肝胆"，使文辞象河水流动般生动。明代造园理论家计成°把造园的境界描述为"虽由人作，宛自天开"。•

○ 17世纪
• 14

诗意的类比手法在中国的传统文学、绘画与造园理论中比比皆是，但在当今以技术世界观为基础的现实主义的城市化运动与城市建筑中很难被实现。均质化的城市景观占据着中国现代化进程的现实表象，但未能提供一种虚构性的文脉使精彩的文化碰撞得以产生。然而，在1980年代新城市运动的早期，在西南山城的建筑学院，有着一群思想活跃的青年教师与学生，他们认真地阅读并思考着博尔赫斯的作品。这其中的领头人物，青年教师汤桦，在1986年日本的一次国际建筑概念设计中获奖，设计的题目是社区的文化广场。十年之后，他在中国的南方城市深圳设计了另一个文化广场并得以建成。在1986年的设计中，汤桦设置了一座断桥来隐语"形而上的桥"。*他建构了一个虚构的建筑迷宫，这个迷宫由一系列记忆的碎片诸如古老城镇的钟塔、月光下在鹅卵石地面的广场中流传的故事、长长的石拱桥引向遥远却又似曾相识的地方等等。•在广场的中央站着一个孤独的人，一个梦想者。钟塔在中国的传统中显得有些异国情调，但它的沉寂的形象引入一种梦中的时间感，就如同我们在意大利画家德·契里克的城市广场作品中所感受到的。断桥与钟塔构成走向神秘世界的途径。整个设计构图让人想起中国的诗意生活景观中那些典型的记忆片断，诸如山水、佛塔、石头拱桥等等。

* 图1

• 15

汤桦建造建筑虚构的理想从来没有被中国城市运动的实用主义实践所阻断。相反，通过理想主义的建筑实践，他把年轻时孤独的阅读与沉思发展成对城市性的一种新认识。在建于1994年的深圳南油广场设计中，他把源自于不同区域文化的几何元素及建筑形体作为隐喻的形象，诸如水平的圆形、垂直的钟塔、带有强烈阴影的廊道、象征性的门廊、曲折的走廊、长而陡峭的阶梯等等。连接沿街面与广场地面的一段狭长而陡的阶梯穿过一段建筑形体中挖出的门廊，不由得让人想起山城重庆的码头意象。这些隐喻的形象被组合进几何形的交融与对立运动中，暗示着道家太极的运动与和谐。广场本身是一个露天剧场，同时也是一个室内剧场的入口。

evokes the paradigmatic memory of the Chinese poetic landscape of water, mountains, a pagoda and a stone arched bridge.

Tang Hua's dedication to constructing architectural fiction was not held up by the prevalent pragmatic practice in Chinese urbanism. To the contrary, he developed his solitary revelry into a new understanding of urbanity through architectural practice. In the Nanyou Cultural Plaza in Shenzhen, built in 1994, the geometrical elements and architectural forms originating from different cultures were used as metaphoric images, such as the horizontal circle, vertical clock tower, shadowed colonnades, gateways, zigzagging corridors and long stairways. These mnemonic images were assembled with a geometrical play of fusion and contrast, which implied the movement of the Daoist diagram of *taiji*, the ultimate harmony. The plaza, working as an open-air theatre, is also the entrance of an interior theatre. The theatricality of the building, incorporating the sky and the distant mountain landscapes, provides an evocative stage for urban life.

As an intellectual architect, Tang Hua observed the paradox between the increasing accumulation of wealth in modern civilisation and the loss of our spiritual home. In his designs, he explored the possibility that spirituality could be retrieved through composing a mythical architectural text. On the built cultural plaza, he wrote: 'This design presents an unnamable mystic tone. Its morphology acts as a metaphor of the cosmos while reflecting the historical sentiment narrated by mythologies and sayings. When these aspects are embodied in architecture, the human existence and the absent divinity begin to be revealed between the earth and the sky.'• More than a decade later, the plaza is now surrounded by homogeneous high-rise buildings and is often covered with kitschy commercial advertisements. However, as he explored in the fictional cultural plaza two decades ago, this built cultural plaza remains as a 'primary gate' for citizens to enter into a spiritual dream.

• 16

Observing the unavoidable standardisation of housing in a universal way of living, Paul Ricoeur points out the conflict between universal civilisation, which is based on the technical worldview, and national culture, which depends upon 'the ethical and mythical nucleus of mankind.' He describes this mythical nucleus as 'the awakened dream of a historical group.'• Through reading Borges and other translations in the early 1980s, Tang Hua and his colleagues acted as a historical group in Ricoeur's sense. They attempted both the mythical and mystic dimensions of human existence, which led them to explore the poetic truth of architecture. It is the ability to keep the dream awakened that continues to be a challenge for Chinese architects. Tang Hua's architecture allows the inhabitant to be a labyrinth wanderer, who is sharply conscious of his own spiritual solitude as well as his longing for poetics. His built project of an office building in Jiangbei of Chongqing peacefully and respectfully stands on a steep slope near the northern end of the

• 17

整个广场的戏剧性场所感将蓝天与远处的山景包容进来，为城市生活提供一个想象的舞台。

作为一名建筑的思索者，汤桦观察到现代文明中的财富积累与失去的精神家园之间的悖论，通过建构神化意义的建筑文本来寻求挽回精神性的可能。对于他自己而言，这个设计呈现着一种无名的神秘情调，其建筑形象隐喻着宇宙的存在，反映着神化与传说故事中的历史感。当这些方方面面被组合到建筑中时，人的存在与神的缺席便在天与地之间被宣示出来。[16] 如今的南油广场周围已是高楼林立，广场的建筑立面上覆盖着各类商业广告，那段穿过建筑的狭长而陡峭的阶梯成为小商贩喜爱的场所，就连二楼的穿透的连廊也成为民工夏日休憩的好去处。从二楼的连廊往下看，猛然意识到广场的中心处由地砖铺砌的迷宫图案。如同汤桦二十年前虚构的文化广场所憧憬的那样，这座建成的文化广场成为市民进入精神梦想的原初之门。

在全球化物质生活日益趋同的今天，人类的居住不可避免地走向标准化，从而造成建立在技术世界观之上的普世居住方式与依靠人类的道德及神秘内核的地区文化之间的冲突。利科将这种神秘内核描述成"特定的历史团体所拥有的惊觉的梦想"。[17] 在20世纪80年代，通过阅读博尔赫斯和其他的翻译文本，汤桦及其同事正是如同利科所言的历史团体那样展开着现实中的梦想，他们探索着人的存在的神话及神秘的维向，寻求着建筑的诗意的真理。对这种梦想的维持正是当今中国建筑师面临的巨大挑战和应该承担的道德责任。汤桦的建筑让人们成为迷宫的孤寂穿行者，在光与景的交织瞬间清新地感知当代人的心灵的孤独和对诗意生活的热爱。他最近建成的重庆江北区招商局办公楼，安静而谦逊地伫立在嘉陵江北桥头的陡坡上，两个垂直交错的方盒子空间内包含着生动的光影变化，悬挑的结构与自然落坡的纤细柱子勾起当地的吊脚楼记忆。* 在浮躁而喧闹的城市化背景中，这座建筑显得异常的孤独与神秘，这与旁边的歌剧院那夸张而炫耀的"国际性"姿态形成宣明的对照。似是而非的是，汤桦建筑的孤独感并不表现与城市记忆的决裂，而是对记忆的延续和对历史感的眷恋。在江北办公楼与市歌剧院之间坐落着已停运的嘉陵江载客索道的机器房。这座建于1980年代的方盒子建筑现在成了城市遗弃的孤儿。把汤桦的建筑与索道机器房做个比较，我们可以感受到他的设计对现代主义建筑及其生活的那个时代的记忆识读。这种记忆的缠绵是对自己

* 图2

bridge over the Jialing River. The two vertically intersected boxes of the building contain a vivid play of light and shadows. The cantilevered structure and the slender columns falling naturally on the slope recall the memory of the local vernacular houses called 'hanging feet houses'.* With the background of the bustling and noisy city, this building appears lonely and mystic and forms an ironic contrast with the extravagant gesture of the municipal opera house. Paradoxically, the solitude of Tang Hua's architecture does not dissociate from the city's memory, yet rather continues it in a preservation of historicity. Between Tang Hua's building and the opera house is the abandoned machine house which used to run the memorable cable car crossing over the Jialing River for the past three decades. Comparing Tang Hua's building with the machine house, we can sense his reflection on modernism transforms the historical age into a lived memory. This mnemonic lingering demonstrates his love for the indigenous culture and land and reminds us of Jean-Paul Sartre's theory on the solitude of love. Sartre stated that lovers always sought for solitude, but such a solitary being was fundamentally for others. The deeper he falls into love, the more he loses himself, the more he is thrown back to his own responsibilities, and hence the lover's perpetual worrying and caring.*

* FIGURE 2

• 18

In the infinite labyrinth, there exists the constantly repeated movement of forking and converging that does not lead to a final goal. The activity of walking through confusing paths is itself the intention of the labyrinth. In both the fictional labyrinth and the garden as a labyrinth, there is something that asks to be understood in what it intends. This meaning does not simply lie in what we immediately see and read, but rather, it rests upon an intricate interplay of showing and concealing, 'a mystery as familiar as it is unexplained, of a light which, illuminating the rest, remains at its source in obscurity.'* The labyrinth requires us to learn how to perceive the labyrinth as it is and how to rise above the universal leveling process in which we cease to notice anything. Such a poetic resistance will lead to Daoist ultimate harmony of deep and distant ethics.

• 19

Tang Hua continues to interpret Borges' works through his recent library project on the new campus of the Sichuan Institute of Fine Arts in Chongqing. In his essay 'The Library of Babel,' Borges imagined a library as the cosmos, constituted of an infinite number of hexagonal towers among which were huge airshafts.* The interior layout of each tower was universalised; along the wall of each side there were five long bookshelves, which could reach as high as the ceiling. There was a long narrow corridor between two towers. Against each sidewall of the corridor there was one closet: one for a restroom, another for sleeping standing-up. There was a spiral stairway in the corridor for vertical transition and a mirror, which implied infinity. In each tower there were two lights which were dim but consistent. This library was infinite and the hexagon symbolised the absolute ideal space. All the people in the library were acciden-

• 20

土生土长文化的热爱。这不由让我们想起让·保罗·萨特关于爱的孤独的论述。他说,恋人总是寻求孤独,但是这种孤独的存在根本上是为他人的存在,爱得越深,越失去自我,也越意识到自己作为人的存在的责任,这让他感到不安。●   ● 18

    在无限的迷宫中,存在着持续的不断分叉与聚合的重复运动,这种运动不导向任何最终的目标。穿过迷茫的路径这一行为本身就是迷宫的意象。在虚构的迷宫与作为迷宫的花园中,存在着某种东西,它要求如其想是地被理解。这种理解方式并不直接地寓于我们的所看和所读,而是寓于出场与不出场的错综复杂的形体与意义的互动。这正如法国现象学家莫里斯·梅洛·庞蒂所描绘的"一种既熟悉又不可解释的神秘,一种照亮了其余而自身又处于晦暗中的光芒"。●迷宫   ● 19
要求我们学会如其所是地感知迷宫,学会超越那些阻挡我们感知的大同标准模式。如此的诗意的抵抗可以导向道家的玄德境界。

    汤桦在最近建成的四川美术学院新校区图书馆设计中,持续着对博尔赫斯文本的记忆与阅读。在其《巴贝尔图书馆》的故事中,●博尔赫斯想象着作为宇宙的图书馆,它由无穷的   ● 20
六边形塔楼组成,塔楼之间是巨大的空气竖井。每个六边形塔楼中的室内布局是统一的,每边布五个长长的书架,书架的高度几乎就是楼层的高度。塔楼与塔楼之间由狭小的廊道联系,廊道两边分别放置着一间壁柜,一间用作站式睡觉,另一间用作盥洗室。廊道中还有垂直联系的螺旋楼梯,以及一面镜子,镜子的反射暗示着无限。每个六边形塔楼中有两盏灯,光线昏暗但持续。这座图书馆是无限的,六边形标志着绝对的理想空间。在图书馆中的芸芸众生都是偶然性的产物,但是作为宇宙的图书馆本身只可能是神的作品。这座图书馆收藏着所有的书,但没有任何两本相同的书。所有的读者都隐隐觉得自己是某种神秘宝藏的拥有者,在这里寻找着自己的护身符,坚信在书架的某个角落藏着珍贵的书籍,虽然它们又是如此难以被找到。读者往往迷信在某座塔楼的某个书架上放着一本概括其他一切藏书内容的一本书,通过寻找这本书,就可以接近神本身。在这个寻找的过程中,每个人接近的程度不同,有些则完全失败,但即使是失败,这也是神的图书馆的最好明证。这座图书馆的布局是无限且循环的,如果有人想穿越整座建筑,几个世纪之后,他会发现自己又回到原点。而这种不断的重复往返,正是神为人建立的秩序。博尔赫斯由此感叹道:"这个优美的。希望使我的孤   ○ 被神所赋予的
独变得愉悦起来。"

tal beings, but the library as the cosmos could only be the product of the eternal divine. The library collection included all books from all over the world with no two books being identical. All library visitors felt as if they were the holders of a secret treasure and looked for their vindications. They believed that some precious books were hidden in a corner, though it was so hard to find them. The visitors also believed that on a certain bookshelf there was a book that outlined the contents of all other books, and they could approach divinity by looking for this book. In this searching process, each person approached divinity to different degrees. Some completely failed, yet that failure could be the best proof for the library of the divine. The organisation of the library was infinitely circling, when traveling through the entire library, a person would be returned to the starting point after centuries. This endless circulation was the very order that the divine established for humans. Borges thus stated that this beautiful hope given by the divine made one's solitude joyful. Interpreting his own library design, Tang Hua wrote:

> Borges imagined the heaven should look like a library; this fascinates me. I like this metaphor and think through it, which makes me hold a special feeling towards the library, which is mystic and divine…My library is a succinct body standing on a hillock and surrounded by agricultural fields, just like a local vernacular brick kiln on the extensive land or a countryside church. The significance of the library is highlighted through a contrast with the medium and small buildings scattered on the campus. The incorporation of genius loci and the carefully preserved agricultural landscape expresses great respect for the deep depth of historicity.•*

• 21
* FIGURE 3

Borges' sinophilia was prominent in his fiction 'The Garden of Forking Paths' and was influenced by Franz Kafka. In the essay 'Kafka and His Precursors,' Borges tried to find those literary ancestors who influenced Kafka's writings.• After careful study, Borges became convinced that one such a literary ancestor was Han Yu,° a Chinese prose writer of the Tang Dynasty. He thought that Kafka's writing tone was very much like Han Yu's. They both shared the writing style of revealing mystery through plain expression. In Chinese history, Han Yu is well known for his writings 'full of strangeness and bewildering; the meaning reveals itself through a moderate and hidden approach.'• Han's poems and prose are good at observing abnormal phenomena while seeking in depth principles. Borges cited Han's short essay 'Deciphering the Unicorn'° as proof for his hypothesis; and concluded that each writer in fact had his own literary ancestors, and the writer's works in turn influenced posterity's view of the past and therefore the future. In his parable 'The Great Wall and the Tower of Babel,' Kafka mentioned that

• 22
° eighth century

• 23
° Huo Lin Jie

在阐释自己的图书馆设计时，汤桦写道：

> "他○还说过，他猜想天堂就应该是图书馆的模样。○'这'令人神往……'我'十分喜欢，也十分沉迷其中。同时对图书馆也就有了一种特殊的感觉，神秘而神圣。"

○ 博尔赫斯
○ 如大地上的事物，如神话般亘古而单纯。

他又写道：

> "'这个设计的'图书馆以一种简洁的形式屹立于山地与田野之中，如同大地上的砖窑，也像乡村的教堂，与校园已形成的中小体量分散布局的建筑物形成对比，凸现其重要地位及象征意义。具有地域精神的建筑形式与刻意保留和设计的农业景观形成有意味的关系，表达对历史纵深的悠久文明的敬意。"●*

● 21
* 图3

　　博尔赫斯的中国情结在其小说《交叉路径的花园》中彰显出来，而他又受德语作家卡夫卡的中国情结的影响。在《卡夫卡及其先辈们》这篇短文中，博尔赫斯试图找出那些曾经影响卡夫卡写作的先辈们。●在经过一番分析之后，博尔赫斯确信影响卡夫卡的先辈之一正是中国唐代散文家韩愈。○他觉得卡夫卡的写作格调○非常像韩愈，两者的文风都显得秘而不宣。韩愈的文风"万怪惶惑，而抑遏掩蔽，不使自露"。●他的诗文善于观察怪异的现象，追求深奥的道理。博尔赫斯引用了韩愈的短篇散文《获麟解》来证明其观点。博尔赫斯总结道，其实每一位作家都有影响自己的先辈们，而作家的作品又影响着我们后人对过去的理解，也影响着未来。在其短文《长城与巴贝尔塔》中，卡夫卡谈到很久以前有位学者写了一本书，把中国的长城与《圣经》中的巴贝尔塔作比较，试图证明巴贝尔塔的倒塌不是因为众所周知的神发怒的原因，而是因为塔基不牢。但这位学者的用意并不是要呼吁提高打造基础的建筑技术，而是为了提出中国的长城将是历史上第一个新巴贝尔塔的安全塔基。这个假设似乎也合乎逻辑：先造墙，后有塔。卡夫卡随即发出一系列疑问，长城的墙基并非圆形，甚至连四分之一圆也不到，它又如何可以成为巴贝尔圆塔的塔基？也许，这位学者只是作形而上的暗喻。但

● 22
○ 8世纪
○ 而非形式
● 23

once upon a time, a scholar wrote a book, comparing the Great Wall and the biblical Tower of Babel and attempting to prove the collapse of the tower was not because God was offended, but rather due to the unstable foundation of the tower. But as analysed by Kafka, this scholar's intention was not to improve building technology but rather to argue that the Great Wall would be the solid foundation for a new Tower of Babel. This hypothesis sounds logical: the wall has to be built before the tower can be constructed. But Kafka raised questions on this hypothesis: since the Great Wall was not in a circle and not even a quarter of the circle, how could it be the foundation for a new Tower of Babel? Perhaps, that scholar was using a metaphor. Compared with a metaphysical metaphor, why not actually build a physical wall? If there was such a real intention of building the wall, why did the scholar not describe at least one building plan? Kafka thus sighed that man always had numerous crazy ideas and, like that scholar, many people wanted to contribute to the same objective. However, human ideas in nature are as changeable as dust. Whenever we want to define an idea, it will break up the definition. The situation of building the wall is just like that, so is human nature.• Kafka's conclusion hints at the collapse of the • 24
Tower of Babel but expresses his deep impression of the magical project of the Great Wall.

If as stated in 'The Garden of Forking Paths' that 'the book and labyrinth are one and the same' and Borges' labyrinthine plot implies a Chinese garden, his 'The Library of Babel' can be interpreted as an intermingling of the Tower of Babel and the Chinese garden. Perhaps, Tang Hua's library provides an opportunity for the audience to stroll about within the human order established by the divine and, as stated by Borges, to be delighted in his own spiritual solitude.

The architectural cases in biblical scripture such as the Tower of Babel, Noah's Ark and the Temple of Solomon occupied a significant place in Western architectural theories during the Renaissance to the 18th century. Many architects and scholars were dedicated to the reconstruction and interpretation of the Temple of Solomon, including the 16th century Spanish Jesuit Juan Bautista Villalpando, the late-17th century Spanish architectural theorist Juan Caramuel y Lobkowitz and the French architect Claude Perrault, the early-18th century Austrian architect and historian Fischer von Erlach and the British scientist Isaac Newton. Pérez-Gómez puts forward that these scholars attempted to retrieve the primordial architectural image and meaning through textual interpretation and figural construction. They all shuttled between the biblical scriptures and Vitruvius' *Ten Books on Architecture*. Villalpando especially endeavoured to reconstruct all the physical details of the Temple of Solomon in the biblical texts and check how they could match Vitruvius' theory of proportion of classic architecture so as to establish the cosmic order of the Renaissance age. Only Newton's reconstruction of the Temple no longer referred to Vitruvis' tradition, but

相比于形而上，为什么不造一圈实实在在的墙而更有说服力呢？如果真的要造墙，为什么不在书中描述一下哪怕是很模糊的方案呢？卡夫卡于是论述到，人总是有许多疯狂的想法，如同这位学者那样，许许多多的人都想为同一个目标而贡献力量。但是，人的想法在本性上如同沙尘般易变；每当我们试图束缚自己的想法，想法就会立马将束缚撕裂，造墙如此，人的本质亦然。卡夫卡的结尾显然是暗指巴贝尔塔的倒塌，同时也让我们意识到中国人众志成城造长城的壮举确实让他思绪万千。 • 24

如果回想博尔赫斯在《交叉路径的花园》中的名言"书与迷宫是同一的"，而他的迷宫背景又暗指中国园林，那么他的《巴贝尔图书馆》可以说是巴贝尔塔与中国园林寓义的交叉。也许，汤桦的图书馆正是为读者提供一个在神为人建立的秩序中重复往返的机会，并如博尔赫斯所言，使人的孤独变得愉悦起来。

《圣经》中的建筑诸如巴贝尔塔、诺亚方舟、所罗门神庙等在文艺复兴至18世纪的西方建筑理论中占据重要的地位。许多建筑师与学者都致力于对所罗门神庙的重构与诠释研究，其中著名的有16世纪末西班牙耶稣士建筑理论家包蒂斯塔·维拉潘多，17世纪后期的西班牙建筑理论家胡安·卡拉姆·洛布科维茨与法国建筑师克洛德·珀乌尔，18世纪初的奥地利建筑师与历史学家冯·厄尔拉赫与英国科学家牛顿等。佩雷兹·戈麦兹指出，这些学者前赴后继地通过文本诠释与图构，试图寻找原初的建筑形象与意义，他们都不约而同地穿梭于《圣经》与维特鲁威的《建筑十书》之间。尤其是维拉潘多的重构试图把《圣经》中所罗门神庙的细节与维特鲁威的古典建筑比例相吻合，从而建立文艺复兴时期建筑的宇宙秩序。只有牛顿的重构不再关注维特鲁威的传统，但仍然维系着传统宇宙学的秩序观。1679年耶稣会罗马学院的德国多产学者阿塔纳修斯·克尔切发表了专著《巴贝尔塔》，通过复杂的数学计算，他试图证明建造巴贝尔塔的想法在技术上是不现实的，即使要达到月球的高度，这座塔也会使地球脱离其原有的宇宙中心位置而导致宇宙的毁灭。克尔切对巴贝尔塔的诠释表现了巴洛克时代理性与数学的精神，同时也反映了人类日益增长的控制世界的能力。克尔切之前发表的大量专著包括《诺亚方舟》、《普遍性的音乐制作》与《中国纪念物图解》。在前两部著作中，他通过建筑与音乐中的和谐比例分析寻求神建立的宇宙秩序，而关于中 • 25

• 26

• 27

his research still maintained the traditional cosmic order.* In 1679, the German prolific professor Athanasius Kircher in the Jesuit College of Rome published a book entitled *The Tower of Babel*. Through complicated mathematic calculations, he attempted to prove that the idea of building the Tower of Babel was technically unfeasible. If the tower could reach the height of the moon, the tower would make the earth veer from its cosmic centre and cause the collapse of the cosmos.* Kircher's interpretation of the Tower of Babel demonstrated the rational and mathematical spirit of the Baroque age and reflected the increasing human power of controlling the world. Other previous works by Kircher include *The Noah's Ark*, *The Universal Making of Music*, and *The Illustrated Monuments in China*.* In the first two books, he sought divine order through exploring architectural and musical harmonies. In the third book, he exhibited a machine box for making music, which contained numerous musical fragments.* Each stored musical unit responds to a specific musical note and scale. After preliminarily arrangment of the order of certain stored units, a music layman can make music through the machine box. The shape of this music box recalls Noah's Ark, and its mystic mechanism reminds us of the Renaissance scholar Giulio Camillo's project Theatre of Memory.*

Kircher has been acclaimed by modern scholarship as the 'last man who knows everything' of the Baroque age. His many books were stored in the 18th century Jesuit libraries in Beijing, which owned over 5,000 European books of the Renaissance to the 18th century. These books acted as the intellectual sources for the Jesuits to design the Western garden within the Garden of Round Brightness. The anonymous scholar in Kafka's 'The Great Wall and the Tower of Babel' appears to be Kircher. In a foreword of his 'The Garden of Forking Parths,' Borges also mentioned his great respect to an anonymous scholar who was said to be able to summarise the contents of a 500 page encyclopaedia.* This scholar with super memory is perhaps, once again, Kircher. The revealed connection among these three authors of different historical ages brings to light the fundamental meanings of architecture emerging from cultural encounters. The mystic relationship of books, labyrinths and the spiritual ark composed by their works suggests the hopeful direction of comparative architecture. Based on such a cross-cultural implication, we can deeply understand the image of Tang Hua's library located on a campus full of agricultural landscapes. Perhaps, this building alludes to a lost spiritual home in a chaotic contemporary culture.

国的论述奠定了他的汉学家地位。在《普遍性的音乐制作》中，他展示了一个音乐制作的方盒子机器，里面存满了音乐碎片。*每个存放的单位对应着特定的音符与音阶，在对存放单位作一定的选择与排列之后，一个不懂音乐的人可以借助这个机器盒子制作出音乐。这个盒子的形体让人想起诺亚方舟，而其神秘的功能又让我们回忆起文艺复兴时期朱利奥·卡米洛的《记忆的剧场》。•

* 图 4

• 28

克尔切被誉为巴洛克时代"什么都知道的最后一人"，他的许多著作被收藏在 18 世纪北京的耶稣会教堂图书馆。这些图书馆拥有五千册的欧洲文艺复兴到 18 世纪的原版书籍，其中包括大量的建筑与园林书籍，它们成为耶稣士设计圆明园西洋楼的智库。卡夫卡在《长城与巴贝尔塔》中提到的那位匿名学者也许正是克尔切。博尔赫斯在为《交叉路径的花园》所写的一段前言中，曾讲到对一位学者的神往，据说那位学者能概述一部想象中有五百页的百科全书的全部内容。•这位超记忆的学者也许又是克尔切。这三位不同历史时间的作者之间的渊源关系，呈现出中西交叉文化中的建筑根本意义的生成。他们的作品中所展现的书、迷宫、精神的方舟之间的神秘关系，也许暗示着比较建筑的发展方向。从这个交叉文化的结论看去，我们可以更深层地理解汤桦设计的图书馆坐落在田园景色般校园中的形象含义，也许它正隐喻着当代文化的迷茫中失去的精神家园。

• 29

FIGURES 插图

1

2

3

4

1 1986年汤桦设计的瓦屋顶居住小区文化广场。
   日本建筑概念设计竞赛获奖作品。
   图片提供：汤桦建筑师事务所
   The Cultural Plaza of a Tiled-Roof Neighbourhood, designed by Tang Hua, 1986.
   Award-winning project in an architectural conceptual design competition in Japan.
   Photo supply: Tang Hua
2 2009年汤桦设计的重庆江北区招商局办公楼。
   图片提供：汤桦建筑师事务所
   Chongqing Jiangbei office building, designed by Tang Hua, 2009.
   Photo supply: Tang Hua
3 汤桦设计的四川美术学院新校区图书馆。
   摄：徐浪
   Library in the new campus of Sichuan Institute of Fine Arts, designed by Tang Hua.
   Photo: Xu Lang
4 制作音乐的方舟。
   Music-making Ark.
   Athanasius Kircher, *Musurgia universalis*, Rome, 1650.

NOTES 注解

1. Paul Ricoeur, *History and Truth*, Charles A. Kelbley tr., Evanston, US-IL: Northwestern University Press, 1965: 23.
2. Ibid., 283.
3. Jorge Luis Borges, 'The Garden of Forking Paths', *Labyrinths: Selected Stories and Other Writings*, Donald A. Yates and James E. Irby ed., New York: New Directions Book, 1964.
4. Ibid., 25.
5. Hui Zou, 'Perspective Jing: The Depth of Architectural Representation in a European-Chinese Garden Encounter', *Chora, Vol. 6: Intervals in the Philosophy of Architecture*. Alberto Pérez-Gómez and Stephen Parcell ed., Montreal: McGill-Queen's University Press, 2011: 287.
6. Hui Zou, 'The Chinese Garden and the Concept of the Line Method', *A Jesuit Garden in Beijing and Early Modern Chinese Culture*, West Lafayette, US-IN: Purdue University Press, 2011: chapter 5.
7. Hui Zou, 'The Philosophical Encounter Embodied by the Yuanming Yuan', *Journal of Environmental Philosophy*, volume 7, issue 1, 2010: 47–61.
8. André Breton. 'Ascendant Sign', *Surrealist Painters and Poets: An Anthology*, Mary Ann Caws ed., Cambridge, US-MA: The MIT Press, 2002: 135.
9. Dalibor Vesely, *Architecture in the Age of Divided Representation: The Question of Creativity in the Shadow of Production*, Cambridge, US-MA: The MIT Press, 2004: 344.
10. Alberto Pérez-Gómez, 'The Myth of Daedalus', *AA Files*, issue 10, 1985.
11. Alberto Pérez-Gómez, *Built upon Love: Architectural Longing after Ethics and Aesthetics*, Cambridge, US-MA: The MIT Press, 2006: 144.
12. Tang Mingbang ed., *Annotation of Zhouyi*, Beijing: Zhonghua Book Company, 1997: 22, 23.
13. Zhou Zhenfu, 'Bixing 36', *The Literary Mind and the Carving of Dragons: Modern Interpretation*, Beijing: Zhonghua Book Company, 2006.
14. Chen Zhi, *Annotation of Yuanye*, Beijing: China Architectural Industrial Press, 1999: 51.
15. Tang Hua, The Community Centre of a Tiled-Roof Neighbourhood', *World Architecture*, issue 2, 1987: 58.
16. Tang Hua, 'Loneliness: Meaning of *Nanyou* Culture Plaza of Shenzhen', *Architect*, issue 2, 1996: 96.
17. Ricoeur, 1, 274–76, 280.
18. Jean-Paul Sartre, *Being and Nothingness*, Hazel E. Barness tr., New York: Washington Square Press, 1992: 491.
19. Maurice Merleau-Ponty, 'The Intertwining—The Chiasm', *The Visible and the Invisible*, Alphonso Lingis tr., Evanston, US-IL: Northwestern University Press, 1968: 130.
20. Jorge Luis Borges, 'The Library of Babel', *Labyrinths*, Donald A. Yates and James E. Irby ed., New York: New Directions Publishing, 1962.
21. Tang Hua, 'Design Notes on Babel Library', *Architect*, issue 5, 2010.
22. Jorge Luis Borges, 'Kafka and His Precursors', *Labyrinths*, Donald A. Yates and James E. Irby ed., New York: New Directions Book, 1964.
23. Su Xun, 'Shang Ouyang neihan diyi shu'. Qtd. in Chen Wangheng, *History of Chinese Classical Aesthetics*, Changsha: Hunan Educational Publishing House, 1998: 506.
24. Franz Kafka, 'The Great Wall and the Tower of Babel', *Parables and Paradoxes*, New York: Schoken Books, 1958.
25. Alberto Pérez-Gómez, 'Juan Bautista Villapando's Divine Model in Architectural Theory', *Chora, Vol. 3: Intervals in the Philosophy of Architecture*, Alberto Pérez-Gómez and Stephen Parcell ed., Montreal: McGill-Queen's University Press, 1999.
26. Athanasius Kircher, *Turris Babel*, 1679.
27. Athanasius Kircher, *Arca Noe*, Self-published, 1675.
28. Francis A. Yates, 'Renaissance Memory: The Memory Theatre of Giulio Camillo', *The Art of Memory*, The University of Chicago Press, 1974: chapter 6.
29. Jorge Luis Borges, 'The Garden of Forking Paths', *Collected Fictions*, Andrew Hurley tr., New York: Penguin, 1998: 67.

12. 唐明邦编《周易评注》，北京：中华书局，1997年，22、23页。
13. 周振甫《文心雕龙今译》，北京：中华书局，2006年，"比兴第三十六"。
14. 陈植《园冶注释》，北京：中国建筑工业出版社，1999年，51页。
15. 汤桦"瓦屋顶居住小区活动中心"《世界建筑》，第2期，1987年，58页。
16. 汤桦"孤寂：深圳南油文化广场释义"《建筑师》，第2期，1996年，96页。

21. 汤桦"营造笔记之巴别图书馆（又译巴贝尔）"《建筑师》，第5期，2010年。
23. 苏洵"上欧阳内翰第一书"引于陈望衡《中国古典美学史》，长沙：湖南教育出版社，1998年，506页。

BLACK

黑

葛明
Ge Ming

TRANSLATOR
Yu Changhui

翻译
于长会

Since 2003, I have been trying to find a special structure, which can be used to think about the boundary of architecture. For instance, how could it help us think about the meaning of 'I' in architecture? How could it help us think about establishing a new social contract through architecture? With regard to these thoughts, three keywords are essential: structure, 'I' and social contract.

从 2003 年以来，我一直在寻找一种特殊的"结构体"，用它作为思考建筑学边界的形式。

譬如它如何帮助思考建筑学中"我"的意义，又如何去建立社会契约。

其关键词为三个：结构体，我，社会契约。

I. SOCIAL CONTRACT
I.I JOHN HEJDUK — SOCIAL CONTRACT OF LETTERS

I have been in the world of John Hejduk for a long time, especially immersed in those well-known dialogues between Peter Eisenman and him:

Eisenman asked:
'John, is this a house? Why doesn't it have any entrance?'

Hejduk replied:
'because the house didn't invite you.'

Here, Hejduk's structure is seemingly personified.

Hejduk went on deducing this form and extending it to all objects. In the Construction Diary, he mentioned an Italian booklet. He stated: 'These books are too flimsy. I'm quite worried about their health. The booklets acted as personal letters made even more private by being made to appear public. Although they were sent out throughout the world, they were expected to be returned.' So he decided 'I would return them in a special private way, they would go on a return journey but accompanied.'* It should be noted that, in such a relationship, 'architecture', 'I' and 'language' have linked together, and form a special *chronotope* and a distinctive *social contract*.

• 1

The concept chronotope comes from M.M. Bakhtin. It is employed to discuss Fyodor Dostoevsky's novels and is often used in his narratives. Many people tend to use chronotope to discuss Hejduk, attribute his works to a special language and even indicate the attribution of the language in the works.

This is key in understanding Hejduk. To convey the *contract*, Hejduk endowed poetry, drawings and structure with an integrated form.* I name it *Nail Architecture*. The drawings have traces of poetry, and the structure contains traces of drawings. All traces show a clue, and the clue is left by the 'I'. For instance, there is a huge fake nail in the structure. It is a need for the drawings which are supplements to the poetry.* Poetry, drawings and structure are always linked to one another. Here poetry is a particular language with a certain form, which is different from daily language — narrative language.

* FIGURE 1

* FIGURE 2

This is duality of Hejduk: on the one hand, he linked poetry, drawings and structure together while keeping them in a separated way; on the other hand, he narrated them in diary and integrated them by means of chronotope, which seemed to return to a kind of linguistics, and to be characterised with 'I'.

一、社会契约
（1）海杜克 —— 信件的社会契约

很长一段时间，我沉浸在海杜克的世界里，尤其那段脍炙人口的对话：

  埃森曼对他说：
  "约翰，这是房子么？却没有入口。"

  海杜克回答说：
  "那是因为你没有得到邀请。"

  在这里，海杜克的结构体拟人化了。
  海杜克继续推演这种形式，以至于一切物品。在《建造日记》中，他提到了一种意大利小册子，他说："这些书这么薄，总让我担心它们的健康。"他说："它们像私人的信件，试图公开却又更加的隐秘。虽然它传遍了世界，但是却又盼望回归。"所以，他决定"归还它们，让它们踏上归途，但是带着伙伴。"* 注意，在这样的关系中，"建筑"、"我"与"语言"连缀在了一起，构成了一个特别的"时空体"，并形成了特有的"社会契约"。 • 1
  时空体这一概念来自巴赫金，用以讨论陀思妥耶夫斯基的小说。它出现在叙事中。很多人用时空体这一词语来讨论海杜克，将其作品归结于一种特殊的语言，指出了它们的语言归属。
  这是理解海杜克的关键。为了传达"契约"关系，海杜克赋予诗歌、图纸、结构体以一致的形式。* 我称之为"铆钉建筑学"。图纸上残留着诗歌的形式，结构体上残留着图的痕迹，显现出一条线索。这一线索都由"我"所遗留。比如结构物上有巨大的铆钉，铆钉是假的，是图纸上需要画它，而图纸又是对诗歌的补充。* 三者连续地维系着这条线索。此时的诗歌是一种专指有形式感的语言，不同于通常所说的语言 —— 叙事的语言。 * 图1

  * 图2

  海杜克一面让诗歌、图纸、结构体三者来回穿梭，保持区别；另一面在叙事的时候把它们汇聚在日记中，建立起时空体把它们统一起来，一切似乎回到了语言，并带有"我"的特征。这是两面的海杜克。

In this way, it points to a definite life journey. This is quite special in modern society. It's such a society that 'the sense of worship and reverence for Death dies away and, the dead is a zero, and even they don't have any value.'• I call it *Architecture of Letter*. It seems like 'returning home'. It seems especially like the returning journey of Odysseus in ancient Greece. Odysseus left and returned, which forms a completeness.

• 2

However, 'the dead is a zero, and even they don't have any value', which implies more 'leaving home' than 'returning home'. It returns to *modernity*.

I.2.

MODERNITY — SOCIAL CONTRACT OF CHANGE

What is modernity? Kafka's work is a representation of it. Walter Benjamin makes it very clear in *Illuminations* 'On Kafka':

> Kafka's work is an ellipse with foci that are far apart and are determined, on the one hand, by mystical experience and,° on the other, by the experience of the modern big-city dweller. In speaking of the experience of the big-city dweller, I have a variety of things in mind. On the one hand, I think of the modern citizen who knows that he is at the mercy of a vast machinery of officialdom whose functioning is directed by authorities that remain nebulous to the executive organs, let alone to the people they deal with.° When I refer to the modern big-city dweller, I am speaking also of the contemporary of today's physicists. If one reads the following passage from Eddington's *The Nature of the Physical World*, one can virtually hear Kafka speak.
>
> I am standing on the threshold about to enter a room. It is a complicated business. In the first place I must shove against an atmosphere pressing with a force of fourteen pounds on every square inch of my body. I must make sure of landing on a plank travelling at twenty miles a second round the sun — a fraction of a second too early or too late, the plank would be miles away. I must do this whilst hanging from a round planet head outward into space, and with a wind of aether blowing at no one knows how many miles a second through every interstice of my body. The plank has no solidity of substance. To step on it is like stepping on a swarm of flies hits me and gives a boost up again; I fall again and am knocked upwards by another fly; and so on. I may hope that the net result will be that I remain about steady; but if unfortunately I should slip through the floor or be boosted too violently up to the ceiling, the occurrence would be, not a violation

○ in particular, the experience of tradition

○ It is known that one level of meaning in the novels, particularly in *The Trial*, is encompassed by this.

这一方式指向了一种特定的生命之旅，在"对死亡的崇拜和敬畏减弱了"，"死者总是一个零，毫无价值可言"* 的现代社会中显得非常特别。我把它称为信件的建筑学，有"回家"的感觉，尤其像古希腊的奥德赛回归之旅一样，出去又回来，有一种完整性，"completeness"。

• 2

但"死者总是一个零，毫无价值可言"毕竟更多地指向"离家"，而不是"回家"，它面向对现代性的讨论。

一（2）现代性 —— 变化的社会契约

现代性是什么？卡夫卡的书充分地表现了现代性，本雅明《启迪》中的《论卡夫卡》有这样一段：

> "卡夫卡的作品像一个圆心分得很开的椭圆；这两个圆心一个被神秘体验支配着，○ 一个被现代大城市居民的体验支配着。至于现代大城市居民的体验，我有许多想法，首先，我想到的是现代市民清楚自己是听由一架巨大的官僚机器摆布的，这架机器由权威操纵着，而这个权威即使对于那些执行器官来说也在云里雾里，而对于那些它们要对付的人们来说就更模糊不清了。○ 当我说到现代大都市居民时，我想到也是当今物理学家们的同龄人。如果你读爱丁顿的《物理世界的本质》中的下面几段，你真可以听见卡夫卡的声音：
>
> > 我站在门槛上正待进入一间屋子。这真是一件复杂的事情。首先我必须推开大气，它正以每平方英寸十四磅的力量在压迫着我，我还得吃准是否踏着这块以每秒二十四英里围绕太阳运行的木板上，只要倏忽之差它就远在数英里开外了。我在干这些的时候，其实是悬吊在一个圆形的星球上，头朝着太空，星际的大风正以每秒不知多少英里的速度穿过我身体的缝隙，我脚下这块木板没有任何质地上的坚固性，踏在它上面就如同踏在一群苍蝇上面。我不会跌下去吗？如果我冒险这样干的话，

○ 尤其是传统的体验

○ 卡夫卡小说，尤其是《审判》里的某一层意义与此紧密相关，这是为人熟知的。

of the laws of Nature, but a rare coincidence...

Verily, it is easier for a camel to pass through the eye of a needle than for a scientific man to pass through a door. And whether the door be barn door or church door it might be wiser that he should consent to be an ordinary man and walk in rather than wait till all the difficulties involved in a really scientific ingress are resolved.• • 3

Walter Benjamin is wise. He believes that the above is Kafka's unique style. It is quite easier for an ordinary man to push a door, but the truth tends to ignore the real reality°—it is ridiculous but controls you in a proper way. This is Kafka-like feeling. In contrast, *3 Standard Stoppages*° by Marcel Duchamp seems much more visualised than Kafka's language, and lacks imagination.

○ '...that reality of ours which realises itself theoretically...in modern physics, and practically in the technology of modern warfare.'
○ 1913—14

Modernity mixes fables and realities. With that, I cannot feel free to apply the architecture of messenger.

1.3.
THE ARCHITECTURE OF PHOTOGRAPHY

Roland Barthes discussed photography in his work *La Chambre claire.*° He suggested that photography provides such a fact that it does not necessarily say 'what is no longer', but only and for certain 'what has been'. It is a new juncture where death enters into daily life.

○ 1980

> Contemporary with the withdrawal of rites, photography may correspond to the intrusion, in our modern society, of a symbolic Death, outside of religion, outside of ritual, a kind of abrupt dive into literal Death. Life/Death: the paradigm is reduced to a simple click...

Roland Barthes also stated:

> The horror is this: nothing to say about the death of whom I love most...her photograph, which I contemplate...the only 'thought' I can have is that at the end of this first death, my own death is inscribed; between the two, nothing more than waiting.•

• 4

It is an era in which death has no meaning. Then a 'flat' death appears, and a new social contract appears.

The death is momentary. Now I'm trying to find a kind of structure which is same with photograph. I'm looking for an architecture of photography not an architecture of messenger.

It is an architecture in correspondence with modernity and an architecture of 'death'.

某个苍蝇会碰到我,让我再次升起来;我再跌下去,再被另一个苍蝇踢回来,如此往复。我或可以希望最终的结果是我仍在原地一动未动;但如果我不幸跌到地板下面去或是猛地被推升到房顶的话,这并不违反自然规律,只是一个不太常见的偶然巧合罢了……

不错,骆驼穿过针眼要比科学家走过门洞容易得多。但明智一些的办法还是乐于做一个普通人,不管是谷仓门还是教堂门,径直往里走,而不是等待有关真正科学的一切难题都被人解决。" •

● 3

本雅明非常睿智,发现这段话深刻地带有卡夫卡风格。作为一个普通人推开门是如此轻易,而真相又往往是忽视了真正的"现实"。——它荒诞却又准确地控制着你。这就是卡夫卡式的感觉,相比之下,杜尚的《三把尺子》○比卡夫卡的语言太直观,少了想象力。

○ 本雅明指出这种"现实"理论上是现代物理学,实际上是现代军火技术。
○ 1913—1914 年

现代性的感觉即一个像寓言却又是现实的状况。有了这些,我就无法心安理得地运用像信使一般的建筑学。

一(3) 摄影的建筑学

罗兰·巴特在他的《明室》○中讨论了摄影,他指出摄影提供了这样一种事实,它表达的不是"已然不是",而是"曾经如是",它是死亡进入日常生活的一个新的节点。"作为和礼仪衰退出现在同一时代的摄影,在我们这个现代社会里,回应的可能就是这种死亡对社会的侵入。这种死亡是一种非象征性的,宗教之外的,礼仪之外的,是一种突然陷入字面意义上的死亡,生/死:这个聚合体在快门'咔嚓'一声时出现了。"

○ 1980 年

罗兰·巴特还说:"可怕的是:对我最爱的人的死,无话可说……当'我'在凝视她的照片的时候,在这第一个死亡之后,我自己的死亡也登记在案了;在这两个死亡之间除了等待,一无所有。" •

● 4

这就是在一个死亡没有意义的时代,一种"平"的死亡出现了,一种新的社会契约形式出现了。

II.
STRUCTURE

I persist in looking for a structure, which is not an imitation of the structure that Hejduk linked to poetry or drawings.

Everyone has his or her own way to deal with the issue.

I start with an object and make it in parallel with language, not back to language.

II.1.
THE AGE OF MECHANICAL REPRODUCTION

This structure attempts to reflect making in correspondence with photography. As we all know, making is a way for us to make contact with the world. We are still living in the 'age of mechanical reproduction', the age codified by Walter Benjamin. Photography is the representation of this age.

In my view, the study of modern art should be based on a research of photography. For instance, the research would be more profound if Picasso's works could be based on photography.

Machinery substitutes hands in the 'age of mechanical reproduction', as machinery is characterised as high efficiency. Photography substitutes the stereotyped printing and becomes one of the fastest reproduced technologies. And photography is the pioneer who frees our hands from the most important technical function and it has transferred this kind of function to the eyes for people looking at the camera lens. Is there any machine functioning at higher speed than the eyes? As eyes can capture an object more quickly than hands, photography can greatly quicken the process of reproducing the image, and it can even keep the same pace of words. For example, 'a film operator shooting a scene in the studio captures the images at the speed of an actor's speech.'• Hence, photography is • 5 the fastest means of reproduction and the best form of it.

The previous arts and crafts existed for rites and were endowed with handmade charm. In the age of reproduction, the crafts are substituted along with the disappearance of their aura. Because of this, the unique friction between the world and us in the age of the handcrafted becomes controllable in the age of mechanical reproduction. Yet after the emergence of photography, eyes become a way to contact the object and the friction disappears. Therefore photography is a means of reproduction or mechanical production with no friction. However, the conflict between photography and friction may still remain.

It is in this manner that the Cranbrook Academy of Art's work *Necessary Friction*,° in the Architecture Department, is presented. ° 1990 The work presents three ways of making in three forms. The first form is that, when the hand touches the chair, it relies on friction;* * FIGURE 3 the second form is a paradoxical use of friction–if the desk doesn't

这种死亡是瞬间的，我开始寻找像照片一样的结构体形式。
寻找摄影一样的建筑学，而不是信使一样的建筑学。
这是对应于现代性的一种建筑学，是思考"死亡"的建筑学。

二、结构体

我坚持要寻找一种结构体，它不需要模仿海杜克那样再加上诗歌或图纸形成一种循环的形式。
每个人的方式不一样。
我从物体开始。物体和语言平行而不是回到语言。

（1）机械复制时代

这种结构体形式试图反思制作。以呼应摄影术，而制作是我们接触世界的方式。我们依然生活在本雅明所说的"机械复制的时代"，这对制作来说是根本性的，摄影是这一时代的表征。　　○ making

　　我的一个观点是：现代艺术研究应该建立在讨论摄影的基础。比如毕加索的作品如果能用平行于摄影的方式来讨论更容易深入。

　　所谓机械复制时代就是用机械来取消手，因为这样快。摄影替代了平板印刷成为最快的复制技术之一。它第一次把手从最重要的工艺功能中解脱出来，并把这一功能移交给了往镜头看的眼睛。还有比眼睛更快的机器吗？由于眼睛比手能更快地捕获对象，所以大大加快了图像复制的工序，甚至能够同说话步调一致。比如"摄影棚中拍摄场景的摄影师，他能够像演员说话一样，拍下各种影像。"● 所以摄影是最快的复制手段，是最好的复制形式。　　● 5

　　以前的艺术品、工艺品都是为了礼仪而存在的，会带有一种手工的东西而有灵韵。进入复制时代后，手工取消，伴随而来的灵韵消失了。正因为这样，我们原本和世界的关系在手工艺时代有特别的摩擦力，到了机械复制时代变成了可控的摩擦力。摄影出现以后，使用眼睛接触物体，摩擦力消失了，所以摄影是一种不需要摩擦力的复制方式或机械生产方式，但两者交战。的可能依然保持着。　　○ engagement

　　正如匡溪建筑系的作品《必要的摩擦力》。中所呈现的。　　○ 1990 年
在这个作品中，以三种形式表示了三种制作方式。第一种即手摸椅子，依赖于摩擦力；* 第二种是悖论的使用摩擦力；○*　　
* 图 3
○ 凳子如果没有摩擦力，其实是无法推动的。
* 图 4

have friction, it actually can't be pushed.* The third form is that, friction disappears, but a square-secluded symptom-like visual conflict appears in this structure. Under such circumstances, the visual sense of a man sitting on the chair is limited by an iron board into a narrow horizontal visual slot.* This complicated installation, which adjusts the visual focus, reminds us of a shooting installation used for keeping a person still during a long period of exposure in early years of photography. Here all the devices surpass the camera itself. There is an aperture facing the eyes, but it is black. A man can see nothing. He is separated from what he sees, just as he is in a special black room, which, in fact, is a special method of reproduction. The friction at this time disappears, but engagement remains, engagement between the body and the world appears in a ghostly way.* Dan Hoffman started to explore the notion that engagement after photography is eventually to become a means of reproduction. This exploration is more profound than research on engagement between the body and the world. It is worth exploring as the Motor City in Detroit becomes gradually deserted. For engagement, the exploration is critical. It is not just the representation of it.

    Criticism is a kind of self-awareness. Criticism is also an essential symbol of modern and contemporary works. *Kitsch* that Andy Warhol referred to is also a kind of criticism of criticism. Certainly, an exploration of mechanical reproduction and the use of industrial materials is also an essential symbol of modern and contemporary works.

* FIGURE 4

* FIGURE 5

• 6

II.2.
FULLNESS — TEMPORALITY

The above section discussed the relevance of photography to technology. Another feature of photography is representation. Photograph has a peculiar feature, which is an image of special fullness. It appears in the presence of the observer without any gaps. Representation of photography is different from the previous arts. It takes time into consideration.

    In accordance with photography, our structure also attempts to rethink the way in which time stays. Reproduction, mentioned earlier, is related to temporality, and so is representation. As the picture is in fullness, it appears in the presence of the observer without any gaps. The observer is not aware of the frame of picture while watching the picture. The picture remains still. When you are facing the picture alone and being stuffed with 'fullness', you have no way to exit this encounter. Barthes stated: 'I am alone with it, in front of it…there is no escape. I suffer, motionless. Cruel, sterile deficiency: I cannot transform my grief, I cannot let my gaze drift.'*

    In the photograph, the stillness of time appears in an extremely abnormal way. Time seems to be frozen. Why is the photograph in fullness? Because you cannot escape, and the picture is motionless

• 7

第三种是摩擦力消失，广场幽闭症式的视觉冲击出现在这一结构体中，其中坐在椅子上的人的视觉被铁板限制在一条狭小的水平视槽中。* 这种调节视觉焦点的复杂装置令人回想起摄影术的早年——在曝光的长时间内用于保持人静态的装置。在这里所有的手段超越了照相机本身。对着眼睛有一个孔，但是黑的，什么都看不见，隔了一个东西再看就像在特殊的黑室。这其实就是特殊的复制形式，摩擦力此时没有了。但"engagement"照样存在，身体与世界的交战以幽灵般的形式出现。霍夫曼就是在摄影成为复制最终手段的情形下来讨论交战。这比普通的身体交战问题讨论得更深入。因为面对日渐荒芜的底特律汽车城，更需要讨论这一问题，这样对交战的讨论才有批判性，不是简单的再现。

  有无批判力是现代作品和当代作品的基本标志，批判就是有自觉能力。即使是沃霍尔的媚俗○也是对批判的一种批判。

  当然，能否对机械复制讨论，能否使用工业生产材料也同样是其标志。

* 图 5

• 6

○ Kitsch

二（2）满——时间性

以上是摄影与技术相关而有的特点，另一个特点是其再现性。照片有一个特别之处，它是一种特别"满"的图像，它没有任何空隙地出现在观者眼前。它的再现与以往艺术形式不同，直接对时间进行了思考。

  因此，这一结构体还试图反思时间停留的方式以呼应摄影术。复制是时间性的考虑，而再现也是对时间性的考虑。因为照片特别满，它没有任何空隙地出现在观者眼前。看的时候不觉得有边框，它截止了。当你单独面对照片的时候，被"满"填充着，没办法找到出口。罗兰·巴特对着这种没有出口的情形描述："我难受，一动不动。这种无所作为是令人痛苦的；我不能排解我的忧伤，我不能使我的目光游离。"•

  在照片中，时间的静止不动以一种极端的畸形形式出现。"时间"好像被堵塞住了。之所以满，是你无法逃离，是因为看的一刹那时间被停住了，你的视点没办法放大缩小，它就是这么满。不管照片如何现代，与我们的日常生活如何贴近，却没有办法改变它停滞的实质。因此它不仅从来不是一种记忆，它让你恍惚，而且还阻断了记忆，成为有碍记忆的东西。

• 7

while you are watching it. So what you see cannot be enlarged or reduced. After all what you see is in fullness. No matter how up-to-date the picture is and how close it is to our daily life, we have no way to change its factual stagnation. Hence, time is not a kind of memory. It blocks the memory. It is a stuff that hinders us to memorise something.

Gerhard Richter put photograph as the object and made it vague in his paintings. It is a representation of the state of: 'what has been but now what is no longer' and it embodies the stagnation. Here stagnation is regarded as the emergence of a kind of oil painting. It is even the photograph of a photograph.*

* FIGURE 6

These are the essentials of photography—temporality, reproduction and stagnation. Letter is also characteristic of temporality, sent to or back in a complete way. It is waiting to be received. So it has desires. It is private but looking forward to being public. A letter is written by someone but conveyed in an implicit way; whereas, a photograph has no desires. A photograph blocks your memory and keeps you at a standstill and even in its fullness. You have no way to imagine or find an exit. You look but you cannot 'see'. It does not lie in 'seeing' but it restricts you from seeing.

III. WORKS
III. I PHOTOGRAPH, NOT OBJECT

In 2010, the topic of the work of conceptual architecture at Southeast University is *Day & Night/Day/Night* and the large black objects.*

* FIGURE 7

The work mainly consists of two large black objects 3m by 3m by 3m in dimension. Besides their size, they do not have much in common. They appear in the main road of the campus or in the front of Zhongdayuan Building without any reason. They don't possess any artificial relations with each other.

A large black object is framed by a black windowpane. It slowly moves along the main road of the campus. There is no starting point and terminal point. The black object blocks people's views. The surrounding world unceasingly appears but disappears as the moving of the large black object. Sometimes it reflects the light, so the surrounding world is unceasingly lightened and darkened.* Another large black object is framed by a black iron sheet. It is an object of five dimensions with one dimension concaved from six dimensions. It occasionally appears in different places on the campus. The concave feature of this black object unceasingly faces the front of the library, the front of the assembly hall, and the front of the School of Architecture. Seemingly, it wants to swallow the exterior space and all the people who are facing it.*

* FIGURE 8–11

* FIGURE 12–14

Here the black objects become a place with all eyes gathered. But they won't let you gaze at them. They form a 'full' and mobile frame of a photograph with the greenness of the trees. That is to say, the

里希特的绘画以照片为对象,他将照片模糊化,是对"曾经如是,已经不是"的表现,是对停滞的表现,将去看照片时出现的停滞作为一种油画出现。它是一种照片的照片。*   * 图6

这个就是摄影术的要点:时间性,复制和停滞。

信是有时间性的,完整的漂移,等待的,有欲望的,私密的又期待着公开,自己写了又兜了圈等等。而照片是没有欲望的,让记忆阻断的,让你停滞的,是满的。你没办法想什么,让你没有出口没有想象,有一种视而不见的效果。它不在"看"上,它让你看不见。

三、作品
(1) 照片,非物体

2010年东南大学概念建筑设计的题目是"白天和黑夜/白天/黑夜"和"大黑块"。*   * 图7

作品最终呈现的是两个3×3×3米的大黑块。它们除了具有同样的尺寸,并没有太多的相同。它们莫名地出现在校园的中央大道上或中大院前,并没有刻意的联系。

一个大黑块被黑玻璃包裹,它非常缓慢地在中央大道上平移着,没有起点也没有终点。黑块阻断了来来往往人的视线,同时在这种缓慢的移动下,周围的世界不断地被大黑块抹除又恢复。由于它有时会反光,所以周围的世界又会不断地点亮又熄灭。* 另一个大黑块,由黑铁皮包裹,是由六面体一面凹进而成的五面体。它偶尔出现在校园不同的地点,它的凹口不断面向图书馆前、大礼堂前、建筑学院,它似乎要将外部的空间和面对着它的人吸入到它的一片黑之中。*   * 图8至11   * 图12至14

在这里,黑块成了目光聚集的地方,但它们无法使你凝视,它们与满树的绿色一起形成了一张"满满"的飘忽不定的照片。也可以说它是为树做的作品。这里必须有树,让人天天在校园中看到树,然后让一个黑块出现,让满校园的树成为一张特殊的照片。绿的颜色甚至像是在黑白照片上染的色一样,不是本有的色,也因此黑块似乎失去了物体的特性。*   * 图15

black objects are a production for the trees. Here the trees are a necessity. They are objects observed in the campus everyday. When a black object appears, all the trees in the whole campus become a special photograph. The greenness is rather like the green used to dye black and white pictures. It is not the innate colour. Because of this, the black object loses its specific property as an object.*

* FIGURE 15

III.2.
PHOTOGRAPH, AND OBJECT

The title *Day & Night/Day/Night* and large black objects not only refer to the photograph, but also to the appearance of the photograph.

When you approach the black object, you can see another photograph, which is similar to the French photographer Eugène Atget's work. He was down and out at his very beginning. Later most of his works were purchased by Man Ray, and he achieved a great reputation. He shot a large amount of his early pictures in Paris at the turn of the century. For instance, in *No. 63 Entrance Hall of the Inner Toule Street*,* we can see two people are standing in the glass door of a restaurant and the shadow of the bank of the Seine from the glass door, as well as the shooting installation of Atget. I think the black object is just like the old photograph taken by Atget. Benjamin commented:

* FIGURE 16

> the picture taken by Atget is like a scene of committing crimes. This scene is desolate and uninhabited. People want to provide kind of evidence to take this scene... The photograph becomes the authoritative evidence of documenting historical phenomena. It also implies a political meaning.•

• 8

It indicates that, the criminal scene is the final destination of the photograph. Certainly, the black object could be related to everything, including mother's photograph.

That also reminds me of Richter's famous series of works *18 October, 1977*.°* The reason why his work pictures a photograph, and even the evidence of a crime, is that it introduces the case of murder and et cetera to itself in an explicit way.

○ 1988
* FIGURE 17

In this way, the large black object gets away from the photograph and becomes evidence on the spot. Because it is a photograph, the black object is not an object, even the greenness of the trees seems to be dyed; also, the black object is a sort of evidence, and so it becomes an object. But the property of the object has changed.

Why do I insist on talking about a structure? Because it is more than enough. The structure forms an object and becomes a part of a photograph, then the property of the object disappears; but the object could become a special object with the emergence of evidence at a crime scene.

三 (2) 照片，且物体

"白天和黑夜／白天／黑夜"和大黑块不只是要说明是照片，还要说明是什么样的照片。

当你靠近大玻璃黑块的时候，又可以看到另外一张照片。这类似法国摄影家尤金·阿杰°的作品，他很潦倒。作品后来由于大量被曼·雷°购买，才享有很大的名声。他拍摄了大量世纪之交的巴黎的旧景。大黑块好像阿杰拍摄的巴黎旧景照。比如在《图尔内街63号门厅》*中，我们可以看到有两个人站在餐馆的玻璃门中，还可以从门玻璃上看到塞纳河堤的影子，以及阿杰的摄影装置。

° Eugene Atget
° Man Ray
* 图 16

本雅明曾说阿杰的照片拍得像犯罪场面，"犯罪场面是荒芜人迹的；人们把它拍下来是为了建立证据。"当然无意中拍下来的照片后来也可能成为犯罪的证据。于是"照片成为历史现象的权威证据，并获得了一种隐蔽的政治含义。"•这句话表明，犯罪场面是照片的一种最终走向。当然，它可以对待一切的事情，对待妈妈的照片也可以如此。

• 8

这让我联想起里希特著名的《1977年10月18日》°这组作品。*之所以说里希特的作品是照片的照片，又是犯罪证物般的照片，就是因其直截了当地运用了谋杀案作为素材等等。

° 1988 年
* 图 17

这样大黑块就从照片中脱身出来，成为现场的一个证物。

因为是一张照片，黑块就不再是一个物体，连树的绿色也像是被染过的；又因为是一个证据，而成为一个物体。此时物体的特征已经发生了变化。

为何坚持使用一个结构体，因其无需更多，结构体构成一个物体，成为照片中的一部分，其物体的特性没有了；物体又会通过如犯罪现场的证物而出现又成为特殊的物体。

因此这里不需要通过诗歌、绘画、结构体之间来回穿梭的方式，不需要每次都以"我"通过想象作为中介。此处的照片无需想象，"我"有时是需要的，有时也可以没有，因为它对每个人都发生，因为它是有力的犯罪证据，所以对每个人都有意味。

Hence, the connections among poetry, paintings and structure are not necessary, and 'I' and imagination as two media for the photograph here are not necessary either. A photograph has various meanings to everyone, for the photograph happens to everyone and it forms a kind of convincing evidence of a crime. Therefore, what I present is quite different from Hejduk. The photograph has some relevance to the structure with photography as a medium. Whether 'I' exists or not is not necessary. 'I' and other 'I' establish another social contract and block the imagination.

III.3
'I'

Here, the large black object also reflects another specific property of photograph. Roland Barthes stated:

> The photograph is literally an emanation of the referent. From a real body, which was there, proceed radiations which ultimately touch me, who am here…the photograph of the missing being…will touch me like the delayed rays of a star. A sort of umbilical cord links the body of the photographed thing to my gaze…
>
> the photographed body touches me with its own rays and not with a superadded light…
>
> colour is a coating…is an artifice, a cosmetic…• • 9

If we want the photograph to be animated, we should get rid of some unnecessary colours, and we should patiently wait for the innate light of the photograph.

This is related to earlier photography. At that time, photography was a kind of chemical reaction, the property of which was formed by radiation. Photography with a digital camera can not be discussed in this way. As the so-called *post-photography* time has come, it is the death of photography. Photography would not be recognised as such if there were no clicks or chemical reactions. Nevertheless, I persist in using photography, though the era of post-photography exists in concurrence.

There are two notable words in Roland Barthes's work *La Chambre claire*: *Stadium* and *Punctum*.

因此这和海杜克的办法不同，照片和结构体通过摄影术连缀，需要"我"又不需有"我"，"我"和别的"我"便构成了另一种社会契约，且其阻断了想象。

三（3）"我"

在这里，大黑块还反映了照片的另一特性。罗兰·巴特曾经指出，摄影是拍摄对象身上散发出来的放射物，然后触及在这里的我。所以那个已消失物体的照片触及我，如同一颗星发出的光延迟了一段时间之后才触及到我一样。一种脐带式的联系连接着被拍摄的物体和我的目光。因此，照片是以其自身的光线触及我，而不是以外加的光线触及我。也因此，照片上的色彩成了外在的东西，是假发，是脂粉。如果要获得照片的生命力，就要去掉一些不必要的色彩，如果要获得照片的生命力，就要耐心地等待物体本身的光线。

● 9

这与早期的摄影有关，因为那时候摄影是一种化学反应，通过感光形成这种特性。数码拍照已经不能用这种方式讨论，已经进入所谓后摄影——摄影的死亡，没有"咔嚓"一声没有化学反应就不是摄影。

我坚持使用摄影术。○

○ 尽管有后摄影

罗兰·巴特的《明室》中有两个词："Stadium"和"Punctum"。

"Punctum"一词有刺伤，小孔，小斑点的意思，还有被针扎了一下的意思。照片上的"Punctum"是一种偶然的东西，它只对特殊的人有用，正是这种偶然的东西"刺痛"我。就像罗兰·巴特从照片的小女孩身上突然发现了他的母亲一样。但摄影是复制的东西，这便有一种悖论，没有想象的复制却能刺痛"我"，跟"我"连缀。

灵韵消失了，让你无法有出口，怎么却又能刺痛我？这是摄影中真正诡异的地方。

"黑"色的物体使周围世界变成了一帧照片，而"黑"正是刺伤我的"Punctum"。

Punctum also means sting, speck, cut, little hole and a cast of a dice. A photograph's punctum is the accident which pricks me. This feeling is like that of when Roland Barthes suddenly saw his mother when looking at the photograph of a little girl. But photography is a kind of reproduction. Subsequently a paradox emerges: the reproduction with no imagination can prick me and establish a relation with me. Aura disappears. You have no way to escape. How could it prick me? How strange the photography is!

The 'black object' makes the surrounding world become a frame of a picture and the 'black' is the punctum of a photograph which pricks me. 'Blackness' tries to present its own light of the past objects and this light needs a period of time to reach us. As for a different 'I', time-lapse is also accidental, indefinite, and hard to wait for. There are too many 'I' on the campus. I have no idea when I will meet the blackness and get pricked. So the photograph-like black object and countless 'I' link together. The light of the black object either could touch 'me' quickly or not touch 'me' forever just as destiny ties people together, which is so indefinite and hard to wait for.

Just like the title of the work, *Day & Night/Day/Night*, time-lapse has constructed a specific space.

Relevant to countless 'I', it is not a simple commensuration. Like the film *Wings of Desire*° directed by Wim Wenders, it is just a commensuration from an angel's angle.   ° 1987

IV.

EPILOGUE

The object is in correspondence with the structure. Language is a need in parallel with the object. How can we keep them in parallel?

...

"黑"试图提示过去的存在物拥有自身的光线，而这种光线会延时到达我们。延时对于不同的"我"又是偶然的，不确定的，无法等待的。校园中的那么多的"我"不知道什么时候会碰见它，被它刺痛。如此一来，像摄影一般的黑块和无数的"我"连接，其光线像某种缘分一样，可能永远也无法到达某个"我"，有的却能很快地到达某个"我"，不确定而无法等待。

因此，正如作品的名字，白天和黑夜／白天／黑夜，其构成了特定的空间。

与无数的"我"有关，这不是简单的通约关系，如文德斯的《柏林苍穹下》，只是用一个天使视角的通约。

四、余言

跟结构体对应的是物体。

与物体平行需要语言，又如何平行呢？

……

---

1. John Hejduk, 'Diary Constructions', *Perspecta*, volume 23, Cambridge, US-MA: The MIT Press, 1987: 78—91.
2. Nicholas Mirzoeff, *An Introduction to Visual Culture*, New York: Routaledge, 1999.
3. Walter Benjamin, *Illuminations*, New York: Shocken Books, 1969: 141—142.
4. Roland Barthes, *Camera Lucida Reflections on Photography*, New York: Hill and Wang, 1981: 92—94.
5. *Ibid.* 3, 219.
6. Dan Hoffman, *Architecture Studio: Cranbrook Academy of Art, 1986—93*, New York: Rizzoli International Publications, 1994: 71—79.
7. *Ibid.* 4, 89—91.
8. *Ibid.* 3, 226
9. *Ibid.* 4, 80—81, 89—91.

2. 尼古拉斯·米尔佐夫《视觉文化导论》（倪伟译），南京：凤凰出版传媒集团、江苏人民出版社，2006年，93页。
3. 汉娜·阿伦特编《启迪——本雅明文选》（张旭东等译），北京：生活·读书·新知三联书店，2008年，151—152页。
4. 罗兰·巴特《明室》（赵克非译，译文略有调整），北京：中国人民大学出版社，2011年，124页。
5. 同注3，233、219页。
7. 同注4，120页。
8. 同注3，243页。
9. 同注4，108—109页。

FIGURES

1

2

3

4

5

6

1 约翰·海杜克《调整的基础》。
John Hejduk, *Adjusting Foundation*, Kim Shkapich ed., New York: The Monacelli Press, 1995.

2 《主体／客体》。
费城艺术大学的光厅，1987 年。
Subject/Object.
The Great Hall of the University of the Arts in Philadelphia, 1987.

3 必要的摩擦力：第一把椅子。
The fisrt chair in Necessary Frictions.
Architecture Studio: Cranbrook Academy of Art, 1986—93, New York: Rizzoli International Publications, 1994.

4 必要的摩擦力：第二把椅子。
The second chair in Necessary Frictions.

5 必要的摩擦力：第三把椅子。
The third chair in Necessary Frictions.

6 《鲁迪叔叔》。
格哈德·里希特，1965 年作品。
Uncle Rudi.
Gerhard Richter, 1965.

插图

7

7 《白天和黑夜／白天／黑夜》。
概念建筑作品，2010 年。
摄：蒋梦麟
Day & Night/Day/Night.
Conceptual Architecture, 2010.
Photo: Jiang Mengling

8

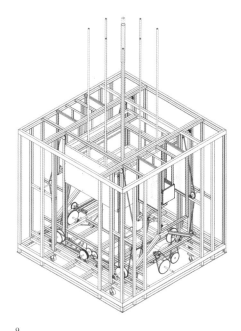

9

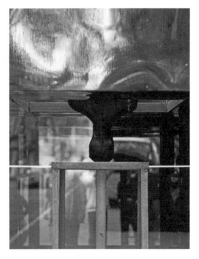

10

11

8 《白天和黑夜／白天／黑夜》
概念建筑作品，2010 年
摄：蒋梦麟
Day & Night/Day/Night
Conceptual Architecture, 2010
Photo: Jiang Mengling

9 《白天和黑夜／白天／黑夜，大黑块》
概念建筑作品，2010 年
绘：2010 概念建筑组学生
Day & Night/Day/Night, Black
Conceptual Architecture, 2010
Sketch: students of Conceptual Architecture 2010

10 《白天和黑夜／白天／黑夜，大黑块》
概念建筑作品，2010 年
摄：蒋梦麟
Day & Night/Day/Night, Black
Conceptual Architecture, 2010
Photo: Jiang Mengling

11 《白天和黑夜／白天／黑夜，大黑块》
概念建筑作品，2010 年
摄：蒋梦麟
Day & Night/Day/Night, Black
Conceptual Architecture, 2010
Photo: Jiang Mengling

葛明
GE MING

12

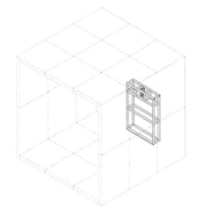

13

14

12　《白天和黑夜／白天／黑夜，空谷回音》。
　　概念建筑作品，2010 年。
　　摄：蒋梦麟
　　Day & Night/Day/Night, Echo Black.
　　Conceptual Architecture, 2010.
　　Photo: Jiang Mengling
13　《白天和黑夜／白天／黑夜，空谷回音》。
　　概念建筑作品，2010 年。
　　绘：2010 概念建筑组学生。
　　Day & Night/Day/Night, Echo Black.
　　Conceptual Architecture, 2010.
　　Sketch: students of Conceptual Architecture 2010.
14　《白天和黑夜／白天／黑夜，空谷回音》。
　　概念建筑作品，2010 年。
　　摄：蒋梦麟
　　Day & Night/Day/Night, Echo Black.
　　Conceptual Architecture, 2010.
　　Photo: Jiang Mengling

15

16

17

15 《白天和黑夜／白天／黑夜，大黑块》。
概念建筑作品，2010 年。
摄：蒋梦麟
Day & Night/Day/Night, Black.
Conceptual Architecture, 2010.
Photo: Jiang Mengling

16 尤金·阿杰《图尔内街 63 号的门厅》，1908 年。
Eugène Atget, *Tambour, 63 quai de la Tournelle*, 1908.

17 格哈德·里希特《1977 年 10 月 18 日》组照，1908 年。
Gerhard Richter, *October 18, 1977*, 1988.

DISCUSSIONS AT THE SYMPOSIUM MENTAL SPACE

会议讨论
心理空间

TIME
29th May, 2011

时间
5月29日

PLACE
1st Floor, East Hall
The Main Auditorium
SEU

地点
礼东二楼报告厅
东南大学

In addition to participants' translation, other works are by Li Hua, Shen Wen & Yu Changhui.

在参与者的翻译之外，其他的工作是由李华、沈雯和于长会完成。

THE SCENES OF THE DISCUSSION

SCENE I
PROJECTION

SCENE III
A NARRATIVE STRUCTURE OF CROSS-CULTURAL ARCHITECTURE

SCENE II
THE FIGURE AS HOME

SCENE IV
BLACK

讨论场景表

场景壹
《投射》

场景贰
《象与家》

场景叁
《交叉文化建筑的情节结构》

场景肆
《黑》

THE PERSONS OF THE DISCUSSION

Mr. Shiqiao Li ················· SHIQIAO
Mr. Dongyang Liu ················· DONGYANG
Mr. Mark Cousins ················· COUSINS
Mr. David Leatherbarrow ········· LEATHERBARROW
Mr. Hui Zou ················· HUI
Mr. Andong Lu ················· ANDONG
Mrs. Wei Chen ················· WEI
Mrs. Yongyi Lu ················· YONGYI
Mr. Ming Ge ················· MING

出场人员表

李士桥 ················· 李士桥
刘东洋 ················· 刘东洋
马克·卡森斯 ················· 卡森斯
戴维·莱瑟巴罗 ················· 莱瑟巴罗
邹晖 ················· 邹晖
鲁安东 ················· 鲁安东
陈薇 ················· 陈薇
卢永毅 ················· 卢永毅
葛明 ················· 葛明

场景壹
《投射》
入场

李士桥 谢谢您,马克。让我们现在开始提问吧。

刘东洋 谢谢,马克。您的讲座就像一场心理疗治,多少令人欣慰的是知道在挣扎里还会获得快感。的确,起码在我自己的实践里,这真是个问题。我弗洛伊德的书读得不多,不知道弗洛伊德是否曾经详述过我们都受到哪些类型的限制。似乎种类很多,有甜蜜也有酸楚。像总被要求去改图,肯定会带来焦虑。而有些焦虑则来自更深的伦理挑战。例如,1980年代和1990年代早期,那时的中国建大马路宽马路并不仅仅是为了解决小汽车交通问题,那时还没有老百姓可以买得起小汽车,规划建设大马路是把它也当成一种现代的标志。可我在加拿大的交通课上,老师特别地告诉我们,仅靠一两条最宽的马路是解决不了交通问题的,道路窄些但是路网密些会好很多。关键是要有像样的公共捷运服务。于是,我在中国的规划项目里,就经常要跟当局在道路宽度上吵架。有时,我妥协,即使我知道那样的规划将来肯定有问题。您刚说过,对于这样的焦虑是没法子医治的,怎么办,这就更让我焦虑了。

卡森斯 我认为您应该将列宁最后的文章之一"宁肯少些,但要好些"给规划院的官员看。这篇文章非常有趣。虽然是列宁的著作,却相当保守。您提到的道路的例子使我们看到这一问题包含了很多城市议题。从某种方面来看,我们的敌人既不是所讨论的地形学问题,也不是所讨论的运动问题。我们的敌人其实是"流线"。不知为何,有关流线问题的历史变得本末倒置了。让我举一个例子。在"AA",非常幸运,我们的校舍得到了扩充。因此相较于以前在贝德福德广场只有一栋建筑,我们现在在有五栋,它们彼此相邻。去年,来了一些自称为"空间规划师"的人。他们来做研究,为建筑师设计这五栋建筑的使用方案做准备。坦白地讲,学生和职员都多少受到了空间规划公司的冒犯。所有他们做的不过是估量哪些人会去建筑的哪些地方。一周后我问他们,"你们想要搞清楚的是什么?""这个么,"他们说,"很明显,我们必须从建筑中的流线问题开始。"我说:"为什么?"这位女士有点不高兴,答道,"因为我们就是这么开始做的。"在她心里,人是先知道流线,才开始走动的。我说"流线在这里并不是问题。"我想说的是,这些18世纪的建筑确实不是一个航站楼,尽管当他们完成设计时,也许感觉上会有点像。这些空间规划师是走廊的先锋。他们视生活为一道走廊中漫长的风景。事实上,在"AA",我们也有防火规范的检察员。他们慷慨地声称不会对我们固执地执行新规定,因为我们处于转型阶段。那些公文几乎就是弗兰兹·卡夫卡的作品。我们正迎向"无菌走廊"的政策。困难在于如何理解这些事情产生的机制,又如何找到抵抗的办法。在"AA"有一扇大门,为了满足防火规范,我们不得不在门上放置一个丑陋的标识"消防出口",好像生怕你会认不出这是扇门。

莱瑟巴罗 我做了些观察,我花太多时间在上面而难以明确地用一个问题涵盖。但它们或许是有用的观察。令我非常感兴趣并希望了解更多的是,对于主体与客体开放性的区分。而我想起您说的第二点,即主体的经济维度,这让我想到开放的两极对立需要有所论证。我愿意理解为,那部分论证有意思地支持了我们讨论的地形学问题,即地形学的历史特征。我想重提主体的考古学与主体量化的无意识之间的两极对立,并没有离题太远。我相信这种两极分化与我们关于焦虑的伦理维度的开放性讨论有关,因为从某种程度来看,焦虑,表面上也是某种外在于主体的东西,即我们介入其他的人或事之中,而这些人或事加剧了我们的焦虑。如果是那样的话,我们需要重新考虑开放性的两极分化。我同意您的看法,对于镜像或镜像化了的主体,地形条件不仅仅包括将过去视为现在,而且更重要的是将现在视为对未来的投入。于是外在性成为主体的一部分,成为构筑中的一部分。外在性不是帮助我们克服焦虑,而是与焦虑同行,只要它被控制在自身范围内,并向各种可能性开放而不受限于过去的阐释。

卡森斯 当然。对弗洛伊德来说,这样说当然不是为了继续制造主客体之间的分离,或所谓的个体与文化之间的分离,或从另一角度来看,内部与外部之间的分离。作为一种相互重叠的系统,弗洛伊德对地形学上的无意识运行方式的表述过于繁琐。只有通过对他人的认同,尤其是最初对母亲的认同,人类才会从幼儿成长为主体。这保留了一种关系即简单的认同,或者说是自我认同模式。弗洛伊德将为"*Je est un autre*●"的说法感到相当高兴。因此,当然不存在清楚的划分。非弗洛伊德派会说,的确在某种程度上,每个人都有其无意识的、与地域无关的却属于这个人的东西。是的,我同意主体最终都牵系在与对他人的依赖及与他人的关联上。的确,你可以说主体不过是对他人的回应而已。我认为至此今天上午的讨论重要的是,是否存在新的实践形式,即被我大而化之地称为数字化或参数化的方式。我不关心计算机在合成形式的数据运用中的细节,我更感兴趣的是论证自己的试验,看看它能否触及和提出这样的问题:这种情况是不是正在发生?从普遍意义上来说这是新事物吗?以及结果如何?

● 我即他者

莱瑟巴罗　以我的经验，我看到当代各种类型的实践中，存在大量的天真的乐观主义，这些实践恰恰将伦理主体的讨论排除在外。他们让设计依数据的优劣来发展，自我生成。对我而言，无论在伦敦、费城、洛杉矶也许抑或上海，情况都差不多，有很多无名的代理机构，他们毫不关心其设计产品是否能给人带来愉悦。

场景壹
落幕

场景贰
《象与家》
入场

邹晖　你列举的一些当代艺术家的作品都表现出强烈的讽喻现象。我想问一下这个现象与中国当代文化的本质联系。这种讽喻的批判性未来发展将会如何？它是否只是针对中国某个历史时期的暂时的策略性方式？

李士桥　从我乐观的视角来看，我曾经认为那是暂时的。但这种东西似乎以不同形式在不同时间反复出现。我认为，为了文化的健康发展，我们需要一种批判的维度；未必是希腊式的批判，可能是某种其他的方式。我用这些艺术家为例，是因为在建筑师的作品中我的确看不到多少实例。我希望未来能在建筑师的作品中，看到与这些艺术家的作品相提并论的表现。最重要的是将这样的讨论发展下去，使其不再总是自满地重新宣称找到"家"了。在英文语境下，乡愁这个词听起来几乎像种疾病，它暗示了要回家的某种极度异常的状况。在中国，我相信回家不意味着乡愁；作为一个病理学状态的乡愁，过去不存在、现在不存在、将来也不会存在，因为事实上当你就在家中时是无家可归的。因此我们该如何看待在家中并同时能维持批判性？

莱瑟巴罗　我对这个问题没有答案。但是对此，我的确有些看法。从静止的石头开始，想起您所说的翻译的经历，和我看到的这次论坛中两种语言之间有趣的碰撞，这个问题的提出听上去顺理成章。当政治理论作家汉娜·阿伦特在第二次世界大战后回到德国时，她被问及如何看待自己的国家。她回答道，"我所熟知的那个国家已经不在了"。问话者说，"完全不在了？"她说"不，语言还在。"我的看法是翻译或卡森斯所说的阐释恰恰构成了一切的起点。举另一个例子，在古代，希腊妇女负责传授语言，这通常借由歌曲传授。以这种方式学习的语言在公共领域中说与他人。也就是说，公共意味着发声，通过语言确立自己。因此，翻译对于涉入公共显得至关重要，它让人超越家庭而进入公共交流之中，突破了家庭谈话的范畴。我相信这与亨特所说的公共言论是相关的。我愿意称其为哲学般的修辞。但这不是事实，而是某些特定情况下的事实。我不知道我是否能想清楚甚至超越一个基本的前提，即如果我们努力表达所思所想，我们便是彼此相关的。毫无疑问那是一种快乐，在人群中重新思考一个人自认为在家中已经知晓的东西的快乐。因此在我看来仍然能从所有那些言论中学到不少东西。而自足性这一我认为的家的现代概念把这种对世界的表达遮蔽了。对于斯维勒·费恩来说，家是自然界的一部分。不可能从家中分离，何况家还是语言和文化的容放地。因此家与城市无法分离。在西方，我们认为，我们的私人生活如此充裕，以致我们认为我们不需要城市。我不确定中国的情况是否不同。我能理解你对语言、文字、它们的形象潜质以及对诠释的要求的优美论述，我相信关键在翻译之中。我认为那是我们应当拥有的惬意。在我看来，我们今天不同语言混合交流中的碰撞是十分有趣的。

李士桥　谢谢。也许我可以对此作点回应，如果需要的话。我想强调语言是第一家园，语言的策略中包含了一种建构家园和家常性的方式。在某种程度上，词义和字母之间的关系与词义和汉字图形单元之间的关系是不同的。这里我试图提出的是在语言学上通过形象王国建立的一种永恒家园的条件。这恰恰与具有公共概念的条件形成了鲜明的对照。这是一种政治生活的概念，这种政治生活对亚里士多德来说是根本性的，当他说冲突出现时，我们效忠的不是家中的朋友而是这种公共生活。对我来说，这是我们思考中文与西方语言间的翻译时一个根本性的议题，它比第一眼反应出来的东西要重要得多。

鲁安东　我想问您一个小的技术问题：为何您使用"象"这个字而不是其他汉文中更为权威性的字、例如"文"。●"文"这个字是来源于纹理。我觉得在这个字之间有不小的区别。"象"是一个按照某种规则的成形过程的结果，而"文"，纹理，则不同——它暗示着"象"之间的关系，以及"象"和它所施与的对象之间的关系。在我看来，你放映的那幅刻着"象"的峭壁的照片更像是某种纹理。它很像是一种赋予力量的过程。我一直有一种猜测，中国园林很像是一种对地形或者自然施加纹理的过程。园林是一种带有纹理的自然。但是纹理对"象"和形象如何陈述之间的不确定的关系非常重要。因此我好奇你为何选择"象"这个字，毕竟这是一个汉文中不太常用的字。

● 纹饰或者文学

李士桥　我的意图是从最简单的单元开始。我认为"文"比"象"要略微复杂一些。"文"由许多"象"组成。当你开始谈论"文"时，这已经远比我的目标复杂了。"象"在某种程度上使我可以同时涉及文学的层面和视觉的层面，因为这是中文的关键所在：即文学性与视觉性的结合，它从来没有设想过文学与视觉之间的差异。

卡森斯　这让我想到了两点。你将中国园林视为景观纹身的观点将使你在加州大受欢迎，却可能在中国不受待见。就你视书法为图像艺术的顶峰来说，我认为拓展这一观点的方式之一是研究英文和中文书法之间的差异，不是关于图而更多地是关于底的差异，甚至是页面的差异。对你来说，书法的底与绘画的底之间是否存在着差别？我的意思是说在绘画中，"底"某种程度上被体验为空间。我不确定在中国书法中您会如何描述"底"，因为其显然不同于英文的书写。这也许是拓展你研究的一种方式。

李士桥　同时设想一种阴和阳的关系，一直是落笔于纸上的策略。所以，它已经是"图"的一部分；而"底"也是"图"的一部分，二者之间不存在"二元对立"。没有一位画家或书法家会说留白不是绘画或书法的组成部分。

场景贰
落幕

场景叁
《交叉文化建筑的情节结构》
入场

鲁安东　我有一些疑问。因为最后看来，你谈到的一些项目似乎在暗示：如果你想要保护或者保存一些记忆，你就开始建造剧场。同时，我对你在中国的文脉中使用"迷宫"这一概念有些疑虑。在我看来，迷宫并非关于迷路，而是关于如何在空间中对迷失的过程加以组织。它是一种空间化的过程。而在中国的文脉中，迷失就像是忽然间被转变为其他方式。那是一种纯粹的迷失，而并非试图对某种东西加以组织，并非试图去加以控制。我认为，之所以迷宫被作为现代建筑的某种起源正是因为它是一种对"组织"的强烈暗喻。我不认为在中国的文字或者图像中有任何类似的说法。其异性实际上体现了某种非常陌生的东西。它反映了欧洲文化在中国文脉中的影响。

邹晖　有必要对"labyrinth"这一概念在文化差异中的含义作进一步理清。在中国传统中与之对应的一个词是"迷宫"。我不认为"迷宫"与西方的"labyrinth"能够完全对等。在我最近关于中国建筑传统的迷宫概念研究中，我试着拾取一些历史的线索，研究与建筑艺术•相关的文本案例。从这些例证中，我们可以找到若干原型来理解迷宫的传统。我追踪迷宫历史的主要目的是寻找戏剧化的空间，尤其是能够承载如 18 世纪文化碰撞的那样的空间。我对"记忆的剧场"概念的比较追踪也是出于同样的目的，此概念源起于 16 世纪威尼斯的卡米洛。该概念所寓意的戏剧性与对历史的积极交往应和着中国建筑传统的主旋律。一种潜在的比较观是我的出发点，它也许有助于打开一个视角来评判什么有价值的已经失去了，并思考如何通过建筑来保护中国文化的核心。

卡森斯　你需要留意说的是哪个层面的记忆。昨天我们谈到了记忆的一个层面。但你所说的，尤其是记忆艺术的部分，它指的是一种记忆的技术。从这个角度来说，我们谈论的是记忆术的艺术。在某种程度上，记忆的剧场可以被称为艺术的剧场，诸如此类。它试图构成一个体系，一方面包含各种知识，同时又可以让个体以某种方式对其自行处理。它与维基百科不同，它能够以某种神秘的方式连接到记忆中的参考资料。

莱瑟巴罗　我想说我更喜欢最后一个建筑，◦我发现，相较于斯蒂文·霍尔的建筑，它能够变成以及表达您所暗示的历史素材方面的东西，而霍尔的建筑在我看来过于夸张、过于清晰、过于聪明了，恰恰与我认为您在寻找的建筑是相悖的。为什么这么说？桥是断的，通向中心的路径尚未完工，我们知道有些东西就在那儿，却还未出现。换句话说，不复存在或尚未出现之物、暗示的气息、对探索的需要、对不完美的容允，这些都是串接最终方案的线索。它们不愿全然显现，因为作为一种文化，我们无法补全他们。事实上，我认为他们的确形成了一个整体，但不是备忘录式的图像集，而要像某种记忆之类的东西，正如之前我所述，它们不是在你所拥有的东西里，而只是在你的渴望之中，渴望它的另一部分，即建筑通过那条线索暗示的那个部分。我将坚信这是对我们所处时代诚实的认识，是对不完美的转译，即使这种感觉需要仔细思量。从方案的其他角度看这种感觉未必成立，这是为什么它招致了如此多的欲望，如同马克在会议开始时所指出的。很明显，一个寂寞脆弱的人仍然在寻找一个可能成形的城镇。我相信这和我们绞尽脑汁所做的事情是相悖的。

刘东洋　我基本能跟得上您讲座的主要线索，不过，不太肯定是否准确捕捉到了您的深意。似乎您将"记忆剧场"视为一种建造行为的原型体验，不管是在西方还是在中国，都具有普遍意义。因此，

● 包括园林
○ 瓦屋顶居住小区文化广场

这个原型体验也可以用来解释为何我们也是在不断通过建筑去寻找记忆，无论是在重庆还是北京？是吗？

邹晖　没有批判的历史视角，●对历史的全面理解是极其困难的。如果我们欲对建筑的现实做出评判并寻求判断的批判性，有必要确立历史的地平线，即在中国经常谈到的历史感。当与历史交往时，总是存在价值的判断。这意味着选择性并需要经由理论的切入来发现根本的问题。为什么我们总是重复讨论历史中的某些建筑，却有意忽略许多其他建筑？通过追寻"记忆的剧场"的历史轨迹，我们能够揭示建筑作为戏剧化空间的根本意义。我在此所举的"记忆的剧场"的案例意欲显示形而上与形而下的交织，这种交织喻示着建筑的形体化含义。这些案例是被笛卡儿的二元论所压制的历史片段与线索。对这些戏剧性碎片的交往全然有别于现代的关于建筑形式的类型学研究。这些历史线索是真实的但又隐而不显。一旦我们通过比较的诠释来接触它们，此进程将有助于思考建筑在面对当今全球化所引发的文化混乱时，是否仍然有贡献于海德格尔意下的诗意的栖居。

场景叁
落幕

场景肆
《黑》
入场

陈薇　我虽然在东南大学工作，但这是我第一次看这个作业的完整表达。我觉得今天的安排很有意思，并且和昨天的内容可能也有些关联。今天上午两个报告给我最大的启发，是说建筑或者景观，能不能在人很困顿的时候提供一个缓释的空间，它是社会和人的一个关系。我觉得东西方在这上面是有差异的。十几年前，在柏林看埃森曼做的博物馆，我觉得他那个建筑蛮戏剧化的。他把东柏林和西柏林在长期分裂时期的那种伤痛，揭示出来了。当时东西德刚刚合并，柏林墙还没有完全拆掉，他的揭示让我感到更痛，一点都没有消解掉，就像刚才葛老师谈到的那张照片。那张照片是静止的，是一种很戏剧化的东西，让你停留在伤痛中间，让你清醒，让你警醒。这可能是整个西方的艺术传统，而中国的艺术，往往是让你在时间的过程中把痛苦悲伤消解掉。比如说电影《活着》，实际上活着已经没什么意思了，但是只要活着就好。主人公的伤痛是很柔软、很绵长的。它会让你觉得生命还是有价值的。这是两种非常不一样的思维方式、空间处理和理解方式。我觉得中国古典园林就是那样一种动态的、流动的、让你消解和释放郁结的一种方式。我有两个问题是针对葛老师做的这个设计的，这个设计有很多学生参与，例如唐静寅。我想问唐静寅，制作这样一个装置，对你心灵上有什么样的升华？它有没有给你的心灵或者情绪一个出口？第二个问题是，你做完了这个装置以后，你觉得在建筑学的教育中增长了哪些知识？或者说，在境界上，有什么样的提升？

卡森斯　首先我必须说我真的很喜欢这个装置。但对我来说，您讲演中的一个问题是作品包含了太多的维度。某种程度上，很难对它做一个简单明确的回应。一方面，您似乎想要将我们通常所说的摄影产生的影响或关于摄影的讨论，比如刺点，转换为现实生活中的三维形式，无论你把这种形式叫什么。我认为这极其有趣。也许只需专注于这一点就够了，因为我认为它已涵盖了很多的东西。我不清楚您谈及里希特绘画的缘由，不大理解里希特的画与您作品的关系。因此在您讲演的最后，我不大搞得清楚您主要的目标是什么。我认为在某种程度上，将力气从"刺点"这一中心主题分离的做法偏复杂了。其次我同样不理解您对白天黑夜的区分。我好奇，这明显与摄影关联，是否之后那些情形的讨论起点。在某种程度上，这与法国导演弗朗索瓦·特吕弗的电影《美国之夜》●相似，其英文译为"日以作夜"。但我不大确定自己理解了白天黑夜的问题。但我的确很享受它。

葛明　《白天和黑夜或白天或黑夜》这一作品主要研究从诗歌到制作。我预先同时设定了一个大黑块和一节诗歌作为起点，而它们之间对应关系的出现是由于出现了"我"。因为"白天和黑夜"，可以扩展为"我的白天和黑夜"，或者是"我的白天或我的黑夜"，这样，"我"就与诗歌联系在一起了。此外，制作也需要"我"的介入，因为必须是"我"跟这个世界的"engagement"，如果没有讨论有无摩擦力的"engagement"，就谈不上制作。这样我们就寻找了诗歌和制作之间的一种联系，这是其一。

其二是关于白天和黑夜或白天或黑夜的解释。满校园的绿树，并没有呈现时间，当校园里出现黑块的时候，突然形成了一组照片，时间的感觉也突然发生了变化，那个绿树也好像黑白照片上染过的色。白天和黑夜或白天或黑夜在这里其实是对照片中所拍事情已经死亡这一事实所做的混淆，照片中呈现的"死亡"就是过去在那而现在又不在那的一个证据。那么白天和黑夜或白天或黑夜实际上是一种假发，它只有在世界作为照片的时候才有意义。如果需要破题，那么必须对它们不是做通常意义上的时间的考虑，才会真正出现时间。

● 如海德格尔所言的"历史性"

● *Nuit Américaine*

最后是关于里希特的画。我特别喜欢里希特的画，里希特的画和安迪·沃霍尔的画是当代艺术世界的两个坐标。我常常想其他的艺术门类有没有可能达到这种画的状态。在做这个作品的时候，我还经常想到德国导演维姆·文德斯拍的《柏林苍穹下》，它主要是用一组组黑白照片构成的一部片子。但这些照片是虚假的，因为假设了天使的存在，才能够让各种黑白的图景同时地出现。那么如何做一个像照片一样的结构体，而不用借助于天使，这是我思考的问题。里希特主要画照片，就是说他直接面向的是"死亡"，这种方式对我有启发。

就这个作品来说，确实还不够集中，似乎主题过多。如果辩解一下，多有一部分是为了教学，如果是我纯粹自己的作品，可能不会那么多。反过来，因为是教学，所以就会有不同的"我"出现，所以我享受这个作品。

卢永毅 葛明老师很多年在孜孜探索设计教学。我理解今天展示的这个课程案例，仍是他要以他的方式引导学生如何在思考、概念和制作之间建立一个桥梁。但我比较习惯于要放在中国现代建筑学发展的历史中去认识他所迈出的这一新的步伐。大家都清楚，回顾我国建筑设计教学的历程，最初，我们采用的是布扎式●的教学，后来我们批判布杂，树立现代建筑的设计原则与方法，当然，现在又反思，我们所经历的现代建筑的发展作为一种形式目标明显，而探索设计方法的形成是不足的。现在葛老师做的这个探索，含有非常明显的意图，就是要超越现代建筑的种种观念和设计方法。我想说，能不能从这样一个历史视角更明确你的教学目标与方法，即，你为设计教学建立了一种什么东西，是我们过去没有的？进而的一个问题是，从布扎到现代主义，比如包豪斯的设计教学思想，甚至到现在一些新的探索，如葛老师的工作，我们总是说受西方的影响非常大，而在我初步印象里，你今天这个将概念和形式建立联系的思维和操作训练，似乎相当西方，你自己怎么看？最后是，不管怎样，从布扎到包豪斯，设计训练仍然立足于建立一种大家都共识的方法，而你现在的方式是不是很个人？如果是这样的话，你是否对设计教学有一种新的认识？你跟学生之间的交流是一种什么样的关系？

葛明 谢谢。我的设计教学现在有三个方向：一是基础教学，二是概念建筑教学，三是园林设计教学，三个一起才更表明我的教学思考。

从本质上来说，我是一个当代主义者，为什么这样说？我有一个参考的坐标——海杜克。在我刚开始设计教学时，非常希望能像他一样来教学生，但后来对他有了一点自己的理解，觉得他的教学主要是一种关于"construction"的教学，他用这个建筑的基本词汇来连缀制图、建造物、诗歌等等，我称之为"铆钉的建筑学"。可以发现在他的制作和制图上都有"construction"的特征，以此完成在不同设计媒介中的翻译。我有点好奇：他在纽约，纽约是当代艺术的大本营之一，当代艺术为什么对他好像没有特别的影响？海杜克的"construction"与他早期的九宫格练习有关，他后期虽然完全不再做这件事情了，但他似乎试图让建筑教育在形式操作的基础上，加入一个特殊的"天使"来加以改进。在我看来，他非常注重各种媒介翻译的方式，而我希望概念建筑能否不完全依赖这些翻译，直接探讨怎么让"当代"进入到设计教学中去。

第二点关于教学方法。我在训练学生的时候，几乎像一个人办了一个非常小的学校，设计虽然只有短短八周，但是训练提前了半年。他们从制作、诗歌、思考一起开始进入这个状态。学生的能力是无止境的，他们从一无所知到能很快进入讨论类似罗兰·巴特《明室》的状态。那我用的方式是什么呢，就是让他们进入制作，进入针对复制的制作。

刘东洋 我插几句，不是问题也不是提问。我觉得这个"概念建筑"的东西如果放在"后现代建筑"的语境里会比较有意思。当后现代建筑走到高度符号化阶段，就快走到穷途末路了。那种建筑的叙述太干瘪。而海杜克的作品却通过诗歌走上另外一条道路。读他的作品介绍很有意思，好像每一页里都有一个小故事，而这个故事往往跟当时的社会没有太大的直接关系，或者说很寓言化。我当时看那些喜欢海杜克的加拿大学生做模型，做得都很诗意。那些模型就介于足尺建筑和虚拟设计之间。我曾问他们的指导老师，这些学生做这些东西有什么用？那位老师回答说，我们培养学生就是从小东西做起，使得你在某种程度上能够熟练地从思想、理念进入操作，再返回来修改设计。这些小东西既具有乌托邦性又具有真实性，在操作层面上，它们是很好的培养学生认识从设计到建造这一过程的手段。

说回葛明老师指导的这些学生作品。我假如什么都不读，没有您的介绍，我直接去看，还是会看到这些东西身上有着某种物体性和仪式感的，那么隆重，像电影里的一个场景。我忽然就想到，这不正是刚刚邹晖老师讲的记忆剧场嘛！●葛老师把"记忆剧场"变成了一个盒子，有了好多层的叠加。也许，像卡森斯刚说的那样，拿掉一些东西，记忆会清晰一点点，当然不是说，清晰到直白。

葛明 做一个小小回应。在我看来，我们的制作跟北美的很多制作不同。他们通常认为概念建筑是一个未完成物或是一个替代品，因此他们的制

● Beaux-Arts

● Memory Theatre

作其实是关于一个大模型的建造，而不是制作。他们也经常提到叙事和制作，但其实多以叙事为主。

刘东洋　整个过程有没有这样一个设计制图的过程，有很好的草图、方案图？

葛明　我讲的制作，是在制作过程中追求像摄影术一样的制作，是对复制技术出现之后进行反思的制作。它与通常的建筑学方式也不太一样，举个例子，我们需要设计图，但有时更重要的是测绘图，因为你做了一半的时候，下一步的精确必须每天测一点再做一点。

此外，我一直坚持以身体来衡量来我们的"结构体"，但不是让"结构体"因身体的尺度而更像一个建筑，而是因为身体的介入，便于思考"我"和这个世界的"engagement"，而这才是重要的。

邹晖　根据葛老师的案例研究，我补充一两点自己的理解。关于海杜克的诗歌、绘画和草图的关系，包括他经常表达的天使主题，我们不能孤立地将它们看成是和阅读者之间的形式阅读关系，而是要进入他的作品的语境和背景中去理解。海杜克这个人本身是充满宗教激情的，你可以从他诗歌的字里行间感悟得出来，包括他画的对某个场所或某个故事理解的草图。他带着心灵去旅行，从与历史的对话中寻找精神的家园。他对文化的理解是通过图示和诗歌解释出来的，而不是把自己关在房子里面想象出某个陌生的构图。思与诗充满了他的想象过程。理解他的作品与历史的关系非常重要。在我的报告里，我试图表达的"情感"因素也可以被理解为建筑师的信仰。如果情感的建筑是可能的话，它就是设计师信仰的体现。你没有这种情感的投入，就没有信仰，这种情感的表述在建筑的方式里是很独特的。也就是说，建筑装置中的身体不是中性的，构造和身体有情感的联系，无论这个情感是批判的还是浪漫的。第二个是葛老师引用的匡溪工作室。同样是在1980年代，与海杜克的库柏联合学院一样，匡溪学生做的装置作品是非常个性化的，很少有团队设计。一个学生要拿到硕士学位，要花一年的时间做这个毕业作品。不管你如何开始，如何进行，最后的作品是必须有相当的理论深度的。没有理论深度的制作是失败的，这让我们想起海德格尔谈到的古希腊概念"techné"，意指诗意的制作。匡溪教师霍夫曼自己的作品也是一样，他仔细地阅读胡塞尔，他的写作文本里充满了对现象学的理解。你会清晰地感到他是通过阅读来打开制作过程的，这个制作的理论语境至关重要。另外，我觉得这个报告中的装置制作过程可以更仔细地考虑如何与现实的反思作用，包括这里边材料细节的决定。就是说，如何使制作过程深入地理论化，以达到对现实的诗意的抵抗。这并非意味着接受某个预置理论，然后花时间制作一个构造，让它作为理论的表象。意义是在过程里面，你怎么记录整个制作过程，包括怎么开始打第一颗铆钉或决定打第一颗铆钉的时候，这些细节都需要同时展开的阅读，沉思与写作来理论化。思与手工关系的最高境界，出现在米开朗琪罗的诗歌中所描述的某种超验的力量支配着工具对材料的令人着迷的加工。

莱瑟巴罗　我将尽量简洁。我觉得最有意思的是社会契约的想法与黑色表皮的结合。关于九宫格——一块黑色的地面和三维的白色方形格子——的一个问题发生在学生操作的时候，其体现的是个人的喜好和偏向，我做学生那会就是如此。这就像曝光外界用的相纸。我想将方案的黑色表皮和中国园林中的池子或湖泊的黑色水面抑或室内的黑色用法以及黑色的人工光做比较。每个例子中表皮都赋予事物特性。我觉得最有意思的是建筑有种容器的感觉，接纳自身之外的事物，黑色内部或黑色表皮能够收纳世界，将其变得可见。我认为这是一件有价值的事情。对此我的感觉是，内部、建筑和表皮进入到了与彼此的契约中，世界能提供什么便收纳什么。在论坛中我们多次谈至此问题：这由年轻的中国学生提出，关于公共，关于共享的问题，关于允许每个人对表皮产生影响、将所有人牵连起来的条件。就此，建筑的沉默引向了某种有影响的力量，无论是一种声音还是一种面向。这就是我对该作品的解读，即社会契约问题的一种希望或潜力。这就像身处机场的体验：你在关口说道"不需报关"。这就是我从黑想到的，不做任何声明，因此产生了一种能力或接纳性。黑既抗拒又接纳。暴露在外却又抗拒外界，这个关于地形学表面的观点，是可能的社会契约的一条线索。这就是我对黑之所想。

场景肆
落幕

SCENE I
PROJECTION

ENTER

SHIQIAO Thank you, Mark. Let us open up for questions.

DONGYANG Thank you, Mark. It's like going through therapy and also a bit comforting to hear that we can get pleasure through this ordeal. Indeed it's very problematic at least from what I know in my own practice. I have read a little Sigmund Freud. I wonder whether Freud ever elaborated on the types of constraints. It seems that there are varieties of constraints, sweet or sour you name it. For example, pressure and anxiety may come from repetitious demand for changing design. Or it can also come from deep down ethical challenges. For example, back in the 1980s and 1990s, bigger and wider roads were regarded in China as a symbol of modernity, and they had been planned and built not just solely to anticipate car traffic, since the majority families then were not able to buy cars. In my studies in Canada though, my transportation teacher had empathetically told us not to rely on one or two wide roads to solve transportation problems, narrower roads but denser road networks appear far better. Of course, a decent public transit service is the key. But in my work here, I have to constantly battle with those planning authorities regarding roads, and I have to surrender or compromise again and again, even when I know a plan like that will be bound to fail. You said that there's no cure for that, which makes me even more anxious now.

COUSINS I think you should give to the planning authorities one of the last essays by Lenin, which is called 'Better, Smaller but Better'. It's very funny. It's by Lenin but is quite conservative. Your example of the roads makes one realise that this problematic covers a lot of urbanism. Somehow our enemy is the one which was neither really discussed under topography, nor what we discussed under the question of movement. Our enemy is really circulation. Somehow the history of circulation came to be one in which the tail is wagging the dog. Let me give an example. At the Architectural Association, we have been very fortunate to expand the premises we have. So instead of one building in Bedford Square, we have about five, all next to each other. Last year, in came people who called themselves 'space planners'. They came in to do research which was preliminary to having an architect propose a way of using these five buildings together. I think the students and staff frankly were somewhat insulted by the firm of space planners. All they seemed to do was to measure who was going where in the building. I asked them after a week, 'what exactly were you trying to understand?' 'Well', they said, 'obviously we have to start with the problem of circulation in the building'. I said, 'Why?' The woman became slightly offended and said, 'Because that's the way we always start.' In her mind you know you start with the circulation and you move on. I said, 'Circulation is not a problem here.' I wanted to make the point that these 18th century buildings were really not an airport, though perhaps when they are finished with it, it may feel like one. They are the avant-garde of the corridor. They see life as one long vista of corridors. Actually we also had fire regulation inspectors at the AA. They said in a very generous spirit that they wouldn't insist on their new regulations yet with us because we belong to a transitional phase. The documents were almost worthy of Franz Kafka. We are moving to a policy of the "sterile corridor". It's very difficult to understand how these things originate and difficult to find a way to oppose them. We have a large front door at the AA, and in obedience to the fire regulations, over the door, we have to put an ugly sign saying 'fire exit', as though you might not realise that it's a door.

LEATHERBARROW I have a couple of observations that would take me a long time, too long to formulate into a question. But they may be useful observations. What I found striking and would like to hear more about is the opening distinction between subject and object. My recollection of the second point — the economic dimension of the subject — suggests to me that the opening polarisation needs to be qualified. And I would think that part of the qualification interestingly bears upon what was said about topography, its historical character. Here I think I'm correct in recalling a polarity between the archaeology of the subject and its economic unconsciousness. I believe this has a link to the opening question about the ethical dimension of our anxiety, because in a sense it is also ostensibly something external to the subject, our involvement with others that adds to our anxiety. If that's the case, we need to think again about that opening polarity. I think you're right in saying that for the mirror or mirrored subject, not only the past as present, but the present as investment in what's to come is part of the topographical condition. So it would be the *externalities* that are part of the subject, part of its constitution. Further, they help us not to overcome but work within that anxiety, in so far as it contains within itself, and open up to possibilities that are not yet foreshadowed by past articulations.

COUSINS Absolutely. To say it, certainly is not to Freud, to continue to reproduce a kind of separation between the subject and the object, or you might say between the individual and the culture, or in another sense the inside and the outside. The way in which those operate as a kind of overlaying system is made redundant by Freud. The little human animal only becomes the subject by virtue of its identification with others, initially with the mother. That remains a relation which is simply a model of identification, self identification. Freud would be quite happy with the formulation of 'Je est un autre', I is another. Therefore there is certainly no clear dividing. Indeed at a certain level it will be, non-Freudian to say, that everybody has their own unconscious, non-localised, but part of the person. Yes. I agree the subject is utterly tied both to a dependence on the other and to the relation with the other. Indeed you might say the subject is nothing but the response to the other. I think what is important at this point, the argument this morning, is whether there is a new form of practice, which I'm loosely calling the digital or parametric. It's not that I'm concerned in detail with the computer which involves a certain use of data in form of incorporation, rather that I am more interested in defending my test on, or my test to be able to get to that point and to ask the question: is this happening? Is this new—I mean new in a generalised sense? And what are the consequences?

LEATHERBARROW Well, certainly in my experience I see a large measure of naïve optimism in the types of contemporary practice that quite happily leave the ethical subject out of the discussion. These practices allow the project to develop by virtue of the data and algorithms that generate solutions on their own. That seems to me fairly typical, whether it's in London or Philadelphia or Los Angeles or perhaps Shanghai, we have nameless agents that bear no responsibilities for the pleasure of their own productivity.

SCENE I
EXEUNT

SCENE II
THE FIGURE AS HOME

ENTER

HUI Some of the contemporary artists' works which you included demonstrate a strong phenomenon of irony. I wonder how this artistic phenomenon can be essentially related to Chinese contemporary culture. Where can this ironic criticality lead us in the future? Is it just a temporary strategic approach necessary for a certain period of history in China?

SHIQIAO In an optimistic frame of mind, I used to think that it is temporary. But this is something that seems to surface again and again in different forms in different times. I think for a healthy development of culture, we need a dimension of criticality; not necessarily of the Greek kind but possibly something else. I use these artists, because I don't really see many examples in architects' works. I hope these amazing works of the artists will find parallel expressions in architecture in the future. What is most important is to keep this discussion going so that it isn't always just a complacent reenactment of returning home. The word nostalgia almost sounds like a disease in English, perhaps in indication of the high degree of strangeness to want to return home. In China, I believe returning home does not mean nostalgia; nostalgia, as a pathological condition, has never existed, does not exist and will not exist, because actually there is no home to return to when you're at home. So how do we think about being at home and being critical?

LEATHERBARROW I don't have an answer for that question. But I do have a comment. It seems sensible, starting with the still stone, and recalling what you said in passing about translation, and what I see as the pleasurable struggle of this conference between two languages. When the political writer Hannah Arendt returned to Germany after the Second World War, she was asked what she thought of her country. She said, 'the country I knew is gone.' The interlocutor said, 'Completely?' She said, 'no, the language survives'. My comment is that translation or what Mark introduced as interpretation is what always constitutes the threshold. To make another comparison: in antiquity Greek women were responsible for teaching language, generally through song. Language learned in this way was then spoken among others in the public realm. This meant to be in public meant to speak,

5月29日
29TH MAY, 2011

to take a stand through language. Accordingly, translation is necessary for entry into the public realm, one transcends domestic into public communication, overcoming the limitations of family conversation. I believe this relates to something Hunt said about public speaking. I would call it rhetoric as philosophy. It's not the truth. It's the truth in the circumstances. I'm not sure that I'm able to think through and beyond the basic premise that we are bound together if we struggle to say what we see and think. Surely that's the pleasure of being among others: to think again about what one thought one knew at home. So in my sense there is still something to be learnt from all those articulations. The expressivity of the world is thrown into shadow by what I would call the modern conception of home as self-sufficient in nature. For Sverre Fehn the home was part of the natural world. There is no way to separate from it and yet it is also the place for language and culture. So it couldn't be detached from the town. In the West, we have reached for private life so much that we think we don't need the town. I'm not sure if that's different here. But my sense is what you said nicely about language, about characters, their figurative potential and their requirements for deciphering, I believe the key is somewhere in translation. I think that's the comfort that we should live in. That's why I understand this struggle with languages today is pretty interesting in the mixed communications.

SHIQIAO Thank you for the comment. Perhaps I could just respond to that, if it needs any response, by stressing that language is the first home and a linguistic strategy contains a way of making home and homeliness. In a way, the relationship between meaning and alphabets is different from that between meaning and a graphic unit in the form of a Chinese character. Here what I try to put forward is a condition, linguistically made possible by the empire of figures, of a perpetual home. And this is put forward in contrast with a different condition where the conception of the public is possible. This is a conception of a political life, the political life that was so fundamentally important to Aristotle when he said that, when there is conflict, our loyalty is not to our friends at home but to this public life. To me it is a fundamental issue when we think about translations between Chinese and Western languages. It's more important than it appears at first glance.

ANDONG I just want to ask you a small technical question. Why did you use the character, 'Figure',• instead of another more authoritative character in Chinese like 'Wen'?° The character Wen originates from pattern. I feel something different between these two terms. The Figure is a sort of outcome of figuration of the chosen plan, and the Wen, the texture, is something different. It implies the relationships between those figures and the relationship between the figure and how it is applied to something. When you show the photograph of the mountain cliff, where the figures are applied, it is actually for me a pattern. It's quite like a process of gaining power. I always have a suspicion that a Chinese garden is quite like a process of giving pattern to topography and to nature. It's nature with pattern on it. But that pattern concerns a lot of the unsettled relationships between figures and how the figures can tell. I wonder why you chose Figure, because the term Figure is not frequently used in Chinese.

SHIQIAO My intention is to begin with the simplest unit. I think Wen is slightly more complex than Xiang. Wen is made of many Xiang. When you begin to talk about Wen, it's already far too complex for my purpose. Xiang in a way allows me to enter into both a literary dimension and a visual dimension because this is what the Chinese language is crucially about: the literary and the visual being blended together. It has never conceived the difference between the literary and the visual.

COUSINS The comment reminds me of two terms. Your idea of the Chinese garden as a form of sort of landscape tattooing, which will make you popular in California but perhaps less popular in China. In terms of your making of calligraphy the apex of the graphic, it seems to me that one way of extending that is to investigate the difference between English calligraphy and Chinese. It is not so much about the figure but about the ground. Even about the page. Is there for you a difference between the ground of calligraphy and the ground of a painting? I mean in the case of painting, the ground is experienced as space in some sense. I'm not sure how you characterise the ground in Chinese calligraphy, because it is obviously different from script. This may be one way to extend your suggestion.

SHIQIAO The strategy of placing things on paper has always had a kind of yin and yang relationship which is simultaneously conceived. It's

• *Xiang*
○ texture or literacy

always already part of the figure; the ground is already part of the figure, and they do not strategise a 'binary opposition'. No self-respecting painter or calligrapher would say that the whiteness is not part of the painting or the calligraphy.

SCENE II
EXEUNT

SCENE III
A NARRATIVE STRUCTURE OF
CROSS-CULTURAL ARCHITECTURE

ENTER

ANDONG I have some doubts, because in the end, some of the projects you have shown seem to suggest to me that: if you want to preserve or keep some memory, you start to build a theatre. I also doubt the use of the term 'labyrinth' in a Chinese context. Labyrinth seems to me not about being lost, but about the organisation of getting lost in space. It is the specialisation of something.• It is a 'lost' just like suddenly being transformed in other ways. That is a pure lost. That's not an attempt of organising something, not an attempt of control. I think the reason why labyrinth was used as a kind of origin of modern architecture is because it is a strong metaphor of organisation. I don't think that we can have any equivalent sayings in Chinese texts and images. The extraordinariness actually shows something very alien. It is evidence of European influence in a Chinese context.

HUI I need to further clarify the concept of labyrinth in cultural differences. If there is a similar term in Chinese tradition, it might be 'migong'.° I don't think the migong can be completely identifiable with the Western concept of labyrinth. In my recent research on the migong in the Chinese building tradition, I attempted to glean some historical clues through studying the literary cases of migong related to the art of building, including gardens. From those cases, we can identify some prototypes to understand the migong in Chinese tradition. My primary intention for tracing the history of migong is to seek the theatrical space as the common ground where cultural encounters can take place as witnessed during the 18th century. So is the 'theatre of memory', a concept borrowed from Giulio Camillo in 16th century Venice. The theatricality and engagement of history implied by this concept resonates well with the main strength of the Chinese building tradition. A sense of comparison is my starting point, which might help open up a perspective for reviewing what value has been lost and how we can preserve the core of Chinese culture through architecture.

COUSINS You have to be careful about what sense of memory you're using. Yesterday we talked about one sense of memory. But in what you are talking about, especially in the art of memory, it is a technique of how to remember. In this sense we talk about an art of memo-techniques. The theatre of memory, in a way, one could call it the theatre of our kind or something. It's about trying to have a structure which on the one hand contains all knowledge and at the same time it let a life somehow manipulate that. It's not like Wikipedia. It's to be able in some mysterious way to connect to references.

LEATHERBARROW I want to say that I rather like the last building and found it can grow into and could express things that you have implied in the historical material much more than Steven Holl's, which I find overstated, over-articulated, too smart, and just the opposite of what I think you are looking for. Why? The broken bridge, the route toward the centre hasn't been accomplished. We know something is there but not yet. The no longer or not yet, the sense of implication, the need for exploration, the acceptance of incompleteness, in other words, they were the strings of the final project. They were reluctant to fully articulate the sentences because we are not capable as a culture of completing them. I actually think they do hold together, but not as memo-techniques, an archive of imagery, but rather memory as something, as I said earlier, not in one's possession; only an orientation of one's longing. It is a desire for that other side that the building seems to suggest by virtue of the string. I would hold on to that—it's a frank recognition of our time, the translation of the incompleteness, if sense that needs to be thought through. The sense isn't guaranteed at the other side of the project, which is why it invites so many desires, as Mark suggested at the opening of the conference. Someone lonely and frail is still looking for a town that might be formed. I believe that's the opposite of what we have done with all of the knowledge.

DONGYANG I think I could follow the main line of your thoughts rather clearly, but I couldn't quite get what you are suggesting for. It seems

• In the Chinese context
○ 迷宫

that you are referring to the Memory Theatre as an archetypal experience of building-act, in a universal sense, both in the West and China. And that is why we build again and again so that the memory can come back through the act of building, whether in Chongqing or Beijing? Do I understand this correctly?

HUI We cannot understand comprehensively the status quo without drawing the critical perspective of history which Martin Heidegger called 'historicity.' If we judge the architectural reality and seek for criticality in judgment, it is necessary to establish a historical horizon which we usually call 'li shi gan' in Chinese.• When we engage in history, value judgment is always involved. That means we have to be selective and need to perceive the fundamental issues through a theoretical approach. Why do we repeatedly talk about some buildings in history while intentionally forgetting many others? Through searching for the historical line of the 'theatre of memory', we can bring to light the fundamental meaning of architecture as theatrical space. The cases of 'theatre of memory' in my presentation all demonstrate, in my view, the interweaving of physical and metaphysical which can be taken for a metaphorical understanding of the embodied meaning of architecture. These cases are the historical moments or clues which are suppressed by the Cartesian dualism. Engagement into these theatrical fragments is absolutely different from the typological study of architectural forms during the modern ages. These clues are real but hidden in history. Once being engaged through a comparative interpretation between East and West, they begin to help us understand what architecture can still offer for poetical dwelling in Heidegger's sense while facing serious cultural confusions in globalisation.

SCENE III
EXEUNT

SCENE IV
BLACK

ENTER

WEI Although I've been working at Southeast University, this is my first time observing the full expression of this project. I think today's arrangement is very interesting. It has something to do with yesterday's discussion. The two lectures delivered this morning are very inspiring, as they both talked about whether a building or landscape can be a space for people to release their minds from a situation of predicament, which refers to a relationship between the society and the individual. I think there is a difference between East and West. More than ten years ago, I visited the museum designed by Peter Eisenman in Berlin, which I think is very dramatic. It was built to show the trauma that eastern and western Berliners had suffered when they were split apart. At that time, East and West Germany were just united, and the Berlin Wall was not yet fully removed. His revelation made me feel more painful, rather than relieved, just as the photograph Mr. Ge Ming showed. The photograph is static, but dramatic. It lets you stay in pain, making you sober, and awakening you. In a sense, this is a traditional approach of Western artforms. But in a Chinese art, the custom would be to let you forget your pain through the passing of time. A good example is a Chinese film named *To Live*.• It implies the fact that there is no point to life but living itself. The pain that the lead role suffers is very soft and enduring, and makes you realise that life is still worthwhile. These are two very different ways in thinking, understanding and organising space. I think a classical Chinese garden sets up a dynamic and fluid platform for people to digest and release their pent-up emotions. I have two questions in respect to the design made by Mr. Ming Ge. The design has attracted many students to be involved with it, including Tang Jingyin. Firstly I would like to ask Miss. Tang, what kind of sublimation did you have to work through working on the installation? Did you feel released? My second question is: what kind of new architectural knowledge and views you gained through making it?

COUSINS I have to start by saying I really enjoyed your talk. But it seems to me one of the problems in your presentation was simply that there was too much going on, too many dimensions of the project. In a way it was difficult to have a single clear response. On the one hand, you seem to want to transfer something that we normally talk about as an effect of photography, or the discussion of photography, like 'punctum', into something as if it were a three-dimensional real life form, whatever it is called. I think that is extremely interesting. Maybe just concentrate on that, because I think it was already very huge idea. I was not was sure about why you talked about Gerhard Richter's paintings. I

• literally, sense of history

• 1994

couldn't really understand what the relevance of Richter is to the project. So by the end of your presentation, I became rather confused what the central objective is. I think in a certain way it's a complexity of the moment that attempts to distract from the central issue of 'punctum'. Then I wasn't clear also about the distinction of day/night and night/day you used. I was wondering if it is a start for the late condition, which is obviously linked with photography. In a sense it would be what the French film director François Truffaut does in a film called *Nuit Américaine*,• which is translated into English as 'Day for Night'. But it wasn't quite clear enough to understand that question of day and night. But I really enjoyed it.

MING The work *Day & Night/Day/Night* mainly focuses on poetry and making. I took a large black object and a poem as a starting point and the corresponding relationship between them arises due to the emergence of 'I'. As 'day and night' can be read as 'my day and night' or 'my day or night', 'I' will then be connected to poetry. In addition, making needs engagement of 'I', engagement of 'I' and the world. If there were no engagements that referred to whether friction existed or not, making would be out of question. Hence we find the link between poetry and making. This is the first point.

The second is about articulation of *Day & Night/Day/Night*. The green trees in the campus did not make time a present consideration. However, when a black object appears in the campus, a frame of pictures, all of a sudden, is formed, which follows an abrupt change of time. The greenness is seemingly like the colour dyed in the black and white pictures. Here *Day & Night/Day/Night* is actually mixed up with the fact that what was recorded in the picture has died. The 'death' in the picture is evidence for what 'it was but it isn't now'. Therefore, in fact, *Day & Night/Day/Night* is the wig. It is nothing but meaningful when the world plays its role as a picture. If it needs interpretation, it should not be interpreted as time in a normal sense. Then time will actually come.

The final point is about Richter's painting. I possess special adoration regarding Richter's painting. Richter's and Andy Warhol's paintings are the most important ones in modern art. I have been thinking about whether other types of art could reach the same level that they achieve. Making of *Day & Night/Day/Night* often reminds me of the film *Wings of Desire*° directed by Wim Wenders. The film is mainly shot in a series of black and white photographs. But the photographs are illusory. It is the assumption of the existence of angels that allows the black and white scenes to appear simultaneously. The question is how to make a photography-like structure without the help of angels. This is what I thought about. And I got the inspiration from the way in which Richter draws pictures, i.e. the death that he faces directly.

In terms of the work of *Day & Night/Day/Night*, I agree that it seems to contain too many themes. If I have to make an excuse for it, it was made mainly for the purpose of teaching. If it was for my own work, it would be much more focused. However, it is out of teaching that a different 'I' could appear. And I do enjoy it.

YONGYI Mr. Ge Ming has been teaching design studio for many years. My understanding of his presentation is that he wants to guide the students to build a bridge between ideas and making in his own way. But I tend to see this as a new step in the development of Chinese modern architecture. As we all know, looking back to the development of architectural design teaching in China, in the very beginning, we adopted the Beaux-Arts system. Then we criticised the Beaux-Arts system and established the design principles and methods of modern architecture. Now when we rethink the whole process we have gone through, it seems that we take modern architecture more as a target for form-making than a way to explore design methods. What Mr. Ming Ge does have is a very clear goal, so to speak, to go beyond various concepts and methods of modern architecture. My first question is, can you say more about your teaching aim and method from a historical perspective? What have you created for design teaching that we did not have in the past? The second question is, from the Beaux-Arts to modernism, from Bauhaus ideas on design teaching to some new exploration of nowadays as Mr. Ge Ming did, we always say that we have been under the influence of Western ideas. And when you make a link between the concepts and form making, my first impression is that the idea and training you showed are quite 'Western'. What do you think of this? The last question is, architectural training in either the École des Beaux-Arts or Bauhaus seems to set up a universal design method. Is your approach very personal? How do you communicate with your students?

MING Thank you very much. My design teaching has three directions: basic training; conceptual architecture; and garden design. A com-

• American Night, 1973
o 1987

bination of the three shows my thoughts in design-teaching more clearly.

In essence, I consider myself a contemporary person. Why say this? A best example for me is John Hejduk. When I started design teaching, I had a wish to teach students in the way as he did. As time went on, I gained my own understanding of him. I think his teaching concentrates more on construction. He links the basic vocabulary of architecture with design, building, poetry etc., which I call 'Nail Architecture'. We can actually characterise his making and drawing by construction, through which different design media can be translated. I once wondered why contemporary art did not seem to make any effect upon him as he lived in New York, one of centres of contemporary art. Hejduk's construction has something to do with nine squares in his early teaching experience. Although he did not do it any more in his later time teaching, it seems that he attempted to make architectural education better on the basis of the operating form through introducing a peculiar 'angel'. In my opinion, he paid great attention to the methods of translation between different media. But I hope conceptual architecture can be sort of independent of these translations, and directly touch upon the question of how to bring 'contemporary' into design teaching.

As to my teaching method, it is almost like that I myself set up a small school when I start to train my students. Although the time for the design studio is only eight weeks, the actual training starts six months in advance. Then students get into the design after learning making, poetry and thinking together. Students' capacity is unlimited. Although they know nothing at the beginning, they can quickly get into a sort of state to discuss Roland Barthes' *La Chambre claire*. How do I teach them? Drive them into making, making that focuses on reproduction.

DONGYANG May I make a few comments, which are neither questions nor questionings? I tend to place 'Conceptual Architecture' within the 1980s postmodern context of North American architecture. I find it intriguing, particularly when seeing the dead end that pomo architecture ran into during that time with all its desiccated Façadism. Hejduk's projects proceed along a different path. His projects can be read almost like fictions, hardly related to the immediate world, or better phrased, being rather allegorical. Most students whom I knew in Canada, who followed Hejduk, often produced wonderful and poetic models. Those models were standing right between full-scale real buildings and virtual designs. Out of curiosity, I once asked their mentor why they devoted such energy to making models. He replied, 'We would rather teach students to fulfil their architectural dreams in smaller things. Models are one of them. Through model making, students get into the good habit of constantly shuttling in-between thinking and making. In addition, these smaller things being both utopian and real, are at a practical scale to get a design built.'

And back again to the students' installations instructed by Ge Ming. Even without your presentation or preliminary reading, even if I encountered them for the first time, their sense of being objects and rituals is obvious. In a way, they would remind me of some scenes within a film, because of their ceremonious characteristics. They are all of a sudden becoming a testimony to what Professor Zou Hui has just talked about — build-act as a Memory Theatre. Professor Ge Ming turns Memory Theatre into boxes. Perhaps, their memory would be all the more legible if they didn't have too many layers as Mark has suggested. But not to a degree of being too literal.

MING May I say something quickly? In my view, our making is quite different from those in North America. To a North American, it is usually believed that conceptual architecture should be an uncompleted object or a substitution. In this way, their making is, in fact, the construction of a large-scale model instead of making. They often mention narrative and making, but actually they pay more attention to narrative.

DONGYANG Did the making of these boxes contain such a process, which required a solid drawing mechanism, from sketches to working drawings?

MING The making that I want to present today is a method following the way of photography in its process. It is a reflection of the emergence of the technology of reproduction. It is quite different from normal ways of architectural design. For example, design drawings are necessary, but sometimes survey is more important. As we are in a midway stage of making, the accuracy of the next step must depend upon what we have made. So every day we make a survey, and then do a bit of making.

In addition, I have insisted that the body is used to measure the 'structure'. It is not to make the 'structure' more like a building because of the scale of body, but more important to think of 'I' and our engagement in the world because of the intervention of body.

HUI Following Ge Ming's case studies in his presentation, I want to supplement a few points on architectural installation. Regarding the interactive relationship of poetry, paintings and drawings in John Hejduk's design works, including his frequent topic of angels, we cannot infer them into a relationship between the writer and audience on mere forms, but rather enter the semantic depth of his works. He is a person full of religious passion, which is demonstrated in his poetry and drawings related to a certain place and narrative. He travelled with all his heart, seeking the spiritual home through dialoging with history. His understanding of cultures is interpreted through his works, rather than acting as a hermit creating illusionary strange compositions. Critical thinking and poetics flow through his imagination, and to understand the innate relationship between his works and the revived history is crucial. In my presentation, the topic of 'emotion' can also be understood as the architect's faith. If emotional architecture is possible, it will be exactly the embodiment of the designer's faith. The projection of emotion into the world will lead to the individual's faith, and the expression of emotion through architecture is always unique. In this sense, the body in/of an installation is never neutral; there should be an emotional relationship between the construct and human body, whether the emotion is critical or romantic. The second reference is the Cranbrook Studio. Like Cooper Union under the leadership of Hejduk, the students' works at Cranbrook during the 1980s-90s were extremely individualised without popular team design. A student spent the whole year on his/her graduation work. No matter how a student completes the process, the final work must reach an expected theoretical depth without which the work would be a failure. This reminds us of the ancient Greek concept 'techne', which means poetic making, analysed by Heidegger. The former teacher at Cranbrook, Dan Hoffman, followed the same theoretical approach and read deeply into Edmund Husserl. His written texts engage in phenomenology and opened up his constructing. Understanding the philosophical background of his approach is crucial for understanding his installations. I thus feel that in your installation studio, more considerations can be put into how the work will act as a critical reflection on reality. In other words, it is about how to consistently theorise the making process in order to convey the poetical resistance. This does not mean the work is the re-presentation of an a priori idea, rather that meaning emerges within the making process. This requires a parallel movement between constructing and reading, thinking and writing. The ultimate relationship between thinking and crafting has been described in Michelangelo's poetry that a supernatural power actually guides the enchanted movement when the tool engages in the material.

LEATHERBARROW I will try to be brief. What I find most interesting is the combination of the idea of a social contract and the black surfaces. One thing about nine square problem—a black ground plane and three-dimensional white nine square grid—is when students worked at it, as I did when I studied, it exposed all the personal interests and tendencies of each individual. It was like photographic paper for the exposure of external conditions. I want to compare the black surface of the project with the black surface of a pond, or a lake in the Chinese garden, or the blackness of the interior, or black artificial light. In each case, surfaces invite qualification. What I found most interesting is the sense of architecture as a receptacle, for something other than itself, this capacity of the black interior or surface to accept the world, to bring it to visibility. I think it is a very positive thing. My sense of this is that interiors, buildings, and surfaces enter into a contract with one another to take from the world what it is able to give. We have come upon this issue many times in the conference: young Chinese students have asked about public ground, about things shared, about the conditions that would allow everyone to have an influence on the surface, bringing everyone together. In that sense the silence of the building is orientated towards some influential source, whether it's a sound or whether it's a direction. That is how I read this, as a promise or potential for social contract. It is like the experience of being at the airport: you say at the customs desk that you have 'nothing to declare'. That's why I thought of the blackness, nothing to declare, and for that reason, the offering of a capacity or receptivity. The black resists and absorbs. This notion of topographical surfaces, as exposure and resistance is a clue about a possible social contract. That is my sense of the black.

SCENE IV
EXEUNT

5月29日
29TH MAY, 2011

DOCUMENTS

文献

SCHEDULE

WED, 25TH MAY
星期三，5月25日

Shanghai Xian Dai Design Group
上海现代建筑设计集团

10:00 — 11:30 AM
《建筑实践与建筑教育》
马克·卡森斯
ARCHITECTURAL PRACTICE & ARCHITECTURAL EDUCATION
Mark Cousins

FRI, 27TH MAY
星期五，5月27日

主持：龚恺　李华
逸夫科技馆大报告厅，东南大学

Chaired by Gong Kai & Li Hua
YS&TB, Southeast University

10:00 — 10:30 AM
"当代建筑理论论坛"开幕
Opening Ceremony

10:30 — 11:45 AM
《艺术在何处？》
马克·卡森斯
WHERE IS ART?
Mark Cousins

2:00 — 3:15 PM
《风景如画的现代性》
约翰·狄克逊·亨特
THE MODERNITY OF THE PICTURESQUE
John Dixon Hunt

3:30 — 4:45 PM
《时间景观中的推移／移逝》
戴维·莱瑟巴罗
PASSAGES IN THE LANDSCAPE OF TIME
David Leatherbarrow

5:00 — 6:00 PM
"山水展"开幕，陈薇主持
前工院一楼展厅，东南大学

Opening of
the Shanshui Exhibition

Presented by Chen Wei
QGY, Southeast University

YS&TB
Lecture Hall at Yifu Science & Technology Building
逸夫科技馆大报告厅

QGY
Ground Floor, Qian Gong Yuan
前工院一楼展厅

EH/ MA
1st Floor, East Hall of the Main Auditorium
礼东二楼报告厅

日程安排

SAT, 28TH MAY
星期六，5月28日
地形学，王骏阳主持
礼东二楼报告厅，东南大学
Topography
Chaired by Wang Junyang
EH/MA, Southeast University

9:30 — 9:50 AM
《地形学是否建筑？》
戴维·莱瑟巴罗
IS LANDSCAPE ARCHITECTURE?
David Leatherbarrow

10:05 — 10:55 PM
《地形学，历史与土地的伸展／谎言》
约翰·狄克逊·亨特
TOPOGRAPHY, HISTORY & THE LIE OF THE LAND
John Dixon Hunt

11:10 — 11:50 AM
《从我家后窗望见的地形学》
刘东洋
A VIEW OF TOPOGRAPHY FROM MY REAR WINDOW
Liu Dongyang

2:00 — 2:40 PM
《地形与形胜》
陈薇
TERRAIN & SCENERY
Chen Wei

2:50 — 3:30 PM
《从〈烟江叠嶂图〉看中国山水画》
林海钟
MISTY RIVER, LAYERED PEAKS: A CONCEPTIOM OF CHINESE
Lin Haizhong

3:40 — 4:20 PM
《建造一个与自然相似的世界》
王澍
BUILD A WORLD TO RESEMBLE NATURE
Wang Shu

SUN, 29TH MAY
星期日，5月29日
心理空间，李士桥主持
礼东二楼报告厅，东南大学
Mental Space
Chaired by Li Shiqiao
EH/MA, Southeast University

9:00 — 9:50 AM
《投射》
马克·卡森斯
PROJECTION
Mark Cousins

10:05 — 10:45 PM
《象与家》
李士桥
THE FIGURE AS HOME
Li Shiqiao

2:00 — 2:40 PM
《交叉文化建筑的情节结构》
邹晖
A NARRATIVE STRUCTURE OF CROSS-CULTURAL ARCHITECTURE
Zhou Hui

2:50 — 3:30 PM
《黑》
葛明
BLACK
Ge Ming

WHERE IS ART?
Mark Cousins

Thank you for inviting me to give a talk. What I want to talk about is something which is difficult to identify and to define, but I think is important. My second point is an apology in a way, that parts of what I'm going to say depend upon the analysis of certain words in English. One way of introducing my topic, which I entitled 'Where Is Art' is to think that it sounds a ridiculous question. It seems at first to make no more sense than asking 'how tall is poetry' or 'what size is music'. And certainly it is not my intention to suggest the arts, what would be called the fine arts, have a location in the traditional sense. But I do want to argue that the arts each have a form of being located, a kind of technology of location which I'll try to identify.

Before I try to describe the argument of my paper I have to say that in some sense it is a paper that is written as a criticism of something else. Something that has become enormously popular in both the United States and in Europe, which is the enormous celebration of curation, and of the curator. It is almost as if curating has been seen as the very ground on which art traditionally operates. Today is not the time to try to explain this phenomenon. Personally I think it would be useful to think that the rise of the idea of curation almost mirrors the rise of the property boom in America. It is also clearly the period of the rise of a number of international biennales, which increasingly centre themselves upon the issue of the curation as a vehicle to lend some distinctiveness to the biennale. And so the term *exhibition* is used more often than the term *art*. The curator now almost takes on the role of an artist and considers artworks as the materials for the art of curation. This ends up becoming a new bureaucracy of art. But, curation is not my topic today. I want to talk about what curation does not understand, which is that the art object has already gone through a process of being presented in the world and that this process happens before. This is what I'm interested to identify today.

Let me put this in the form of a proposition. Let us assume that what an artist does is in some sense to make a work of art. That, I will be arguing, is not enough. In order to become part of the world, the artwork has to be presented to the world. I put it very clearly so that at least you can remember my argument. It doesn't make me right. Let us say that this process of presentation is the way in which the artwork becomes an art object.

Let me try to clarify this term 'presentation'. It is not a philosophical term. I'm not concerned with the famous philosophical problems of the presence of being. Indeed, I think that the use of philosophical category comes too soon and tends to obscure the actual reality of this process. If we look at the English term presentation, it normally refers to something which is formal and definite. If someone is introduced to someone, it is called a presentation of them. It would normally be under certain prestigious social circumstances, so you would be, for example, 'presented' to the Queen. You don't just get introduced to her, you get presented. And the same occurs in other rather historically based situations. It is not, as it were, that you don't know who the Queen is. You are not surprised that she is this elderly lady. It is kind of an act, a social act. It has a relation towards what some philosophers call a speech act. So if I say I would like to 'present' to you so-and-so, I'm not really conveying knowledge. I'm engaging in a certain act.

艺术在何处？
马克·卡森斯

感谢你们邀请我来做这个演讲。首先，我想说明，我今天要谈的东西是难于界定的，但非常重要。其次，我要抱歉的是，我讲的一部分内容是基于英语中一些特定词汇的分析。我介绍主题的方法是把今天演讲的主题"艺术在何处"当成是一个听上去荒谬的问题。乍一听到这个问题，就像问"诗歌有多高"或者"音乐有多大"一样，不着边际。当然，我想要谈的并不是给艺术○一个传统意义上的位置，而是要探讨每种艺术都有其被定位的方式，我想要辨明的是一种关于定位的技术。

○ 或美术

在我试图描述我的观点之前，我需要说明，这篇文章在某种程度上是对其他一些事的批评。在美国和欧洲，有件事早已大行其道，这就是对于策展和策展人的大肆称颂。策展几乎已经被视为艺术一直运营的基础。今天不是我解释这个现象的时间，但就我个人来说，一个颇有助益的思考方式是，策展观念的兴起恰恰反映了美国房地产繁荣的兴起。显然，这也是大量的国际双年展兴起的时期。国际双年展越来越依托于策展，把策展视为突出双年展独特性的载体。因此，今天，展览这个术语比艺术这个术语更常用。策展人现在基本上扮演了艺术家的角色，将艺术作品看成是策展艺术的材料。其结果是，产生了一种新的艺术官僚体制。不过，策展不是我今天的主题。我想要谈的是策展无法理解的事，即艺术作品已经经历了一个在这个世界上呈现的过程，而这个过程发生在展览之前。这个过程才是我感兴趣并要详细说明的。

先让我们来预设一个前提：假设一个艺术家所做的工作，是制作一件艺术作品。我稍后将会阐明仅此是不够的。为了成为这个世界的一部分，艺术作品必须被呈现到这个世界上。我将它说得如此明确，是希望你们能记住我的论点，倒不是说它最确切地表达了我的意思。我们说，呈现的过程是艺术作品成为艺术物品的方式。●○

● 1
○ art work becomes an art object

在此，我要解释一下"呈现"的含义。●"呈现"不是一个哲学概念。我不关心那些著名的存在之在场的哲学问题。事实上，我认为太早地使用哲学范畴，会遮蔽呈现过程的实际情况。英语中"presentation"○这个词，通常是指某种正式明确的事物。假如某个人被介绍给另一个人，这叫做他们之间的呈现。这通常发生在一些特定的、显赫的社交场合，比如说，你将被"呈现"给女王。你不仅仅是被介绍给她，你要呈现自己。同样的情形发生在另一些很有历史传承的场合。这时，你不会不知道谁是女王。你不会感到惊奇，你知道女王就是个老妇人。这是一种表演，一种社交性质的表演。它有些像一些哲学家称之为的"演说"。因此，如果我说我要呈现给你什么什么，我不是真的想要传达信息，而是在进行某种特定的表演。

● 2
○ 呈现

当这涉及到艺术的问题，我的论点是，各种艺术都有某种特定的方式去

When it comes to the question of art, my argument is that artworks are themselves always presented in a particular way and each art has a different form of presentation. We can understand then presentation as a certain mechanism or a certain technology for transforming the artwork into an art object. Now of course some aspects of this topic are discussed. But it usually assumes that the mechanism of presentation is the same as what we would call the form or the medium of the work. But if that were true, a sculptor would have to reinvent sculpture each time they made a sculpture. Obviously artists understand the rules of the mechanism and therefore produce a work of art that will go through the mechanism.

I apologise for putting this in such an abstract way. So let me now turn to a concrete example. First of all, let me just invite you to look around this hall, just to notice what's in it, and how it is arranged. In some sense, this space would be recognisable at the point of the origin of Greek drama. Originally ancient Greek drama didn't yet have a theatre, but it had a certain and definite arrangement. It had a physical arrangement of the audience. It introduced another category called the chorus. It had an elementary wooden stage which is called the *skēnē*. And then perhaps it had some representations behind, usually of a door. But it also came with, and we must consider this as part of the arrangement, a number of what are normally called conventions. On the stage, they would be actors not in the modern sense. They would be in a way speaking the role of someone who the audience knew well, a hero of the past. The stage created a new space, which permitted dialogue, an exchange of words between the actors and the intervention of the chorus, the idea that actions did not take place on the stage, but that people would appear on stage and report them.

It is very difficult to recapture. It is important to understand the astonishing character of what in respect to drama I will insist is an invention. This invention involves elements which often are not acknowledged in the history of theatre. We tend to assume that the idea of drama is obvious and self-evident and so we neglect what an astonishing invention this was. Just to give one element of the invention is that the stage, the *skēnē* has to be considered as somewhere where you can represent anywhere except the stage itself. It represents anywhere except itself. My argument is designed to establish that this machinery of presentation is more than what people would call the medium or the form.

If we look at this hall, one major difference is, instead of having a representation of a door at the back, a door which Roland Barthes famously describes as this 'tragic door', we have a screen, and you can see that we have reversed the relationship of the inside and outside. In Greek tragedy it is the inside that you didn't see, now it is the outside which you see represented. We can understand that this is an astonishing invention but there's no inventor. I've initially taken Greek theatre as an example of what I'm calling the machinery of the technology of presentation. Now perhaps we can move on from Greek drama, by taking one word we've already met which is the *skēnē* and look at the strange role that it played in the English language by the word 'scene'. Afterwards perhaps you'll understand why perhaps I could call this the scene of art.

Let me present you with some of its uses in English. Here we obviously meet translation problems of the term 'scene'. Sometimes in English it is used to mean a small part of a play. Sometimes it is used in the word 'scenery' to indicate on the stage the representation of something like a building or sometimes a landscape, but the word 'scene' is also used though it changes its spelling. It's also used for the natural, and for what would be thought of as the natural beauty of the landscape. Sometimes the word

呈现它们自身，每种艺术都有不同的呈现形式。我们可以理解，呈现是把艺术作品转变为艺术物品的一种特定机制或技术。现在，当然这个话题的很多方面都在被讨论。但通常的假设是，呈现的机制与我们所称之为作品的形式或媒介是同一种东西。然而，如果真是这样，雕塑家每做一个雕塑，都不得不重新发明雕塑这门艺术。显然艺术家们了解这种机制的规则，并且因此制作出运用了这一机制的艺术作品。

我很抱歉，这样的说法相当抽象。所以，现在我来举一个实在的例子。首先，我请你们来环顾一下这个演讲大厅，注意一下里面都有些什么，它们是如何被安排的。在某种程度上，这个空间可以被认为是起源于古希腊的戏剧。最初，古希腊的戏剧并没有剧场，但是有特定的、明确的安排和组织。它在物质层面上，有一种对观众的安排，并引入了另一种艺术门类——合唱。它有一个简单的木质舞台，被称作"skēnē"。或许，在后面还有一些表现式的布景，通常是一扇门。但这不仅仅是一些物件，而是整个戏剧安排的一部分，与之相伴的，是我们通常所说的。传统或惯例。在舞台上演出的人不是 ○ 戏剧
现代意义上的演员。他们以某种方式讲述观众们熟知的角色，比如，一个历史上的英雄。舞台创造出一种新的空间，使对话、演员之间的语言交流以及合唱班之间互动得以发生。其理念是行动并不在在舞台上发生，而是演员登上舞台，向观众汇报这些行动。

要重温那种场景是很困难的。我坚信，重要的是去理解戏剧那不可思议的品质是一项发明。而这种发明结合了诸多戏剧史上名不见经传的要素。我们总以为戏剧性的概念是显而易见和不证自明的，所以常常忽视这是一项多么令人惊讶的发明。这项发明的一个要素就是舞台，而舞台被看成是可以表现除了舞台以外其他任何地方的一个所在。舞台可以再现任何地方，除了它自身。我在此想要说明的是，这种通过设计确立的呈现机制要远超出人们称之为媒介或形式的东西。

假如我们环顾这个大厅，就会发现它与希腊剧院的一个主要区别：舞台 ○ 主席台
后面没有画一扇门，一扇罗兰·巴特所描述的著名的"悲剧之门"。相反，这里有投影屏，你可以看到我们置换了内外之间的关系。在希腊悲剧中，这里是你看不见的内部，而现在是被呈现在你眼前的外部。在我们所知的范围里，这是一项没有发明者但令人惊讶的发明。前面我以希腊剧场为例，说明了我所说的呈现技术的机制。现在，或许我们可以从希腊戏剧这里继续我们的分析：取用我们已经遇到过的单词"skēnē"，看看它在英语中通过"scene"这个词所扮演的角色。之后或许你会明白为什么我会把这叫做艺术场景。

让我先说说"scene"这个词在英语中的一些用法。显然，我们这里会遇到"scene"这个词在翻译上的问题。有时，在英语里，这个词指一出戏中的一幕。有时它被用作"scenery"。中，来指在舞台上的一栋建筑或一种 ○ 布景
景观，虽然拼写有变化，但使用的是同一个词根。还有一种用法是指自然或者景观的自然美。有时，"scene"这个词意思是指很糟糕的行为，被称作

艺术在何处？
WHERE IS ART?

'scene' means behaving very badly, what is called 'making a scene'. When it is linked to 'scenery' in the theatre, it's obviously linked to words which are close to it, like scenography, which narrowly and traditionally meant the painting of perspective representations on the stage, but later on becomes part of a vague direction within an art form. So in film, you will often have a scenographer. But it is not clear of what the definition of that role is. I discovered to my great pleasure there is a specialised use in the 19th century of a scenite, which is someone who lives as a nomad. It is a dwelling on the move. You can readily see that the term 'scene' is what we might call conceptually a kind of nomadic term. It is the arrangement of something as a kind of mechanism, a kind of identity.

I think that there are central periods in art history and indeed in architecture, where in a sense those media had been crucially related to setting up the scene. I was recently in Mortara in Italy, where you see it time and again, a kind of competition between painting and architecture in a setting, not by designing a building but of creating a kind of scene. Indeed at the Architectural Association School of Architecture, I wish that the new discipline of landscape urbanism recognised that it could learn a great deal from the Renaissance, from someone like Giulio Romano, especially in this question of scene.

The term 'scene' we have seen is something of a nomad. It is as if it moved the crop of art, undermining any philosophical clarity. I would even like to suggest that one of its major functions is to undermine the philosophical certainties of the analysis of art. We might say that the enemy of the term 'scene' is that conceptual couple which are brought as 'reality' and 'representation' in a simple form while people think there is a reality, and there are representations of it, true or false, close or not very close. Discussion of the arts is not really helped at a banquet by the appearance of 'representation' and 'reality' at the table. 'Reality' presents itself as being the absentee landlord of the arts, and representation is the tax collector. When they appear at the table, representation and reality make a great deal of noise in the discussion but the art forms find them very boring and wish they would go home. Indeed the idea of art that it is or ought to be a representation of reality seems to me to have the question the wrong way round. An analysis of the scene, or the 'presentation' or mechanism of the presentation, actually determine or are a model for human behaviour.

Let us return to the example of Greek tragedies. It is clear in a sense that the act at the heart of Greek tragedy is a tension between traditional values and new ways of thinking about them. Everyone is agreed in some sense that the rise of drama in the form of all tragedy is closely connected to the experiment with democratic form in Athens. When I say democratic I really mean the central issues of how to take decisions and how to frame discussions about those decisions. There is no clear way to organise the discussion or indeed suggest the mechanism of decision making. Clearly the tragedies worked through many of the problems that will be based in the assembly, in the political assembly. Indeed it was not uncommon for actors or dramatists also to be famous politicians. We know that famous orators of the fifth century BCE were extremely knowledgeable about Greek drama. So you see here how the arts with this mechanism profoundly influence social life.

Let me give one additional small example, the anthropological research of the French anthropologist in the first half of the 20th century Marcel Mauss. It is called 'The Techniques of the Body'. In his paper, Mauss put forward the evidence which is, he said in about 1910, that young French women started walking entirely differently.

"making a scene"。当这个词联系到剧场中的"scenery"时，很明显它会与"scenography"这样一些拼写相近的词联系在一起。"scenography"最传统和狭义的用法是指仅指舞台上以透视绘制的场景，但后来变成了这种艺术形式中一个模糊的方向。因此，在电影中，经常会有一个"scenographer"。但是对这个角色的定义却并不明确。我很高兴地发现，在19世纪还有一种特别的用法"scenite"，指过着游牧生活的人。一种在移动中的起居生活。你现在应该马上可以看到"scene"这个词在概念上有一些游牧的意味。这指的是将事物组织在一起，形成某种机制、某种特征。

○ 大吵大闹
○ 舞台布景
○ 舞美，配景透视

○ 舞美师

我认为，在艺术史上，也同样在建筑史上，有些重要的时期，艺术手段及媒介是与创建场景紧密地联系在一起。我最近去了意大利的莫尔塔拉，在那里你会常常看到，绘画和处于场景之中的建筑交相辉映，而这种场景是通过创造一种景观而非设计一栋建筑而产生的。事实上，在"AA"，我希望景观都市学这门新学科能够认识到，它可以从文艺复兴、从诸如朱利奥·罗马诺等人那里学到很多东西，尤其是在创建场景这个问题上。

我们已经看到"scene"这个词是指某种游牧的东西。正因如此，它移动了艺术的取景框，破坏了哲学上的清晰。我甚至要说它的主要的功能之一就是破除了对艺术分析的哲学确定性。我们可能会说，"scene"这个词的敌人是一对概念术语——现实和再现。简单地来说，人们通常会认为这有一种现实，那儿有对于这种现实的一些再现，不管是真是假，是接近还是不太接近。如果我们把艺术讨论比作一场晚宴，"再现"和"现实"两位客人出现在桌边对讨论不会有什么真正的帮助。"现实"是作为艺术缺席的领主来呈现他自己的，而"再现"则是税金收取人的角色。当他们出现在桌边时，"再现"和"现实"会在讨论中争吵不休，但是"艺术形式们"会发现他们很无聊，希望他们早点儿回家。事实上，对我来说，讨论"艺术是否应该是对于现实的再现"这个问题好像有点儿南辕北辙。对于场景、呈现或是呈现机制的分析，实际上决定了或者就是人类行为的一个范式。

○ Reality & Representation

让我们再回到希腊悲剧的例子。显然，在某种程度上，希腊悲剧的表演核心就是传统价值观和新的思考方式之间产生的紧张冲突。在某种程度上，大家都会同意，悲剧形式的兴起与雅典的民主模式的试验有紧密的联系。当我说民主时，我实际上是指如何进行决策以及如何界定关于这些决定的讨论等中心议题。这里没有明确的组织讨论或建议决策的机制。很明显，悲剧解决了其中很多问题，这些问题是城邦市民大会、政治集会的基础。事实上，不论是对演员、剧作家还是著名的政治家，这两者都是相通的。我们都知道，公元前5世纪的著名演说家们的希腊戏剧知识是异常渊博的。由此，你可以看到艺术与呈现机制如何深刻地影响了社会生活。

• 3

让我来举一个小例子，这是20世纪前半叶，法国人类学家马塞尔·莫斯作的一个人类学研究。他的这项研究被称作"身体的技艺"。莫斯在他的论文中提出证据称，1910年左右的年轻法国女人的走路方式发生了完全的

His explanation of this change is the initial import of American silent movies. Suddenly a much freer way of walking which was characteristic of the States destroyed the traditional tiny steps that French women were trained to move with. And within two years, according to Mauss, a fundamental transformation of a simple bodily function occurred. I can see that my research has various problems involved in it. I haven't really been able to identify the clearest way of distinguishing between the form, the medium and the mechanism of presentation. But hopefully I should be able to resolve this. It is not going to come by formal definition. It is more being able to show the different functions of the form and difference from the mechanism of the presentation. With that I'm going to conclude.

改变。他的解释是这个变化是源于美国无声电影的引进。突然间，一种美国特色的、自由得多的走路方式，摧毁了法国女人过去被训练的传统小碎步的行走方式。根据莫斯的研究，在两年之内，一个简单的身体机能发生了重大的转变。我能看到我的研究在这方面还有很多问题。我还没有能够找到一个最清晰的方式，去辨别形式、媒介和呈现机制之间的区别。希望我今后能够解决这个问题。我要做的不是给出正式的定义，而是要能够说明形式的不同功能和呈现机制之间的差别。以此，作为我演讲的结束。

李真 译

译者注：
1 在这里，艺术作品"Art Work"强调的是艺术家在个人化的创作阶段的主体性，在这一阶段，艺术品只与艺术家有关系，且仅基于艺术家的主观感知。而艺术物品"Art Object"则强调的是当艺术品脱离创作阶段，进入公共领域后，被视作流通物品的客体性。
2 本文中涉及到的一个主要的英语单词"present"，作为动词，有呈现、汇报之意。由这个动词衍生出来的一对名词"presentation — representation"，通常分别被翻译分"呈现"和"再现"。"presentation — representation"是英语学术讨论中经常会出现的一对概念，它们之间的差别主要体现在它们与"reality"（现实或被表现之物）之间的关系。"presentation"强调以一种经过设计的机制尽可能还原现实或被表现之物的特质，其重心始终在被表现之物上。而"representation"则更多的是一种创造性方式的呈现现实，其重心往往不在于还原或展现原有的现实之物的特质，而强调呈现手法的创造性。
3 古希腊的悲剧是整个西方戏剧的起源，最早的希腊戏剧都是悲剧。演讲者在这里所说的希腊戏剧和希腊悲剧是一回事。

THE MODERNITY OF THE PICTURESQUE
John Dixon Hunt

We coin words and names to refer to and to indicate habits or ideas for which there existed no previous or useful term. Many of these were devised and introduced in relation to the visual arts, and Chinese too has many examples. One of these words coined in English is the term 'picturesque'. The term has enjoyed a long life; yet, more recently, it has succumbed to an ineffectual, bland and largely journalistic usage. Nonetheless, four elements of the picturesque are worth considering:

1. The relation of words and images
2. The role of movement of feet and the movement of mind when responding to landscape
3. The full meaning of 'nature' in landscape painting that insisted upon *human* nature;
4. The role of using landscape depiction and appreciation as a way of defining and authorising a sense of nationhood or local cultural identity

The English word 'picturesque' came into English in 1685 in the book *Painting Illustrated* where William Aglionby noted that some north Italian artists were 'working A la pittoresk, that is boldly'. The phrase emphasised how certain painters drew attention to the bold technique or practice of their *medium* — paint was worked vigorously on the canvas, or the draughtsman or engraver was concerned to use his *burrin* or engraving tool on the plate so that the printed version was also busy and bold. The phrase did not seem to indicate any need other than to point to an artist's technique; it had in fact no connection with landscape, though it certainly was used by artists who depicted landscape.

The term in Chinese that is closest to 'picturesque' is the characters for 'picture-like'. The term appeared very early in the first century CE, and initially referred only to human beauty; only in the early Sung did the term become associated with nature.•   • 1
But even then it never singled out the virtues of roughness, irregularity and variety, that came to characterise much pictorial definition in England.•   • 2

The picturesque was a matter not just of images, but of images *and* words. Paintings in the Renaissance and for centuries afterwards always invited a verbal response to their images, if they had not in the first place relied on words to initiate a subject. Picturesque landscapes, ruins or fragments* stimulated the mind and imagination to   * FIGURE 1
think of what had transpired there, and to do that it necessarily invoked words to explain such things. When the 20th century land artist, Robert Smithson, looked back to the 18th century picturesque theorist Uvedale Price and his Three Essays on the Picturesque, it was because it *explained* in words something in what was seen.

Images were not used alone or by themselves; the verbal accompaniment to the picture was also crucial: it took various forms — maybe simply as a caption for the image, sometimes dating it with time and place, but more often in the form of an accompanying text that explained elements of what the picture displayed, sometimes printed as letterpress, or even as a spoken commentary when looking at an image.• So   • 3
does Li Gefei in his famous essay on famous Luoyang gardens note that everything in

风景如画的现代性
约翰·迪克逊·亨特

对于那些没有既定或有效术语来指称的习俗或思想，我们时常创造一些新的词汇或名称来指代和表述它们。其中许多词汇的发明和引入都与视觉艺术有关。中文中也有许多这样的例子。英语中就有这样一个被创造出来的术语，叫做"风景如画"。这个术语已经沿用了很长时间，近来却沦于一种无意义、空洞、几乎新闻一样的使用方式。尽管如此，风景如画仍然有四个方面的特征是值得我们思考的： ○ Picturesque

1. 文字和图像的关系
2. 身体的运动和思想的运动在景观感受过程中的作用
3. 建立于人造自然基础之上的景观绘画中"自然"的完整含义
4. 景观的描绘与评鉴对界定和确立国家或地方文化特征的作用

1685年，威廉·阿格里昂比在著作《插图》中首次于英语世界使用了"风景如画"这个词。他在书中提到一些意大利北部的艺术家"正致力于一种自由活泼的"A la pittoresk"的创作方式"。这个词用于强调画家对活泼自然的技术和绘画媒介的关注——绘画被狂放地铺陈于画布上，或者用于指称绘图员或刻版师对使用制图或刻版工具以使印刷成品显得稠密和狂放的努力。表面上看，这个词仅仅是指艺术家的某种技术，与景观并无实质上的关联，尽管这种技术的确被描绘景观的艺术家所使用。

中文中最接近"风景如画"的词汇是"如画"。早在公元1世纪，这个词就已经出现，并仅仅用于指称人的美貌。直到北宋，这个词的内涵才扩展到自然的范畴。但即便在北宋，如画这个词也并未意指粗糙、不规则和多样变化的特质。而这些在英格兰却构成了某种图像性定义的特征。 ● 1

● 2

"风景如画"不仅仅是图像的问题，而涉及图像与文字的结合。文艺复兴及其之后的数个世纪中，即便绘画的主题最初不是由文字激发而来，也总会诱发语言对图像的反馈。风景如画的景观、废墟或断壁残垣激发人们思考和想像其中已然消逝的那些东西,*而这种思考和想像必然需要借助文字的解释。20世纪的大地艺术家罗伯特·史密森号召回溯18世纪风景如画理论家尤维达尔·普莱斯及其著作《风景如画三论》，正是因为它以文字解释了所见之物的某些内涵。 * 图1

图像往往不会单独使用，而以各种形式同时呈现的文字也同样重要：它们可能只是图像的标题，有时会注明时间和地点，但更多的时候是与画放在一起的用以解释画中所绘各种元素的相关文字，有时好似信笺，有时是一段画评。正如李格非在其著名的《洛阳园林记》中所说的那样，园林中的万物 ● 3

a garden, be it flowers, pavilions or meandering paths, 'all profound thought'.°   ° all (*display*) profound thought

There is a fundamental collaboration between word and image in China. Sometimes, in fact, the words seem to have been more important than an actual garden or an imagined landscape. Some accounts of gardens and landscapes had no visual element at all that we know of, like Wen Zhengming's 'Record of the Dwelling in the Mountains at Jade Maiden Pool', or the Ming hermit Liu Yuhua's 'Garden that is Not Around' in which his garden only existed in a literary account for which he drew on many earlier texts.• But for the well-known 'Garden of the Unsuccessful Politician' by Wen Zhengming there was an integrated album containing a prose narrative along with 31 poems and 31 distinct images.• So both words and images were a part of a larger understanding.

• 4
• 5

It is a matter of common knowledge how much Chinese landscapes were imaged with characters that draw out the fuller meaning of the image—maybe to say who owned the painting, the place depicted, or maybe by inscribing words on the image that draw out their fullest meaning.* Such words authorised an essential element of picturesque that is common to both Western and Chinese landscape—the theme of what we would call 'associationism'—i.e. what it was that places, or images, triggered in the mind, memory and imagination.

* FIGURE 2 & 3

The picturesque always emphasised *movement*, which is unavailable if we deal simply with photographs or static images: movement of the feet, but also of the mind. In the early 19th century Robert Southey defined the picturesque as a new 'science'° and he said that it derived from a 'course of summer travelling', where exploration was of the essence. This is hard to illustrate, since so many 18th century and indeed modern images show only static or frozen moments.* We go very wrong if we think only of photography, or what we capture in a viewfinder of a camera.* But we might look to a *text* without images like William Gilpin's *Dialogue on the Gardens of Stowe* of 1748, where two people move through this extensive site and talk and discuss as they go. This was Gilpin's very first publication, and he went on to be one of the most prominent authors of picturesque travel in England. We might compare Gilpin's *Dialogue* with the distinction that Chen Congzhou makes between what he calls 'in-position viewing' and 'in-motion viewing'; both modes are needed to achieve a full experience of a garden.

° i.e. a way of knowing

* FIGURE 4
* FIGURE 5

We have lost this essential mode of understanding how the picturesque was captured, experienced, in motion. The modern architectural critic, Peter Collins in his *Changing Ideals in Modern Architecture*, writes of how 'the spectator, as he advances, and as he move away, distinguishes in the distance a thousand objects, at one moment found, at another lost again, offering him delightful spectacles'.• It was also valued by the sculptor Richard Serra in applauding 'deambulatory space and peripatetic vision' in describing the effect of his sculptures as they came to be placed in urban settings.•

• 6
• 7

But movement of the feet, as Gilpin always made clear, promoted a movement of the mind. Two late 18th century theorists of the picturesque—Uvedale Price and Richard Payne Knight, among others—both concentrated on this mental movement when discussing the picturesque. As I understand this in Chinese examples, the doubleness—the exterior and interior° worlds—were rarely disconnected. In Guo Zhuang's gardens on the western lake was a pavilion on a rock called Shang-xin Yue-mu Ting.°

° inward
° pavilion pleasing both eye and mind

Pictures and therefore the picturesque were always about human nature not just nature, though during the later 18th century this aspect was gradually lost sight of, and the landscape itself seemed sufficient. Earlier, though, it followed that even

"都展现了深刻的思想",无论是花木、厅堂或者蜿蜒的路径。

中国的文字和图像之间从根本上是紧密相关的。事实上,文字似乎比一座真实的园林或绘画的景观更为重要。有些园林和景观的记载中根本没有任何视觉元素的存在,例如文徵明的《玉女潭山居记》或者明代隐士刘裕华的《乌有园记》。而后者所记述的园林是作者根据早前的文献想象出来的,仅存在于文字的记述中。但在文徵明著名的《拙政园记》里则包括了一篇园记、31首诗以及31幅独立的册页。于是,文字和图像共同构成了一种更为全面的理解。

• 4
• 5

众所周知,中国山水的想像在很大程度上借助文字来传达更完整的画作内涵。它或许会告诉你画作的主人、画作所描绘的场所,或者于画面之上题写文字来表述最为完整的含义。这样的文字正是中西景观所共通的如画的一个基本要素——我们称之为"关联主义"主题,也就是场所或画面在我们思维、记忆和想像中所激发的东西。

* 图2和3

风景如画总是强调运动,而身体的运动以及思维的运动是我们在单单面对摄影或静止画面时所无法获得的。19世纪早期,罗伯特·骚赛将风景如画定义为一种新型的"科学",并声称风景如画脱胎于一种以探索为核心的"夏日旅行的过程"。这一点很难用图像来说明,因为众多18世纪以及现代图像呈现的都仅仅是静止或凝固的瞬间。我们不应当只是从摄影术或取景器的角度去思考风景如画,也可以研究诸如威廉·吉尔平的《1748年斯托园旅行对话》这种没有图像的文本。文中,两人穿过广阔的场地,一边行走一边交谈和讨论。这是吉尔平出版的第一部著作,随后他著书不断,成为英格兰最为杰出的风景如画游记作家之一。我们可以比较着陈从周的"静观"——"动观"论来看吉尔平的《对话》,可以见到,为了获取完整的园林体验,这两种模式都是必要的。

○ 一种理解世界的方式

* 图4
* 图5

我们已经丧失了理解如何在运动中获取和体验如画性的这种根本模式。现代建筑评论家彼得·柯林斯在其著作《现代建筑设计思想的演变》中写到"观者在他的来去之间辨别出远方大量时显时隐的物体,获取令人愉悦的视觉体验。"雕塑家理查德·塞拉在描述其雕塑作品成为城市装置所产生的效果时也很看重这种"游荡的空间和巡游的视觉"。

○ 1998年

• 6
• 7

然而,吉尔平说得很清楚,身体的运动促进了思维的运动。以18世纪晚期的两位风景如画理论家尤维达尔·普莱斯和理查德·佩恩·奈特为代表,许多人都在讨论如画时专注于这种思维的运动。就我所理解,在中国的园林中,这种外部世界和内部世界的双重性往往是勾联在一起的。西湖边上的郭庄有一座石上亭,叫做"赏心悦目亭",即是同时愉悦了眼与心的亭子。

○ 内心

绘画以及风景如画一直以来关注的都是人造自然,而不仅仅是自然。尽管到18世纪晚期这一点已经逐渐丧失而关注景观自身似乎已经足够。但是在更早些的时候,即便是描绘景色的绘画再现的也是人类如何以及为何属于某个特定场所。画家鲁本斯所描绘的斯蒂恩城堡及他自己的地产景观,从某

* 图6

images of scenery were representations of how and why people belonged in a particular place. A landscape by the painter Peter Paul Rubens of his Steen Castle* and of his own estate was, as it were, a portrait of a person, in this case of Rubens himself. Throughout the 18th century representations of landscape, like those in William Birch's *Les Délices de la Grand Bretagne* in 1791, looked at almost every landscape in Great Britain as an emblem or explanation of a place and those who lived or used it.• That was also relevant to the psychological processes of Price and Knight. I find an intriguing parallel with Chinese paintings of an owner's own property,* called *bie hao tu*—by-name pictures, because the owner's name was given to the picture of that place.

    * FIGURE 6

    • 8

    * FIGURE 7

A modern commentator on the early picturesque wrote that 'travel meant discovery, and discovery meant knowledge. In this respect travel became a 'mental accumulation' of ideas, images, narratives and connections…drawn from spatial as well as historical experiences.'• This is still the case today, though we have forgotten that the picturesque was a means of understanding how the land looked, but was *not* a model for creating one. We have insisted too much on how garden and landscape designs were modelled on painting.° Paintings were essentially a tool for comprehension, for learning how to look, how to seize the substance of a place, to parse what you saw and learn about its components—to ascertain the *genius loci*. For that skill, one would certainly have learnt from looking at carefully composed landscape paintings, and that scrutiny° sharpened the mind in its other inquiries beyond paintings.

    • 9

    ° the actual number of such known examples are relatively few.

    ° that 'tutorial'

The modern picturesque needs two fresh responses: it needs to rethink its original procedures, as I have suggested; and it needs to apply them to contemporary landscapes. We can, certainly, still learn from paintings, but not by mimicking their pictorial formulae as a basis or model for our own contemporary landscapes. But we need, for instance, less Salvator Rosa and more Edward Hopper. We need to rethink our assumptions about the varied and disjunct landscapes that Uvedale Price, quoted by Smithson, relished; we should apply them to what Alan Berger has called 'drosscapes', landscapes—so familiar and ubiquitous these days—that require both the rethinking of our assumptions as well as more practical ways of responding to their redesign. Peter Latz and Partner did that at Duisburg in Germany, with its conversion of a derelict and toxic steel mill complex into a public park.*

    * FIGURE 8

So the picturesque may still be of service today. We always learn about the places where we live or visit by looking and thinking, and these days we rely even more than the 18th century upon a strongly visual culture—films, computers, video games as well as paintings, though perhaps we tend to let the machines do the looking for us. We also learn endlessly by movement—we live in a hugely mobile world; and if, as some people displace themselves often and are endlessly mobile, then a sense of the immediate locality becomes a necessary means of orientating oneself; despite our globalism, locality is immeasurably important. And learning about this sustains our sense of belonging, whether nationally or locally. Which is also why I think we need to expand our grasp of what elements contribute these days to a picturesque mode of looking and learning: as regards both physical nature and human nature.°

    ° the world around us *and* the world within us

种意义上而言都是人物即他自己的肖像画。纵观18世纪的风景画，如威廉·伯奇1791年出版的《大不列颠的欢愉》中的那些画，几乎将英国每一处景观都视为一个地点以及生活于其上或使用着它的人们的象征或诠释。这很符合"Price"和"Knight"所讨论的心理过程。很有意思的是，我在中国绘画中发现了同样的情形。这种描绘主人私产的绘画叫做"别号图"，即用主人的名字来命名描绘这个场所的绘画。

● 8

＊ 图7

一位现代评论家曾这样评论早期的风景如画："旅行意味着发现，而发现意味着知识。在这种意义上，旅行成为源于空间以及历史体验的各种思想、图像、叙事和联系的'内在累积'……"今天的情况依旧如此，尽管我们已经忘却了如画是一种理解世界的方式而不是一种创造世界的模型。我们过于关注园林和景观设计如何模仿绘画。绘画本质上是用于理解的工具，是对如何看，如何捕捉场所实质、分析所见、了解其组成以明确场所精神的能力进行练习的工具。而为了获得这种能力，我们必然要研读经过仔细构画的景观绘画，锻炼精微的观察和诠释，使我们在应对绘画之外的对象时能够思维敏锐。

● 9

○ 我们所知的真实案例其实相对很少。

现代如画必须对两件事作出新的应对。其一，我们需要重新思考我之前提到的如画最初的实现程式；其二，将这些程式应用于当代景观。我们当然还要向绘画学习，但不是通过模仿它们的构图定式，将其作为当代景观的基础或模型。我们需要少一些诸如萨尔瓦多·罗萨绘画的那种浪漫主义，而多一些爱德华·霍普的现实主义。我们需要重新审视自己对史密森认为尤维达尔·普莱斯所赞赏的多样而独立的景观的认识，并应将这些认识应用于现今普遍存在的被阿兰·贝格尔称为废物景观°的景观之中。这既要求我们重新思考我们的认识，也要求我们找到更有效应对再设计的方法。

○ drosscape

彼得·拉茨合伙人工作室在德国杜伊斯堡就做了这样一个案例，将一个废弃有毒的轧钢厂转变成一个公园。＊

＊ 图8

因此，风景如画在今天或许还是有其用武之地的。我们一直通过看和想来理解着我们生活或参观的那些场所。而今天，我们甚至比18世纪的人更倚赖强烈的视觉文化，诸如电影、电脑、影像游戏以及绘画，尽管我们现在可能更倾向于让机器来代替我们完成看的工作。我们也同样不断地通过运动来获得对场所的理解，因为我们生活在一个极具移动性的世界中。有些人甚至经常变动生活地点即不停地移动，那么即时的地方性的获取就成为他们自我定位的必然方式。尽管我们处于一个全球化的时代，地方性却无与伦比地重要。获得地方性巩固了我们对国家或地区的归属感。正因为如此，我才认为我们需要拓展自己的视野，捕捉当今对如画式观看和理解模式有益的元素——这种模式既注重物质自然，°又注重人造自然。°

○ 外部的世界
○ 内心的世界

唐静寅 译

FIGURES

1

2

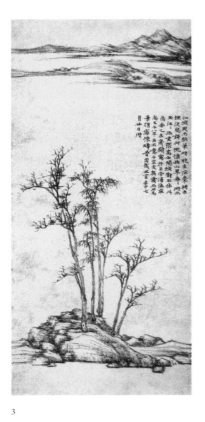

3

4

1 英国维多利亚时代早期拍摄的约克郡修道院遗迹。
   Early Victorian photograph of the ruins of Fountains Abbey, Yorkshire, England.
2 英国汉普敦新森林。
   Scenes of New Forest in Hampshire, England.
   William Gilpin, 'Illustrated notebooks', *Remarks on Forests Scenery*, 1781.
3 倪瓒（元）《渔庄秋霁图》。
   Ni Zan (Yuan Dynasty), *Clearing Autumn at a Fishing Village*.
4 纽约斯坦顿岛的弗莱士科尔斯公园。
   渲染：詹姆斯·科纳景观设计公司绘
   图片提供：纽约市
   Fresh Kills, Staten Island, New York.
   Renderings: James Corner Field Operations
   Photo supply: City of New York

插图

5

6

7

8

5　在尼亚加拉大瀑布边相互照相留念的中国游客们。
　　摄：约翰·狄克逊·亨特
　　Chinese visitors photographing each other at Niagara Falls.
　　Photo: John Dixon Hunt
6　彼特·保罗·鲁本斯《斯蒂恩城堡》。
　　伦敦：国家美术馆
　　Peter Paul Rubens,
　　A View of Het Steen in the Early Morning, c. 1636.
　　London: National Gallery
7　杜琼（明）《友松图》。
　　Du Qiong (Ming Dynasty), *Befriending the Pines*.
8　德国北杜伊斯堡公园。
　　摄：约翰·狄克逊·亨特
　　The Park of Duisburg-Nord, Germany.
　　Photo: John Dixon Hunt

NOTES   注解

1. I am much indebted here to Wai-Kam Ho, "The Literary Concepts of 'picture like' (Ju hua) and 'picture-Idea' (Hua-i) in the relationship between Poetry and Painting', Words and Images: Chinese poetry, calligraphy & painting, Alfreda Murch and Wen Fong ed., New York: Metropolitan Museum of Art, 1991: 362–3.
2. *Ibid.*, 383. Nor, says the author, did it imply pessimism for the ruined and dilapidated.
3. I.E. Diderot on Salon landscape paintings. Elizabeth Carroll Waldron O'Connor, *Diderot as a Critic of Landscape Painting: A Study of the Salons of 1759–1781*, Northampton, US-MA: Smith College, 1960.
4. Stanislaus Fung, 'The Imaginary Garden of Liu Shilong', Terra Nova, volume 9, issue 4, 1997: 15–21. See further in his 'Notes on the Make-do Garden', Utopian Studies, volume 9, issue 1, 1998: 142–8.
5. Andong Lu, 'Deciphering the reclusive landscape: a study of Wen Zhengming's 1533 Album of the Humble Administrator's Garden', Studies in the History of Gardens and Designed Landscapes, volume 31, issue 1, 2011: 40–59.
6. He is discussing the old 18th-century idea of 'parallax', and his quotation is drawn from the architectural writings of Jacques-Germain Soufflot. A more Marxist account is in Slavoj Žižek, The Parallax View, Cambridge, US-MA: The MIT Press, 2006. Another interesting example of a landscape perceptive akin to or similar to parallax is the account of driving in Australia that was cited in George Raitt, 'Ekphrasis and illumination of painting: the end of the road?', Word and image, volume 22, issue 1, 2006: 17.
7. Yve-Alain Bois, 'A picturesque stroll around Clara-Clara', October, volume 29, 1984.
8. Birch's work is examined and illustrated in Emily T. Cooperman and Lea Carson Sherk, William Birch: Picturing the American Scene, Philadelphia: University of Pennsylvania Press, 2010.
9. Christopher Ridgway, 'Rethinking the Picturesque', Sir John Vanbrugh and Landscape Architecture in Baroque England 1690-1730, Christopher Ridgway and Robert Williams ed., Stroud, UK: Sutton Publishing Ltd, 2000: 172—191, especially 182 et. seq.. For a check list of these English travels see G. E. Fussell and C. Goodman, 'Travel and Topography in Seventeen Century England', Transactions of the Bibliographical Society, series 2, issue 10, 1930: 292—311.

1. 这一观点主要来自"Wai-Kam Ho", 参见英注1。
2. 同英注1, 383页。作者还指出, 在北宋, "如画"这个词也并未暗含对废墟和断壁残垣的感伤。
3. 即狄德罗在沙龙一书中对风景画做的评论。
7. 他说的是18世纪提出的"视差"问题, 文字引自雅克·热尔曼·苏夫洛的建筑文章。斯拉沃伊在《视差》(The Parallex View) 一书中有更为马克思主义的言论。另一个接近或类似视差的有趣例子是一段关于在澳大利亚驾车旅行的描述, 这段文字被乔治·赖特尔在一篇有关言说的论文中引用。参见英注7。

PASSAGES IN THE LANDSCAPE OF TIME
David Leatherbarrow

Passages in space are designed most intelligently and experienced most fully when seen as *passages in time*. Such is the conclusion one reaches after considering Fumihiko Maki's design for the Annenberg Public Policy Centre,° recently finished in Philadelphia at the University of Pennsylvania. Maki himself suggested this thesis decades ago. The first page of his first publication ends with the following line: 'any order introduced within the pattern of forces° contributes to a state of dynamic equilibrium—an equilibrium which will change in character *as time passes*.'• Forty five years later in his most recent book he argued the same premise: 'The ideal work is one that accurately expresses—by its modernity—the particular present in which it is constructed, yet is able to transcend that time and continue to exist. Time alone is the final judge of any work of architecture.'• Both statements recall an ancient commonplace: *veritas filia temporis*, truth is the daughter of time.

○ APPC

○ that defines the city and its architecture

• 1

• 2

The history of the thesis is instructive. In the 17th century Francis Bacon argued for the progressive unveiling of truth. Leonardo da Vinci, a century earlier, understood time to be the cradle of invention, as did Aristotle in antiquity. Still earlier, Hippocratic writers described the discovery of medicine as the outcome of extended periods of research. And Plato, in *Laws*, asserted that the truth of any piece of legislation is arrived at through long processes of correction and improvement. *As time passes* order becomes apparent, as does its adequacy to the full reality of our condition. Reflecting on his several seasons of work at the Hillside Terrace in Tokyo, Maki wrote:

> Looking back…the process that led from Hillside Terrace's first to sixth phase suggests not only changes in our notion of public space and the evolution of modernism, but also what I would call the landscape of time.•

• 3

Some settings seek to preserve a time of beginnings through re-enactments. Others participate in a time of continual change, accenting not the sameness of situations, but their difference. Of course time is also known in experiences of spatial depth, the passage between layers of space.

> The singular sense of space…at Hillside Terrace…is the result of a deliberate design approach that has created continuous unfolding sequences of spaces and views, taking advantage of the site's natural topography…giving an impression of substantial depth.•

• 4

All together, these passages indicate three distinct temporalities in architecture: the time of incremental planning and construction, the time of environmental modification, and the time of experience, between moments of movement and rest or approach and arrival, followed by departure and recollection. Let us review these several types of *passage* and see how they help us understand the design and experience of the APPC, as well as essential dimensions of architecture, the landscape, and the city.

In the decades before Maki became involved with the University of Pennsylvania campus something like his thesis of 'incremental planning' seems to have served as

时间景观中的推移／移逝
戴维·莱瑟巴罗

当我们把空间中的推移看作是时间的移逝时,对空间路径的设计才可最明智,对其体验也才最饱满。在细品槙文彦所设计的安纳伯格公共政策中心之后,° 我们便可得出上述结论。数十年前,槙文彦自己也曾提出过这样的观点。在他第一本书的首页末尾,他这样写道:"那些对城市和建筑起决定作用的力量所产生的任何秩序,都有助于形成一个动态的平衡,并且,这种平衡的具体特性会随时间的推移而发生改变。"• 45 年后,槙文彦再次表明了相同的观点:"理想的作品以它的现代性精确地表达它建造时的状况,但它同时也能超越那一时刻而持续地存在。能够对任何建筑作出最终评定的,唯有时间。"• 以上两句话都令人想起一句古语:"真理是时间之女。"°

    ° APPC

    • 1

    • 2
    ° *veritas filia temporis*

历史上也有很多人论述过这一主题,17 世纪,弗朗西斯·培根认为真理是逐渐被揭示的。此前一个世纪,列奥纳多·达·芬奇更认为时间是发明创造的摇篮,更早的亚里士多德也是这么认为。而更为久远的《希波拉底》的作者把药物的发现描述为研究扩展时期的成果。柏拉图在《律令》一书中声称:任何一篇法律都是通过长时间的更正和改良才获得它的真谛。秩序因时间的流逝而凸显,就像只有时间才能呈现全面而真实的境况。在回顾他在代官山住宅项目持续多年的工作时,槙文彦这么写道:

> "回望代官山住宅项目从第一期到第六期的整个过程,它反映的不只是我们对于何为公共空间在概念理解上的改变,以及现代主义本身的演变,更有我所称的时间性景观。"•

    • 3

某些环境通过不断重修来保持最初那一刻的形象。另一些环境则参与到持续变化的时间中来,不再强调环境的一致,而是它们的差异。当然时间性也能在对空间深度体验——穿梭于空间的不同层次——中来感知。

> "代官山住宅项目中独特的空间感,便源自一种有意识采取的设计方法,它利用基地自然地形的特征,创造出连续展开的空间与视线序列,创造出一种具有显著深度感的空间印象。"•

    • 4

总之,这些空间的推移° 显示了建筑中三种独特的时间性:渐进式的设计与建造当中的时间性,环境嬗变中的时间性,以及体验当中的时间性——在游动与休憩之间,在接近与到达之间,以及其后的离去,还有念想。让我们重温这几种类型的空间推移,看看它们如何帮助我们理解安纳伯格公共政策

    ° 时间的移逝

the basis for the development of the buildings in the APPC's immediate vicinity.

Given its age and division into schools, the campus as a whole can be understood as a series of additions and alterations—some to buildings, some to ensembles, and still others to the city. The design of the better projects on the campus not only responded to changes in the encompassing urban pattern but also disclosed new possibilities. The entry pavilion added some years ago to the Annenberg School, for example, established a new court on the street side and gave better definition to a large plaza on the campus side. In the end, that project's urbanism was more successful than its architecture. Nevertheless, the APPC seeks to be similarly responsive and anticipatory: the placement of its main public room, the 'agora,' relatively deep in the urban interior intends to develop the public potential of mid-block walkways, articulating urban depth by multiplying the layers of space that can be shared.* Incremental planning of this kind views existing topography as both an occasion and a prompt. The chronicle of its development will unfold over decades, as did the urban richness of Hillside Terrace.

    * FIGURE 1

A very different sense of time is apparent on the building's façade, the adjustable elements of which allow it to record daily and seasonal change. Two basic parts make up the façade: a curtain wall which mediates the weather and a system of birch panels behind the glazing which allows for more local modulation of air and light, admitting air and casting shadows, performing what might be called breathing and gnomonic functions. Each layer has operable elements: most of the birch panels slide sideways like vertical louvers while the horizontal vent windows open outward.* In addition to localising comfort control and recording the times of day and season, these movements also animate the façade, indicating changing patterns of use. Here, iconography has operational or performative substance: images are prompted by environmental conditions, sometimes having visible effect, sometimes not.

    * FIGURE 2

Temporality reveals spatiality in still other ways in this building, in the configuration of parts that invite and reward movement through its several settings. Consider the façade once again. At the same time that their changing positions indicate patterns of use within the building, the glazing and panels also expose and conceal different depths of interior space. Various aspects of offices can be seen, also deeper passageways, shared spaces, and even parts of buildings beyond this one.* Because some aspects of these settings are seen and others are hidden full understanding of any one of them is only intimated, which is to say delayed. Some parts are shown but their counterparts are concealed. This is because overlapping results in occlusion, occlusion that has been increased because the layers of the façade have been multiplied and its parts allowed to shift. And what the façade inaugurates, the plan and section elaborate, for passage into and through the building encounters openings and concealings that initiate and prolong spatial discovery. Content is made apparent and kept out of reach, awareness excited, appreciation postponed. Occlusion is not the only technique of spatial adumbration; corner entries, diagonal movements, and divided pathways serve the same function.

    * FIGURE 3

To see this clearly consider the foreshadowings that pace ones passage from the campus-side entry to the main public spaces. The free-standing column at the entry corner splits both the view and approach along two lines, one that accelerates past the gridded façade toward the crossroads in the distance and another that is slowed by the shadows that shelter passage into the reception hall.* Similar devices° were put to use at Hillside Terrace to structure approach sequences. In the Philadelphia project, two paths also diverge from the reception desk, one toward the guest lounge on the

    * FIGURE 4
    ° corner column, diverging paths and contrasting lighting

戴维·莱瑟巴罗
DAVID LEATHERBARROW

中心的设计和体验，以及建筑、景观与城市的一些根本维度。

在槙文彦与宾夕法尼亚大学校园结缘前的数十年，他的一些观点，譬如"渐进式的设计"理论，似乎已经成为安纳伯格公共政策中心周边建筑发展的基本模式。由于校园的悠久历史和各个学院的分散布置，整个校园可以被看做是由一系列的加建与改建而形成——这些变动有些是针对建筑，有些是针对建筑群，还有些则针对城市。在校园中，理想的建筑设计不仅要回应身处其中的城市形态的改变，也要为今后提供和展现新的可能性。例如，多年前安纳伯格学院加建的一个入口大厅，就在临街一侧创造了一个新的庭院，同时也更好地限定了位于校园一侧的大广场。结果这个项目在城市层面比在建筑层面更为成功。安纳伯格公共政策中心的设计，也在寻求类似的对于环境的因应以及对于未来的提示：它的主要公共空间°的位置相对深入城市内部，意图激发位于地块中间的那些步道的公共性潜力，并通过增加共享空间的层次来清晰地表现城市深度。*这种渐进式的扩建规划视现有地形环境既是机遇也是条件，其发展过程将会持续数十年之久，就像代官山住宅项目所呈现出的城市多样性一样。

° 可称其为"广场"

* 图1

建筑的立面则呈现出另一意义上的时间，可调节的建筑构件同时记录着时间和季节的变幻。两种基本构件组成了建筑的立面：幕墙，用来调节气候；玻璃后面的桦木板系统，用来调节局部的空气和光线，让空气流通并投射影子，行使着类似于呼吸和时钟的功能。立面的每一层都有可调节的构件：大部分桦木板能像垂直百叶一样向两侧滑动，而水平的通风窗则可以向外开启。*除了按具体要求来调节舒适度，并记录时间和季节的变幻之外，这些变化也使立面更加生动有趣，并暗示出内部使用模式的变更。此时，建筑的形象其实已经带有了操作或效能的实质，这些形象被环境条件所激发，有时产生视觉上的效果，有时则不然。

* 图2

在这一建筑中，瞬时性还以另一种方式展现空间性，这表现在建筑各部分的配置关系中，它们通过不同的组合来邀请身体的运动并且也回报这种运动。让我们再次回到立面。玻璃窗和桦木板的位置变化在暗示建筑内部使用模式的同时，也呈现并隐匿了内部空间的不同深度。我们可以看到办公室的各个角度，还有后面的过道、共享空间，甚至可以透过去看到外面其他建筑的局部。*由于这些位置一部分可见而另一些部分不可见，所以对任一构件的完整理解都只能是暗示性的，也就是说是延迟的。有些部分得以呈现而另一些则被隐藏。这是因为重叠导致了视线遮挡，并且由于立面层次的增加，以及立面上各构件的可变性，而更为增加了视线遮挡。立面所开启的这种特征，由平面和剖面进一步推进，因为进入或通过建筑会遇到各种开口和隐匿之物，它们启动并延长了空间的体验。空间展现在眼前，却又非身体所能及，然后是兴奋莫名的知晓，最后是姗姗来迟的欣赏。视线遮挡不是空间暗示的唯一手法：角部入口，对角流线，以及分叉的路径也都具有同样的作用。

* 图3

为了使这一点更为清楚，我们可以来看看从校园一侧的入口到主要公共

façade's opposite corner and a second down the passage toward the *agora*. A run of columns in this latter passage also splits movement,° but farther along these paths rejoin, exactly where the threshold to the *agora* concentrates a range of spatial opportunities: the first treads of a stair on the right; direct entry to the *agora*, between those treads and another path-splitting column; a wider entry and view to the exterior, on the other side of the column; and a diagonal approach to the space that normally acts as the stage, at the far left.* Together with these obvious indications are several others that are less apparent but no less indicative: the wide balcony above, that roofs the deepened doorstep; the reflections of the front façade mirrored on the glass doors; and the fragment of an aperture behind the rising stair, an aperture that will—upon entry—expand the space of the *agora* into the Annenberg entry court and the street beyond.* Thus, the threshold offers a cluster of opportunities that implies a structure of involvements, involvements that will be enjoyed well beyond this single setting. The first—the involvements—becomes apparent at the moment of arrival, the second —the opportunities—once movement forward takes its first interior step. And additional steps are greeted by yet other ensembles of suggestions, which are followed by still others as one progresses through the building's public spaces.

    What kind of spatiality do we have here? And what kind of time does it elaborate? Elements that abut one another in this building seem to insist on preserving their distinctness, as if the whole were governed by a principle of disconnection. This insistence also makes them° unwilling to suffer some overarching or integrative framework.° Juxtaposition among these elements and settings leads to disjunction not coordination, on the thesis that adjacency can be estranging.* Maki has suggested that clusters of incompatibles typify the topography of the modern city, Tokyo in particular. That premise makes the building the city's epitome, perhaps also its accusation.•

    The *agora* assembles individuals. Can it be seen as an effort to embody the democratic principles the APPC as an educational institution seeks to promote? The counter principle seems clear: settings that hold themselves tightly together often do so by keeping others out, as is the case when 'community' building assumes a politics of exclusion. Local disharmony, by contrast, allows linkages to more distant territories. How? Within the *agora*, or the 'family room' at the base of the atrium, elements that seem indifferent to one another in a single setting align themselves with others nearby, the way siblings free themselves from familial ties in order to enter into social and professional friendships. Surface layering starts the clock of spatial passage while the clustering of heterogeneous volumes rewinds it. Permeable edges serve the same purpose, extending ones orientation outward and endlessly. Prolongations of this sort are especially clear in the corner stairways, which couple vertical movement with horizontal awareness. The extent of this awareness is very wide, which is to say urban.

    Spatial order such as this might be criticised for being incidental or episodic. That would be fair, I suppose, were there not complimentary strategies for indicating self-sameness and stability. One is the repetition of key elements, a few of which have been named already: the corner entries, the path-splitting columns, the repetitive intervals of the glazing grid, and so on. Self-sameness is also shown in the simplicity of the building's primary massing: the symmetrical façades, front and back, as well as the simple volumes for the main public spaces, the *agora* and the family room. Furthermore, those two large interiors exercise magnetic pull over neighbouring offices and passageways, leaving no doubt about the orientation and primary order of the entire ensemble. While much of the building argues for movement, change, and temporal unfolding, its

○ toward the glazed seminar room on the left, and the service facilities on the right

* FIGURE 5

* FIGURE 6

○ the single column, each of the balconies, the stair, the various apertures, etc.

○ except, of course, on the building's exterior

* FIGURE 7

• 5

空间的进入路径上的那些预示。位于角部入口的独立柱把我们的视线和路径分为两个方向：其一是快速通过格栅状的立面指向远处的交叉路口，其二则是沿着被阴影遮蔽的通道缓缓步入接待大厅。* 类似的策略。在代官山住宅项目中被用于组织路径序列。在费城项目中，从位于接待处的桌子开始路径再次一分为二，一条通往位于沿道路立面另一角的休息厅，另一条沿过道向西通向室内的"广场"。而过道中伫立的一排柱子又将路径分成两条，。但这些路径在远处又重新汇合，并在"室内广场"的入口汇集了通向一系列空间的可能性：右边是楼梯的第一组踏步；在那些踏步和另一根让路径分叉的柱子之间则可直接进入大厅；在柱子的另一侧，则是较宽的出入口和视野，通向室外；而一条对角线路径通向远处左边的空间，通常作为舞台使用。* 在这些明显的空间指示以外，还有另一些不太明显但依然具有指示作用的元素，如通道上方的宽挑台，它遮盖了向内伸出的入口空间；还有反射在玻璃门上的前立面的影子；以及位于向上梯段后面的窗洞的局部，一旦进入"广场"，便会感到这个开口便将空间延伸至安纳伯格学院前的入口庭院，以及更远处的街道。* 因此，这个将进未进之处提供了一系列的可能性，而这些可能性又暗示了内部的空间结构，它们远远超越了在入口处所能看到的单一视景。在我们到达的那一刻，首先可以见到各种空间上的可能性；当起步进入，内部空间结构则开始显现。随着向里深入，穿过建筑的公共"广场"，不同的空间场景渐次呈现，变幻无穷。

　　在这里，我们拥有的是怎样的空间性？它又呈现出怎样的时间性？看上去，这个建筑中相互毗邻的元素都坚持保留它们的差异性，仿佛整个建筑都被一种分离的原则所控制。这种坚持也使这些构件。不愿意去承受某些总体性或整合性的框架关系。构件与环境之间的同时存在导致的是分裂而非协调，也就是说，所谓的相邻也可能恰恰是疏离。* 槙文彦曾经指出，由不协调之物构成的群体是现代城市状况。的典型特征。这一看法使得建筑既是城市的缩影，可能同时也是城市的罪状。●

　　"广场"是不同个体的聚集。它是否能被视为是一种努力，以体现作为一个教育机构的安纳伯格公共政策中心所寻求提倡的民主原则？但与之相反的原则似乎也很清楚：那些把自身紧紧联系在一起的整体，往往通过排除其他要素来达到这一点，就如"社区"建筑所采取的排他性策略。与此相反，局部的不协调使这些构件可能与相对遥远的领域建立起联系。但如何实现这一点呢？在"广场"内部，或说是中庭底部的"家庭室"中，那些在这一总体环境中看上去彼此无甚关系的构件，却与不远处的其他构件结合在一起，这就有如手足兄弟脱离家庭纽带踏入社会，或建立起职业上的关系。于是，建筑的表皮开启了移逝之钟，但那些异质性的构件却似乎又要反方向而动。通透的边缘界面也为着同样的目的，把人引向一个无尽延伸的外部。这种延伸在结合了垂直运动和水平感知的角部楼梯处表现得尤其清楚。这种感知的范围十分广阔，而有了某种城市性。

* 图4
○ 角部的柱子、分叉的路径和光线的对比

○ 分别通向左侧敞亮的研讨室，和右侧的服务用房

* 图5

* 图6

○ 独立柱子、每个阳台、楼梯、各种开口等
○ 当然，建筑的外部除外
* 图7
○ 特别是东京
● 5

时间景观中的推移／移逝
PASSAGES IN THE LANDSCAPE OF TIME
373

repetitions, simple massing, and stable orientation assert sameness and fixity, as if its fluctuations were nothing more than different ways of revealing its constancy.

Maki has written that in Japanese culture ideas of depth are often connected to notions of inwardness.• Both share a sense of what is out of reach. In the examples I have adduced, we have seen that layers of space, discreet settings, and architectural elements can be recessive and withdrawing—even if they expose aspects of themselves. On this account, what is partially present can also keep itself remote. I have tried to show that spatial order of this showing/concealing kind is contingent on temporal structures: the settings 'behind' any one we now see were experienced some time 'before,' while those 'beyond' this one are yet to be experienced. One way to describe the present—this present moment—is to characterise it as a former future. Grasping that conception means accepting each temporal moment's individuality and comparability. If the comparative sense of the present as a former future can be granted, we take the next step and see the past as a recent present. Similarly, the present can be seen as an impending past, a moment that is just about to leave us. This would mean noticing the way it edges its way into the time that once was. Likewise, the future would be a present to come, and a past to follow. All of this suggests that each moment in time indicates its own individuality and a set of implications—the 'vectors' to which I referred earlier. Just as vestiges await their own transformation, anticipation is nurtured by recollection. A city's past sets the stage for the revelation of what is yet to come, a new avenue, garden, or neighbourhood in a certain part of the city. What is yet to come through project making arises by virtue of awareness of what it has been. The 'has been' and 'yet to come' are the two-part framework for understanding the present. At any given moment places and times have linkages to distended conditions that are *present as hidden*, which is to say recalled and anticipated, seen partially and implied. The key point is that the work's recessive aspects allow it to *give more than it shows*, to yield content that exceeds what is seen at any moment as well as what might be expected or remembered. Work such as this joins step with urban and natural reality: transcending what has been designed into those two sources of the project's renewal.*

• 6

* FIGURE 8

类似这样的空间秩序，有可能因为其偶然性或片段性而遭受批评。不过这样的批评实在有失公允，因为建筑师同时也采用了平衡性的策略，以暗示其一致性和稳定性：一种途径是通过关键构件的重复，这一点上文已经提到了一些，如角部入口、让路径分叉的柱子、幕墙窗框的重复等等。此外，建筑基本体量的简单性也表达了一致性，其前后立面对称，主要公共空间°体量简洁。此外，这两个主要的室内空间还产生强大的吸引力，影响着与之相邻的办公室与过道，使得整个空间的方向性和基本秩序都明晰肯定。虽然这一建筑主要表达了游动、变换、及其在时间向度中的渐次展开，但是，它无言的重复，简单的体量，以及当你身在其中时所感受到的稳定的位向，无不确认了这个建筑中的一致性与固定性。似乎，所有的波动与变换，都只不过是揭示其不变与连续的不同方式而已。

○ "广场"和"家庭室"

　　槙文彦曾这样写道，在日本文化中，深度的概念常常与内向性相关。•两者皆有一种不可触及的感觉。在我所引述的例子中，我们也已经看见空间的不同层次、整体环境的平淡内敛，以及退隐的建筑构件。°由于这个原因，部分的呈现倒能让它本身显得疏远。前面我已试图能够说明，这种或呈现或隐匿的空间秩序的达成，其实取决于某种时间性的结构：位于我们现在所见"后面"的场景，在"之前"的某一时刻已经被体验，而这"之后"的场景仍在等待着我们。描述"当下"°的一种方式是把它表述为"过去的将来"。要理解这一概念，意味着要承认每一个瞬间的各自特征以及彼此之间的相容性。如果我们能够认可这种把"现在"视作一种"过去的将来"的理解方式，那么接下来也就能把"过去"理解成最近的"现在"。类似地，"现在"可以被理解成最近的"过去"，一个即将离我们而去的瞬间。这意味着注意到它缓缓进入过去时间的方式。同样地，"将来"是即将到来的"现在"，和即将远去的"过去"。所有这一切都表明绵延之时间中的任一瞬间，都呈现出它自身的特征和一整套含义——即我早先提到的那些"向量"。就好像遗迹等待着自身的转变，预期也因回忆而饱满。城市的"过去"有如一个舞台，上演着将要呈现"将来"，那些崭新的林荫道、花园，或是城市里某个地方的邻里社区。那些通过新的计划而产生的，其实源自于对已有条件的认识。"已经存在"和"将要到来"是理解现在的两分结构。在任何时刻，场所和时间都与更大范围的环境相联系，这种环境以一种"隐匿地存在"的方式来呈现，也就是记忆与期待，你只能见到它的部分，并且还是通过间接的暗示。关键在于，建筑中潜藏的部分其实要比它显现出来的透露更多的信息，它释放的内容也超出我们在任一时刻的所见，也超出我们的预期和记忆。这样的项目结合了城市环境和自然现实：它超越了对于这两个因素的人工化介入——而这样的两个因素又正是项目自身得以更新的源泉。*

• 6

○ 即使它们也呈现出自身的部分样貌

○ 现在这一时刻

* 图8

史永高　译
感谢东南大学建筑学院
博士生陈宁在初译阶段的工作

FIGURES

1

2

3

4

5

插图

6

7

8

1—8
槙文彦设计的安纳伯格公共政策中心，2009 年。
摄：戴维·莱瑟巴罗
Annenberg Public Policy Centre, designed by Fumihiko Maki, 2009.
Photo: David Leatherbarrow

NOTES

1 Fumihiko Maki, *Investigations in Collective Form*, St. Louis, US-MO: School of Architecture, Washington University, 1964: 3.
2 Fumihiko Maki, 'Time and Landscape: Collective Form at Hillside Terrace', *Nurturing Dreams: Collected Essays on Architecture and the City*, Cambridge, US-MA: The MIT Press, 2008: 259.
3 *Ibid.*, 74.
4 *Ibid.*
5 Maki has made this point in several essays, in particular see 'The Japanese City and Inner Space', *ibid.* 2, 150. Implicit in my characterisation of this building as the city's 'epitome' is the well-known analogy between the house and the city, to which Maki has also referred in several essays, often citing Leon Battista Alberti. That the analogy should be seen as *non-reversible* — that it is fine to see the house as a small city but not to see the city as a big house — is a point I argue in *Architecture Oriented Otherwise*, New York: Princeton Architectural Press, 2009.
6 Fumihiko Maki, 'The Japanese City and Inner Space', *Nurturing Dreams: Collected Essays on Architecture and the City*, Cambridge, US-MA: The MIT Press, 2008: 150.

注解

5 槙文彦在其多篇文章中都表达过这一观点，特别在其论文"The Japanese City and Inner Space"。在此报告中，我把安纳伯格公共政策中心当做城市的"缩影"的暗示源自著名的在房屋与城市之间进行的类比。槙文彦同样在他的好几篇文章中，通过对莱昂·巴蒂斯塔·阿尔伯蒂的引述来提及这一点。但是我并不认为这种类比是可逆的，即或许我们可以把房屋视为一座小城，但是却不可把一座城市视为一个大房子。我在《Architecture Oriented Otherwise》一书中对于这一点曾经作过论述。

SYMPOSIUM PHOTOS

论坛照片

SYMPOSIUM PHOTOS

At Southeast University Campus,
Nanjing, 27 May 2011.
Photo: Lai Zili

摄于东南大学校园，2011 年 5 月 27 日。
摄影：赖自力

论坛照片

SYMPOSIUM PHOTOS

MARK COUSINS
马克·卡森斯

CHEN WEI
陈薇

DAVID LEATHERBARROW
戴维·莱瑟巴罗

JOHN DIXON HUNT
约翰·狄克逊·亨特

LIU DONGYANG
刘东洋

LIN HAIZHONG
林海钟

论坛照片

WANG SHU
王澍

LI SHIQIAO
李士桥

ZOU HUI
邹晖

GE MING
葛明

WANG JUNYANG
王骏阳

LI HUA
李华

SYMPOSIUM PHOTOS

At Opening Ceremony of the Symposium,
Southeast University, Nanjing, 27 May 2011.
Photo: Lai Zili

摄于东南大学"当代建筑理论论坛"开幕式，2011 年 5 月 27 日。
摄影：赖自力

论坛照片

At A Lecture of the Symposium,
Southeast University, Nanjing, 28 May 2011
Photo: Lai Zili

摄于东南大学"当代建筑理论论坛"会场，2011年5月28日。
摄影：赖自力

At the *Shanshui* Exhibition,
Southeast University, Nanjing, 27 May 2011
Photo: Lai Zili

摄于东南大学"山水展",2011 年 5 月 27 日。
摄影:赖自力

REVIEW

评论

ON THE ARCHITECTURAL UNCANNY

AN INTERVIEW WITH DR. ANTHONY VIDLER

INTERVIEWER
He Weiling

TRANSLATOR
Li Hua
Geng Xinxin

TIME
28th February
2012

SITE
Cooper Union
New York
United States

由《建筑的异样》展开
安东尼·维德勒访谈

访问　　　翻译
贺玮玲　　李　华
　　　　　耿欣欣

时间　　　　　　　地点
2012年2月28日　　库珀联盟
　　　　　　　　　纽约
　　　　　　　　　美国

由《建筑的异样》展开：安东尼·维德勒访谈
ON «THE ARCHITECTURAL UNCANNY»

维德勒：面对这样一个采访的一个难题是，我已经完全不记得很久以前我写了什么，甚至连为什么要写它我也忘记了。

贺玮玲：事实上这是我的第一个问题。我在读《建筑的异样》° 时意识到，它早在 20 前就出版了。可是对我而言，它一点也不过时。相反，它已经成为当代建筑理论的经典之作。作为读者，这本书触动我的一个方面就在于异样是一个十分私密的° 话题。它给我在情感上的触动与我在阅读"崇高"这个主题时的感觉非常相似，是一种个人内心的强烈感受。所以我的第一个问题是，最初您是如何找到异样这一概念的？

° *The Architectural Uncanny*

° intimate

维德勒：一直以来，我都对弗洛伊德感兴趣。我始终认为他的案例研究非常重要，无关于与他们在实践中的对错，而在于他们开启了诠释、文学和美学的新领域。很多年前，大概是 19 世纪 80 年代，我被请去主持一个有关神话故事的会议。普林斯顿大学德国语言与文学系的一位同事建议我看看德国浪漫主义早期恩斯特·特奥多尔·阿马多伊斯·霍夫曼的短篇小说和弗洛伊德对其中一篇小说《睡魔》° 的随笔。而弗洛伊德关于《睡魔》的随笔，毫无疑问是在他对异样感觉分析的框架中写成的。让我感到触动的是霍夫曼和弗洛伊德在思考异样时，都没有仅仅把它看做一种心理感觉，而是将其看做一种源于空间的感受——一种既是情感上也是空间上的感觉。霍夫曼另一篇名为《克来斯泊》° 的小说讲的是一座奇怪房子的建造。这所房子的主人是一名政府公务员，建造时，他不是先计划好门窗的位置，而是先砌墙，然后遇到哪儿就在哪儿开洞。这个故事让我乐不可支，因为整栋房子事实上就是一个人心理状态的隐喻——一所"异样"的房子。

° *The Sandman*

° *Councilor Krespel*

于是，我开始沿着这条线索追寻下去。这条线索的起点是弗洛伊德的随笔《异样》，° 在这篇随笔开始的部分，弗洛伊德列出了德语词"unheimliche"在字典中的标准定义。显然，"Heimliche"的意思是家常的，° 但就字面上，"unheimliche"既有陌生的也有异样的意思。接着，弗洛伊德试图将异样和恐惧、害怕进行区分，异样确实是一种令人不安的感觉，但并不是像纯恐惧那样强烈。他总结到，异样作为一个美学的范畴与埃德蒙·伯克提出的崇高不同。伯克的崇高是来自于对死亡的恐怖的一种惧怕，而对于弗洛伊德而言，异样是细微但却意义重大的情感。

° *Das Unheimliche*

° homely

贺玮玲：更多的是一种日常的感觉。

维德勒：没错。当一个人有一种似曾相识——好像以前经历过的某事——的奇怪感觉时，所迸发出的就是这种感觉。这种奇怪的感觉是你明明在一个地方，却又感到与它的距离和错位。这使我想到，在最近许多试图动摇、"分解"° 现代建筑作品的纯净性的背后隐藏着一个相似的过程。因此，它触动我的是，将其作为一个重要的主题或比喻，从不同的角度看待当代建筑动摇了现代主义抽象概念的表层。

　　于是，我对异样的观念史产生了兴趣，这让我直接追溯到了 19 世纪的文学之中。因此，我的这本书的第一部分讲诉了异样的感觉在文学、哲学和美学中如何体现的历史，而第二部分是由关于当代艺术和建筑的评论文章组成的。当然，建筑自身在本质上不是异样的。但是人在其中所体验的环境可以引发异样的感觉，这就像艾森斯坦电影中蒙太奇所产生的"震撼"效果一样。总的说来，这本书勾画了一种不稳定的情景，其主题延伸到了我的第二本心理分析的著作《扭曲的空间》° 中，《扭曲的空间》探讨的是都市恐惧症的问题，而不是家居场景中的异样。

° de-construct

° Warped Space

　　贺玮玲：与《建筑的异样》相比，《扭曲的空间》是同一主题在更大范围上的扩展。

维德勒：是的。《建筑的异样》和《扭曲的空间》是姊妹篇。《建筑的异样》是关于内部、家居的不稳感，而《扭曲的空间》则与恐怖症有关，包括心理学家、心理分析家和哲学家在现代空间 —— 大都市空间 —— 的出现中，发现的广场恐怖症、幽闭恐怖症等所有的恐怖症反应。

　　贺玮玲：什么是恐怖症？现代生活中的这种焦虑又是什么？

维德勒：我认为它源于过去两个世纪工业和大都市文化快速的转变，和个体在不同时刻感受到异化的那种空间体验的转变。因此，异化的问题，空间恐惧的问题在现代十分普遍。诗人威斯坦·休·奥登在第二次世界大战期间曾写过一出剧，叫《忧虑的时代》。° 恐惧，无论是强烈的恐惧感，还仅仅是个体无法掌控环境的担心，从 19 世纪末开始就已经支配着城市中的日常生活。

° *The Age of Anxiety*

　　贺玮玲：我想恐惧是这本书提出的一个主要问题。

维德勒：是这样的。有门禁的社区就是恐惧的一种迹象，分区制、° 监控摄像机都是恐惧的表现；而恐惧，在当代世界政治的助推器。

° 分区：zoning

　　贺玮玲：所以，您认为异样是您来看待当代社会中恐惧的一种批判性视角吗？

维德勒：异样确实是那样一种感觉，有些事不太对，它无处不在却又说不出来，比起始终存在的恐惧，它更像是萦绕于心头的一种感觉。它不是鬼，而更多的是与社会的不稳定性有关，例如眼下次贷危机中涌现的不稳定性，一下有房子，一下又没有了。显然，这就是当代城市生活动荡本性的"unheimliche"意义。

贺玮玲：好的。您自己有没有过异样的经历？

维德勒：我自己异样的体验来自于第二次世界大战间对伦敦的轰炸。因为住在离伦敦很近的地方，我能听到"V-2"火箭导弹从我们镇上呼啸而过，和父母一起躲避轰炸，等炸弹落地时，被爆炸的冲击波扔到一边。这就是我最初的恐惧。我还有很多呢。○笑

贺玮玲：好的，我就不追根问底了。

维德勒：其实，有个非常了解我的人读了《建筑的异样》以后，发现它就好像是我的自传。

贺玮玲：您不认为很多人都是这样的吗？在某种程度上，我们总是与某些想法紧密相连。

维德勒：是这样的。

贺玮玲：我想换一个话题。在这本书的前言，您写道："从18世纪末开始建筑与异样的概念紧密相连。"为什么这么说？那个时代有什么特别之处酝酿出了这种亲密的联系吗？是因为工业革命还是因为启蒙运动？为什么您认为是在18世纪末异样开始在建筑中兴盛起来？

维德勒：我认为这是因为18世纪末的启蒙运动，在笛卡儿、帕斯卡之后出现的个体意识，个体作为认知存在，其思想源自外部世界，而越来越没有了依赖于宗教规则下的安全感。世俗化和科学思想能够解释这个世界，但却不能使它变得稳定可靠。

贺玮玲：什么意思？

维德勒：受到启蒙的自我是这个世界上孤独的自我，总会受到所有不可控的自然力量——地震、闪电、暴风雨等——的影响。如何使自然灾害，像火山爆发或1740年代里斯本地震变得合理化——如何"驯服"它们？这就是

为什么埃德蒙·伯克将崇高的概念发展成为一个美学的范畴，用以描述这些事件和这些事件的再现，它们的确不美但却非常强烈有力，足以产生审美的影响。从这点来看，异样的概念可以说是崇高感比较微弱的时刻。

  贺玮玲：在《建筑的异样》中，您通过人的体验建立起了物理空间和情感的关系，身体被空间撕裂、改变和触发。那么异样存在于何处？在空间的体验中还是观察者中？您是不是赞同异样的产生于对物理空间的精神和心理的延展中？

维德勒：人们常常问我这样一个问题，特别是读完《建筑的异样》的学生：如何设计一个异样的空间。显然，异样的空间从不是设计出来的。任何空间都可以触发出异样的感觉，这取决于个体的感觉。

  贺玮玲：所以异样不可能是设计目的？

维德勒：没错。它是个体和空间中或事件里某种东西之间的关系产生的效果，这个空间或事件深深地激发了个体的历史无意识。

  贺玮玲：那是一种非常个人化的体验。

维德勒：是的，这的确是一种非常个人化的体验。所以，我所做的是，延伸其概念的使用，将其扩展到当代建筑学中。但是马克·卡森斯说得很好，他将我所做的描绘为一种建筑学的心理分析。°在某种程度上，它是一种隐喻，也是理解建筑学某个特定手法的一种方式，一种解释的模式。

°与弗洛伊德的个体心理分析相对

  贺玮玲：是建筑感受到它还是观者感受到了它？

维德勒：我想说的是，如果建筑是主体，面对它所完成的某种空间体验，建筑将会深深地感受到侵扰。这是一个隐喻。

  贺玮玲：好的。既然我们谈到观者，或者说读者，您对这本书在中国读者间的反应有什么样的预测？

维德勒：我不知道。首先，我不知道现在在中国是如何接受心理分析学的。所以我不知道弗洛伊德在中国是否有什么当代含义。当然，城市的急速转变可能也会激发出一种不稳定感，与20世纪早期在欧洲和美国的城市所经历的相似。

贺玮玲：异样的概念使我想起约翰·海杜克所说的"弦外之音"的概念。在他的定义中，"弦外之音"是那种看似简单直接，但同时又让读者感到并不仅于此的某种事物。那么，这两个概念间有什么关系吗？总的来说，我想看看异样是否在一个概念网之中。

○ otherness

维德勒：我想是有的。我认为这与俄国语言形式主义者所说的"疏离"有关，一种是文本或特定形式的文字所产生的一种距离感。因此，异样存在于一个理论的网络中，有人或许会将它称之为疏离的网络，它将个体和社会的异化联系在一起，将范围更广的政治和社会态度、情感联系在一起。

○ estrangement

贺玮玲：自然中是否存在着异样？

维德勒：近年来，我已经和自然没有什么接触了。

贺玮玲：我对此感到好奇，因为我们一直关注的是文化。这可能与我个人对自然的兴趣有关。

维德勒：是这样啊。只有当人主体在自然中时异样才存在于自然之中。自然可以激发异样，就像自然可以激发崇高，可以激发恐惧一样。但是自然本身无法体验异样。

贺玮玲：我的最后一个问题与未来有关。在未来，您会如何想象异样与互联网，"dubstep"和数字设计的关系？在数字设计中，设计师热衷于与恐怖电影和科幻电影相关的荒诞感觉。

○ 当下流行电子乐的一种风格

○ grotesque

维德勒：我不认为这与异样有任何关系。就像你刚才所说的那是荒诞，而不是异样。荒诞是一种你在视觉上看到的纷扰、不同或困难，但是你并没有那种以前曾经经历过的某事的感觉。数字媒体是相当表面的。它就在那，社会媒体也一样，"你有的就是你所看到的。"

贺玮玲：那么，对21世纪的建筑学生，异样的意义是什么，起的作用是什么？

维德勒：它与任何就空间和潜意识关系的讨论所起的作用相同。它使学生对他们自身的感觉，对他们和空间的关系，对他们自身与其潜藏的恐惧间的关系有所察觉，进而察觉到他们与整个世界的关系。

VIDLER One of the problems with an interview like this is that I've completely forgotten everything I've written from a long time ago and I don't even remember why I wrote it.

> WEILING That'd actually be my first question. I was reading *The Architectural Uncanny* and realised that it was first published twenty years ago. The book didn't seem dated to me. On the contrary, it has become a classic of contemporary architecture theory. One aspect of the book that strikes me as a reader is that the uncanny is a very intimate topic. It gives me a similar emotional charge as when I was reading about sublimity. Something that one feels very strongly inside. So my first question is, how did you arrive at this idea of the uncanny originally?

VIDLER I've always been interested in Freud. I've always found his case studies important. It doesn't matter if they are in practice right or wrong, but rather if they open up areas of interpretation, literary or aesthetic. Many years ago, maybe in the 1980s, I was asked to give a conference on fairy stories. A colleague in the German department at Princeton suggested that I look at the short stories of E. T. A. Hoffmann in the early Romantic period in Germany, as well as the essay by Freud on one of these stories, *The Sandman*. Of course, Freud's essay on *The Sandman*, was framed within his analysis of the feeling of the uncanny. It struck me that Hoffmann and Freud were both thinking about the uncanny not just as a mental sensation but as a sensation that derives from space — an emotional and a spatial sensibility. Another of Hoffmann's stories, called *Councilor Krespel*, involved the construction of a strange house that was being built for an official of the bureaucracy, where instead of planning the position of doors and windows, he first built the walls and then 'found' entries by running up against them. It amused me because the whole house was in fact a metaphor for somebody's mental state — an 'uncanny' house.

Then I started to follow a trail. The trail started from the beginning of Freud's essay *The Uncanny*° where he lists the standard dictionary definitions of the German word 'unheimliche'. 'Heimliche,' of course, means homely, and literally *unheimliche* means unhomely but also 'uncanny'. Then Freud tries to distinguish the feeling of uncanniness from that of terror, or fear, where an uncanny feeling is certainly disturbing, but not as overwhelming as sheer terror. He concludes that the uncanny as an aesthetic category, is therefore distinct from the high sublime, which as Edmund Burke proposed is a fear drawn from the terror of death itself. The uncanny, then, for Freud is a minor, but significant sensibility.

○ *Das Unheimliche*

> WEILING More of an everyday feeling.

VIDLER Yes. It's really a feeling that erupts when one has a strange sense of déjà vu, of something that one has experienced before. That strange feeling that you are in a place but distanced or displaced from that place. It struck me that a similar process lay behind many recent attempts to destabilise or 'deconstruct' the purity of many works of modern architecture. So it struck me as being a very important theme or trope to look at the different ways that contemporary architecture has disturbed the surface of modernist abstraction.

Then I became interested in the history of the idea of the uncanny and this took me all the way back through 19th century literature. Thus the first part of my book tells the history of the way in which the sense of uncanniness is represented in literature, philosophy and aesthetics, while the second half is composed of critical essays on contemporary art and architecture. Architecture itself, of course, is not intrinsically uncanny. Rather, the circumstances in which it is experienced can lead to an uncanny feeling—like the 'shock' effect of montage in a film by Eisenstein. Overall, the book sets a scene of destabilisation, a theme that is traced further in my second psychoanalytical book, *Warped Space*, which looks at questions of urban phobias rather than the domestic scenes of the uncanny.

> WEILING In relation to *The Architectural Uncanny*, maybe *Warped Space* is an extension on a bigger scale.

VIDLER Yes, *Warped Space* is the companion book. *The Architectural Uncanny* is about an interior, domestic feeling of destabilisation, and *Warped Space* is about phobia, agoraphobia, claustrophobia, all the phobic responses that psychologists and psychoanalysts, as well as philosophers, detected in the emergence of modern urban space—the space of metropolis.

> WEILING What is this phobia? What is this anxiety in modern life?

VIDLER Well, I think it comes from the rapid transformations in industrial and metropolitan culture over the past two centuries, transformations in spatial experience from which individuals at various moments feel alienated. Therefore, questions of alienation, questions of spatial fear, are very prevalent in the modern period. The poet W. H. Auden wrote a play during the Second World War which he entitled *The Age of Anxiety*. From the end of the 19th century, fear, whether fear of attack or simply of conditions over which the individual has no control, has governed everyday life in the city.

> WEILING I guess fear may be one of the major questions that this book raises.

VIDLER Yes. Gated communities are an indication of fear. Zoning is an indication of fear. Security cameras are evidence of this fear; contemporary politics throughout the world is fuelled by fear.

> WEILING So would you say the uncanny is a lens through which you critique fear in the contemporary society?

VIDLER The uncanny is indeed that sense of something not quite right, pervasive but intangible—more like a haunting than an ever-present fear. It's not a ghost, but more to do with the instability of society—an instability that came to the fore in the present sub-prime mortgage crisis—now you have a house, now you don't. Certainly it is an unheimliche sense of the unstable nature of contemporary urban life.

> WEILING Right. Have you personally had any uncanny experiences?

VIDLER My foundation uncanny experience was during the bombardments of London in the Second World War. While I lived quite close to London, I heard the V-2 rocket bombs as they crossed over my town, took shelter with my parents, and was often blown over with the force of the blast when they landed. That was my first fear. I've had many more.°

○ Laugh

WEILING Ok, I wouldn't dig into them.

VIDLER In fact, someone who knew me very well, read *The Architectural Uncanny*, and observed that it was as if I had written my autobiography.

WEILING Don't you think that's the case for a lot of people? Somehow we are deeply connected to certain ideas.

VIDLER Yes.

WEILING So, I might change gears here. In the preface of the book you wrote: 'architecture has been intimately linked to the notion of the uncanny since the end of the eighteenth century.' Why is that? What's special about that time period that incubated such intimate linkages? Is it the industrial revolution? Is it the age of enlightenment? Why do you think the end of the 18th century is when the uncanny started to blossom in architecture?

VIDLER I think that's because in the 18th century Enlightenment, following Descartes and Pascal, there emerges a sense of the individual as a perceiving being whose thoughts are derived from the external world, and increasingly without a secure sense of religious order to rely upon. Secularisation and scientific thought were able to explain the world, but not to secure it.

WEILING What do you mean?

VIDLER The enlightened self is a self alone in the world and subject to all the uncontrollable forces of nature—earthquakes, lightning, storms. How could natural disasters such as volcanic eruptions, or, in the 1740s, the Lisbon earthquake, be rationalised—how could they be 'tamed?' This is why Edmund Burke developed the idea of the sublime as an aesthetic category to qualify those events and representations of events that were certainly not beautiful but powerful enough to have an aesthetic effect. And following from this, the idea of the uncanny as, so to speak, a minor moment of the sublime.

WEILING In *The Architectural Uncanny*, you established a relationship between the physical space and emotions through embodied experiences. The body is torn, alerted, and triggered by the space. So where does the uncanny exist? Does it exist in the experience of space, in the viewer? Would you agree that the realisation of the uncanny occurs in a mental or psychological extension of physical space?

VIDLER I've often been asked, especially by students after reading this book: how can an uncanny space be designed? And of course, an uncanny space can't ever be designed. Any space can trigger an uncanny feeling depending on the individual's feelings.

> WEILING So 'uncanny' cannot be a design intention?

VIDLER No. It's an effect of the relationship between the individual and something in the space or in the event that triggers something very deep in that individual's historical un-conscious.

> WEILING It's a very personalised experience.

VIDLER Yes, it's a very personalised experience. So it's a stretch to use the notion in terms of extending it, as I do, to contemporary architecture. But Mark Cousins put it very well when he characterised what I was doing° as a kind of psychoanalysis of architecture. In a way it's a metaphor. In a way it's a way of understanding certain moves in architecture, a mode of interpretation.

° as opposed to Freud's psycho-analysing of the individual

> WEILING Is it the architecture who feels it or the viewer who feels it?

VIDLER What I'm implying is, if architecture was a subject, architecture would feel deeply disturbed in the face of certain spatial experiences that had been done to it. It's a metaphor.

> WEILING Yes. Since we talk about the viewer, the audience basically, do you have any speculations as to how the Chinese audience would respond to this book?

VIDLER I have no idea. First of all, I have no idea of how psychoanalysis is now being received in China. Therefore I have no idea as to whether Freud has any contemporary meaning. Certainly the extraordinary rapidity of the urban transformation might well provoke a sense of destabilisation similar to that experienced in European and American cities in the early 20th century.

> WEILING The idea of uncanny makes me think of the idea of 'otherness' that John Hejduk talked about. In his definition, 'otherness' is something seemingly straightforward but at the same time it is making the audience feel something else is going on. So what's the relationship between these two ideas? Basically, I'm trying to see if the uncanny as an idea has a conceptual matrix around it.

VIDLER I think it does. I think it is related to what the Russian linguistic Formalists called 'estrangement',° a sense that a text or a particular formation of words could create a sense of distance. So the uncanny is within a theoretical network which one might call the network of estrangement, which links to individual and social alienation, which links to much broader political and social attitudes and sensibilities.

° отчуждение

WEILING Does the uncanny exist in nature?

VIDLER I haven't talked to nature recently.

WEILING I was curious because we've been focusing on culture. That may also have to do with my personal interest in nature.

VIDLER Right. The uncanny is in nature only if the human subject is in nature. The uncanny can be provoked by nature. Like the sublime can be provoked by nature, like terror provoked by nature. But nature itself doesn't experience the uncanny.

WEILING So, my last questions would be about the future. Looking forward, how would you imagine the uncanny in relation to the internet, dubstep, and digital design? In digital design, designers are keen on the grotesque sensation in relation to horror movies or sci-fi movies.

VIDLER I don't think that has anything to do with the uncanny. That has to do with what you just said, the grotesque, which is not uncanny. The grotesque is something you visually see as disturbing, different or difficult, but you don't have the sense of something that you've experienced before as returning. The digital media is very literal. It's very there. Social media is very there. 'What you see is what you get.'

WEILING So what role does the uncanny play for architecture students in the 21st century?

VIDLER It plays the same role as any discussion of the relationship between space and the unconscious. It sensitises students to their own feelings, to their own relationships to space, their own relationships to their hidden fears, and therefore to the world in general.

EPILOGUE

尾声

BIOGRAPHY

CHEN WEI
Professor of Architecture and Academic Director of Architecture History and Theory at Southeast University. She also undertakes a number of academic duties, including Vice Chairman of the Architecture History Council of the Architectural Society of China, Vice Director of the Architecture History Council of Science and Technology Academy of China, Vice Director of Chinese Yuanmingyuan Council, and one of the experts of the State Administration of the Cultural Heritage of China. Having been teaching and researching Chinese architectural history, she has published more than 100 articles, one book and a set of CDs on ancient Chinese architecture, and has co-edited 10 other books.

MARK COUSINS
Director of History and Theory Studies at the Architectural Association School of Architecture, Visiting Professor of Architecture at Columbia University, University of Navarre and Southeast University, and cofounder of the London Consortium. He is the author of and a contributor to a number of books, including a book on Michel Foucault (with Athar Hussain), and the introduction to a new translation of a selection of Freud's papers on the unconscious. He has published on the work of Tony Fretton and many other architects and artists.

GE MING
Associate Professor of the School of Architecture at Southeast University, Editor of Journal of Asian Architecture and Building Engineering, and the journal Garden and Architecture. He has participated in several international exhibitions, including one of China's Pavilions, 'Murmur', at the 11th Venice Biennale in 2008. He has completed the work of 'Ru Yuan' Garden, published more than 10 papers, and been invited for academic activities by universities including ETH Zurich.

WEILING HE
Associate Professor at Texas A&M University College of Architecture, where she has been teaching design studios and seminars, while maintaining her architectural practice. Her research field is design theory with particular focus on translations across different forms of art, formal descriptions of space, metaphors of making, diagramming and visual thinking. She has attended a number of international academic activities and exhibitions, published in journals and at conferences, and received several research grants.

JOHN DIXON HUNT
Emeritus Professor of the History and Theory of Landscape in the Department of Landscape Architecture, University of Pennsylvania School of Design; Ex-departmental Chairman and Ex-acting Dean of the School. He is the founding editor both of Studies in the History of Gardens and Designed Landscape and of Word & Image. He has published a number of books, including *Greater Perfections: The Practice of Garden Theory* and *The Picturesque Garden in Europe*.

DAVID LEATHERBARROW
Professor of Architecture and Associate Dean for Academic Affairs at the University of Pennsylvania School of Design, and the winner of the 2012 Provost's Award for Distinguished Ph.D. Teaching and Mentoring. He is primarily known for his contributions to the field of architectural phenomenology, and has published several books, including *The Roots of Architectural Invention: Site, Enclosure, Materials*; *On Weathering: The Life of Buildings in Time* (with Mohsen Mostafavi); *Topographical Stories: Studies in Landscape and Architecture*; *Surface Architecture* (with Mohsen Mostafavi) and *Architecture Oriented Otherwise*.

LI SHIQIAO
Weedon Professor in Asian Architecture, School of Architecture, University of Virginia, where he teaches and researches emerging issues in contemporary Chinese cities. He has taught in London, Singapore and Hong Kong, and lectured over the world. His writings have appeared in high profile international journals, and his books include *The Chinese City, Architecture and Modernization* and *Power and Virtue, Architecture and Intellectual Change in England 1650—1730*.

LIN HAIZHONG
Professor of Chinese Painting at China Academy of Art in Hangzhou, and Vice Director of Research Centre of Painting and Calligraphy in China Academy of Art. His works have been exhibited in a number of art exhibitions in Shanghai, Nanjing, Shenzhen, Chengdu, and New York, and are included in the collections of the National Art Museum of China, Shanghai Art Museum, Art Gallery of Guangzhou, and the British Museum. He is also the author of six books on Chinese landscape painting.

LIU DONGYANG
Freelance writer. Searching in-between disciplines, between building and city, between the thingness of humanity and nature as a process, his long term interest is in the history of a specific site, and the practical wisdom rooted in local people as a possible alternative for tackling our current ecological and cultural crisis, through understanding people's experience of and with their land.

ANTHONY VIDLER
Professor and Dean of the Irwin S. Chanin School of Architecture of The Cooper Union. A historian and critic of modern and contemporary architecture, specialising in French architecture from the Enlightenment to the present, he received the architecture award from the American Academy of Arts and Letters in 2011. His publications include *The Writing of the Walls*, *Claude-Nicolas Ledoux*, *The Architectural Uncanny*, *Warped Space*, *Histories of the Immediate Present: The Invention of Architectural Modernism*, and *The Scenes of the Street and Other Essays*.

WANG SHU
Architect/Professor and Head of the Architecture School at China Academy of Art, cofounder of Amateur Architecture Studio. He has received a number of architecture prizes, including being awarded the position of 2012 Laureate of The Pritzker Architecture Prize. His works have been published and exhibited widely. His most widely referred to works include: Ceramic Houses, Vertical Courtyard Apartments, Ningbo Contemporary Art Museum, Five Scattered Houses, Xiangshan Campus of China Academy of Art, Ningbo History Museum, and the Tengtou Pavilion for the Shanghai Expo 2010.

ZOU HUI
Associate Professor of the University of Florida. He is a former Fellow of Garden and Landscape Studies at Dumbarton Oaks, Harvard University. His areas of research include comparative studies in architecture and garden histories, and architectural philosophy and criticism. He published the English book, *A Jesuit Garden in Beijing and Early Modern Chinese Culture* in 2011, and the Chinese book, *Fragments and Mirroring: A Two-Fold Discourse of Comparative Architecture* in 2012.

人物简介

陈薇
东南大学建筑学院教授、博士生导师和历史与理论学科学术带头人。兼任中国建筑学会建筑史学分会副理事长，中国科学技术学会建筑史专业委员会副主任委员，中国圆明园学会学术专业委员会副主任委员等职。长期从事中国建筑史的研究，发表学术论文80余篇，出版专著一本、专业光盘一函，参编著作10本。

马克·卡森斯
英国建筑联盟学院历史与理论学科主持，美国哥伦比亚大学、西班牙纳瓦拉大学建筑学院和东南大学访问教授，伦敦联合学院创始人之一。他是多本书籍的作者和撰稿人，如《米歇尔·福柯》、弗洛伊德新译文选《论潜意识》的序言等，并曾为多位著名建筑师和艺术家及其作品和展览撰写评论。

葛明
东南大学建筑学院副教授，《亚洲建筑与工程》杂志编委，《园林与建筑》一书的主编之一，发表论文多篇。多次参加国际建筑展，包括设计2008年威尼斯国际建筑双年展中国馆之一 "Murmur"，建成作品还包括 "如园"，曾赴苏黎世联邦高等工业大学等多所高校进行学术交流。

贺玮玲
美国得州农机大学建筑学院副教授，从事设计工作室和研讨会的教学工作和建筑实践。主要研究通过艺术的不同形式，空间的形式表达、制作的隐喻、图解和视觉思维进行解译的设计理论。曾多次参加于世界举办的学术活动、展览，发表学术论文多篇，并获得研究基金多项。

约翰·迪克森·亨特
美国宾夕法尼亚大学设计学院教授（退休），前执行院长和景观建筑系主任。他是杂志《园林和设计景观历史研究》和《词与图像》的创办者，主要著作包括《更加完满：园林理论的实践》、《欧洲风景如画园林》与《建筑的另一种方向》等。

戴维·莱瑟巴罗
美国宾夕法尼亚大学设计学院建筑学教授及负责学术事务的副院长，并因出色的博士生教学和指导而获得2012年度院长奖。其研究方向主要包括建筑如何显现、如何被感知以及地形如何塑造建筑。主要著作有：《建筑创造之源：场地、围合、材料》、《论天气：时间中的建筑生命》、《地形学的故事：景观和建筑研究》、《表面建筑》等。

李士桥
美国弗吉尼亚大学建筑学院教授。曾在英国、新加坡、香港等地任教，和世界各地知名大学举办讲座。当前的研究课题主要集中于当代亚洲城市建筑理论。在国际知名刊物上发表文章多篇，并著有《权力与美德，英国1650—1730的建筑与思想变迁》（英文）、《现代思想中的建筑》、《中国城市》（英文）等专著。

林海钟
中国美术学院国画系教授、博士生导师，中国美术学院书画鉴赏研究中心副主任。作品被中国美术馆、上海美术馆、广州美术馆、伦敦大英博物馆收藏，并著有《山水画技法》、《水墨画教材——山水画的技法》（日本版）、《中国历代名家技法集萃——山水卷石法（上）》、《中国历代名家技法集萃——山水卷树法（下）》、《古画临摹》、《早春图画法》等。

刘东洋
作者长期徘徊在建筑与城市之间、物化的人性与自然的生长之间，总希望触摸土地的逻辑与日常生活的弥散。因此，会从桌子上书本、铅笔、茶杯的关系，一直关注到水井旁的人群以及大尺度空间内河流改道与城市兴衰的交织。概言之，作者关注"基地史"，关注具体人群对于土地的体验、解读和改造，也关注由此带来的可能生态及文化危机，并总想从普通人的智慧中寻找设计对策与答案。

安东尼·维德勒
库珀联盟艾文·钱尼建筑学院教授、院长。现代与当代建筑历史学家、评论家，启蒙运动至今法国建筑研究专家。2011年获美国艺术与文学协会建筑奖。主要著作有：《墙的书写》、《克劳德·尼古拉斯·勒杜》（获约翰·罗素·希区柯克奖）、《扭曲空间》、《建筑异样》、《当下史：创造建筑的现代主义》、《街道场景及其他论文》等。

王澍
中国美术学院建筑艺术学院教授、院长，建筑师，业余建筑工作室的创办人之一，多项国际大奖的获得者，包括2012年普利兹克建筑奖。其作品刊登于世界各类书籍及杂志，在世界各地美术馆、艺术设计中心以及建筑院校展出，并受邀发表演讲。代表作品包括瓷屋、垂直院宅、宁波当代美术馆、五散房、中国美术学院象山校区、宁波历史博物馆、上海世博宁波滕头案例馆等。

邹晖
美国佛罗里达大学建筑系副教授，哈佛大学杜巴顿中心园林史研究员（2001）。其研究领域包括建筑史与园林史的比较研究、建筑哲学及评论。著有《北京的耶稣士花园与中国早期现代文化》（英文）和《碎片与比照：比较建筑学的双重话语》（中文），并在国际知名杂志及中国建筑杂志上发表有关建筑史、园林史、哲学、汉学及中国当代建筑和城市化的论文。

GLOSSARY

	§	地形学	topography 31, 47, 221
《插图》	Painting Illustrated 359	地形	dixing, Topography as terrain or contours 78, 80, 82, 86, 88, 111, 113
《大观》	Grand View 283		
《红楼梦》	Dream of the Red Chamber 283	地质学	geology 105
《获麟解》	Huo Lin Jie, Deciphering a Unicorn 293	地志	dizhi, A local 'gazetteer' 80, 84
		蒂沃里	Tivoli 111
《离骚》	Lisao, The Lament 287	地貌	dimao, Topography as land surface or appearance 78, 80, 86, 88
《诗经》	Shijing, Book of Odes 287		
《投射之模》	Projective Cast 241		
《易经》	Yijing, Book of Change 285		E

A

阿尔多布兰迪尼别墅园	Villa Aldobrandini 123
爱斯特别墅园	Villa D'Este 115, 123
安尼特园	Anet 145

F

废物景观	drosscape 363
分解	de-construct 391
分区制	Zoning 391
风景如画	Picturesque 359
枫丹白露园	Fontainbleau 145
弗朗西斯·培根	Francis Bacon 369

B

bi	比, literally 'compare' 286
彼蒂宫	Pitti 123
波波里花园	Boboli garden 123
波茨坦无忧宫苑	San Souci 145
波维斯城堡台地园	Powis Castle 145
Brief Notes upon Reading the Local History and Gazetteers, Dushi fangyu jiyao 读史方舆纪要 82	
布拉托里诺美第奇别墅园	the Medici villa at Pratolino 115

G

giardino segreto	秘密花园 283
公共广场	public campo 37
古生物学	palaeontology 105

H

海杜克	John Hejduk 337
红衣主教	Cardinal 125
红衣主教里阿里奥	Cardinal Raffaele Riario 129
红衣主教蒙达多	Cardinal Peretti Montalto 131
红衣主教卡纳迪尔·冈伯拉	Cardinal Gianfrancesco Gambara 129
荒诞	Grotesque 394
回归	nostos 247
霍夫曼	Dan Hoffman 337

C

场地手术	Field Operations 35
场所性	Topos 83

D

大地	land 113
大连市城市规划管理条例	Dalian City Planning Administrative Guidelines: References for Application *Dalian shi chengshi guihua guanli tiaoli* 96
地方	topo 47
地貌学	geomorpholoy 105, 106
地貌研究的历史	a history of landform study 105
地文学	Physical Geography, 现在, 多译为自然地理学 85
地文学	physiography 105

I

J

计划	programme 33
家常的	homely 390
建造	construction 337
景观	landscape 113
景观路径	scenic route 241
景色	scape 37

词汇表

K

卡斯特洛别墅园	Villa Castello 133
肯尼思·弗兰姆普敦	Kenneth Frampton 106

L

戴维·莱瑟巴罗	David Leatherbarrow 106
兰特别墅园	Villa Lante 125
历史性	historicity 285
罗马美第奇别墅园	Villa Medeci 115

M

making a scene	大吵大闹 355
曼·雷	Man Ray 319
媚俗	Kitsch 315
描画	graphein 47

N

内摄	introjection 239
尼普顿神	Neptune 123
nostalgia	思乡、怀旧的 247

O

P

presentation	呈现 351

Q

R

人生地理学	Human Geography,现在,多译为人文地理学 85

S

赛维鲁阿尔卡扎尔园	The Alcazarin 145
scenery	布景 353
scenery	舞台布景 355
scenite	游牧生活的人 355
scenographer	舞美师 355
scenography	舞美,配景透视 355
实践	praxis 35
疏离	estrangement 394
私密	intimate 390
Special Geography	特殊地理学 85
孙子兵法	Sunzi bingfa, The Art of War 82

T

痛苦	algia 247
topos	处所 81
投射	projection 239
土地	land 37
托马索济努齐	Tommaso Ghinueei 129

U

V

W

维尼奥拉	Giacomo Barozzi da Vignola 129
无忧宫	Sans Sous 145

X

西北巴格内亚村	Bagnaia 125
弦外之音	otherness 394
现实	Reality 355
小广场	piazzi 37
信仰的体现	embodiment 337
xing	兴,evoking 286
叙事	narrative 337

Y

亚平宁山	Apennines 133
伊恩·伦诺克斯·麦克哈格	Ian Lennox MeHarg 129, 152
异样	Das Unheimliche 390
艺术物品	art object 351
艺术作品	art work 351
尤金·阿杰	Eugene Atget 319
约翰·海杜克	John Hejduk 394
运行	operation 37

Z

再现	Representation 355
制作	making 337
中国风	Chinoiserie 145
自然地理学	physical geography 105

ACKNOWLEDGEMENTS

From the organisation of the Forum of Contemporary Architectural Theories (the Forum) to collating and publishing Architecture Studies (AS), there has been a tremendous amount of effort, mentally and physically, by many individuals. Moreover, as the publication of Architecture Studies is bilingual, the workload doubles. Without all the help and support, the publication of AS is inconceivable. ●To the speakers and the participants of the Forum, thank you for responding to our invitations and sharing your knowledge, insights and thoughts. Only with your contribution, could the Forum and AS generate solid substance to intrigue readers. ●To the translators of *Topographical Stories* and *The Architectural Uncanny*, thank you for providing meaningful subjects for the Forum and AS. ●To the Members of the Academic Committee of the Forum, as well as the Translation Advisors, thank you for your great support which has been vital to secure the planned outcome. Special thanks extend to AA Director Brett Steele, Dean of the School of Architecture of Southeast University Professor Wang Jianguo, Professor Liu Xianjue, Professor Dong Wei, and the President of Shanghai Xian Dai Architectural Design Group Zhang Hua, and Principle Architect of Shanghai Xian Dai Architectural Design Group Shen Di. ● To the dozens of participating volunteers during the Forum, your effort has ensured our progress to date. We would like to particularly mention the following people: Lu Andong, Susan Chai, Shen Yang, Shi Yonggao, Zhu Yuan, Yu Changhui, Tang Jingyin, Geng Xinxin and Zhang Yinan. ●To the friends who helped in publishing AS, in China and overseas, thank you for translating, editing and collating the articles and discussions of the Forum. These enormous yet sometime trivial jobs enabled us to present the knowledge, thoughts and many sparks at the Forum with the best quality attainable. They are Shirley Surya, Huey Ying Hsu, Susan Chai, Lu Andong, Professor Bi Fei, Yu Changhui, Tang Jingyin, Geng Xinxin, Kong Dezhong and Shen Wen. ● To Huang Juzheng and Sun Lian from China Architecture and Building Press, and Twelve Design, thank you for your support and assistance with the publication of AS. ●To Professor Wang Junyang, thank you for your suggestion which benefited the Forum considerably. ●We won't be able to list every person's name here. Nevertheless, to all who played a part in the process, please accept our heartfelt gratitude. ●Driving architectural theory studies and building cultural exchange platforms are not accomplishments that a few individuals can accomplish. They require a long-term, collective effort from many. It has been several years since the initial planning of the Forum. It grew from a thought related to architectural translation to a long-term research project; from an individual's isolated work to an activity which involves many. ●While receiving great support from colleagues, we also come to the realisation of the enormity of the task we have taken on.

We would like to dedicate this publication to everyone who encourages, supports and helps us along the way.

Nanjing, 09 June 2012

致谢

无论是"当代建筑理论论坛"的组织,还是《建筑研究》的编辑,都是一项费心费力的工作。加之会议与出版以中英文双语进行,更是增添了加倍的工作量。难以想象,没有诸多同仁的支持与帮助,这本专辑能够付梓出版。● 感谢各位主题演讲者和与会学者的响应与参与,没有他们,无论是研讨会还是本期的《建筑研究》都无法想象。● 感谢《地形学的故事》和《建筑异样》的译者,没有他们,我们的议题难以得到落实。● 感谢论坛学术委员会委员以及中英文翻译顾问的支持。来自英国建筑联盟学院(AA)校长布雷特·斯蒂尔,东南大学建筑学院王建国院长、刘先觉教授、董卫教授,现代建筑设计集团的张桦总裁和沈迪总建筑师的支持,是整个计划得以实施的保障。● 感谢会议筹备和进行期间,各位参与者的工作。鲁安东、柴舒、沈旸、史永高、朱渊、于长会、唐静寅、耿欣欣、张一楠以及东南大学建筑学院的十数名老师与志愿者的工作,保证了会议的顺利进行。● 感谢在《建筑研究》编辑过程中各位朋友的帮助,他们首先包括文章的译者。中国美术学院的毕斐教授对中国绘画中重要概念的翻译提供了关键的指引。身居海外的雪莉·苏里亚、许蕙滢、鲁安东对部分英文、艾莉克莎·加莱亚对全书英文进行了校对。"AA"的柴舒,和东南大学的于长会、唐静寅、耿欣欣、孔德钟、沈雯担负了许多繁琐的协助性工作。● 感谢中国建筑工业出版社黄居正先生和孙炼女士的支持,以及十二工作室的设计师的帮助。● 感谢王骏阳教授的建议,使论坛的筹划受益颇多。● 篇幅所限,恕我们不能在此一一致谢。● 无论是推动建筑理论的研究,还是搭建文化交流的平台,远非几个人的力量所能达成,它需要长期的集体合作。从论坛最初的筹划到《建筑研究》的出版,历时数年,从一个翻译的想法,逐渐发展成一个长期的研究计划;从一个个体的独立工作,逐渐演化为一项多人参与的活动。● 在这个过程中,我们得到了同行者的支持,更感到了任务的艰巨。

此记,谨献给每一位支持、鼓励和参与了的同仁。

2012 年 6 月 9 日于南京

图书在版编目（CIP）数据

建筑研究 02 ／（英）卡森斯，陈薇主编．— 北京：中国建筑工业出版社，2012.8
ISBN 978-7-112-14609-3

Ⅰ．①建⋯ Ⅱ．①卡⋯ ②陈⋯ Ⅲ．①建筑理论－文集－汉、英 Ⅳ．① TU-0

中国版本图书馆 CIP 数据核字 (2012) 第 197971 号

COLOPHON
尾署

chief editor Mark Cousins Chen Wei	主编 ［英］马克・卡森斯 陈　薇
deputy chief editor Li Hua Ge Ming	执行主编 李　华 葛　明
series editor Sun Lian Huang Juzheng	责任编辑 孙　炼 黄居正
english proofreading Alexa Galea	英语校对 ［英］艾丽莎・加莱亚
assistant editor Yu Changhui Laura Lao	助理校对 于长会 ［英］劳拉
proofreading Li Hua Ge Ming	总校对 李　华 葛　明

建筑研究 02
Architecture Studies 02
主编：［英］马克・卡森斯　陈薇
执行主编：李华　葛明
*
中国建筑工业出版社出版、发行（北京西郊百万庄）
各地新华书店、建筑书店经销
北京方嘉彩色印刷有限责任公司印刷
*
开本：850×1168 毫米　1/16　印张：21　字数：700 千字
2012 年 8 月第一版　2012 年 8 月第一次印刷
定价：128.00 元
ISBN 978-7-112-14609-3
（21917）

版权所有　翻印必究
如有印装质量问题，可寄本社退换
（邮政编码 100037）

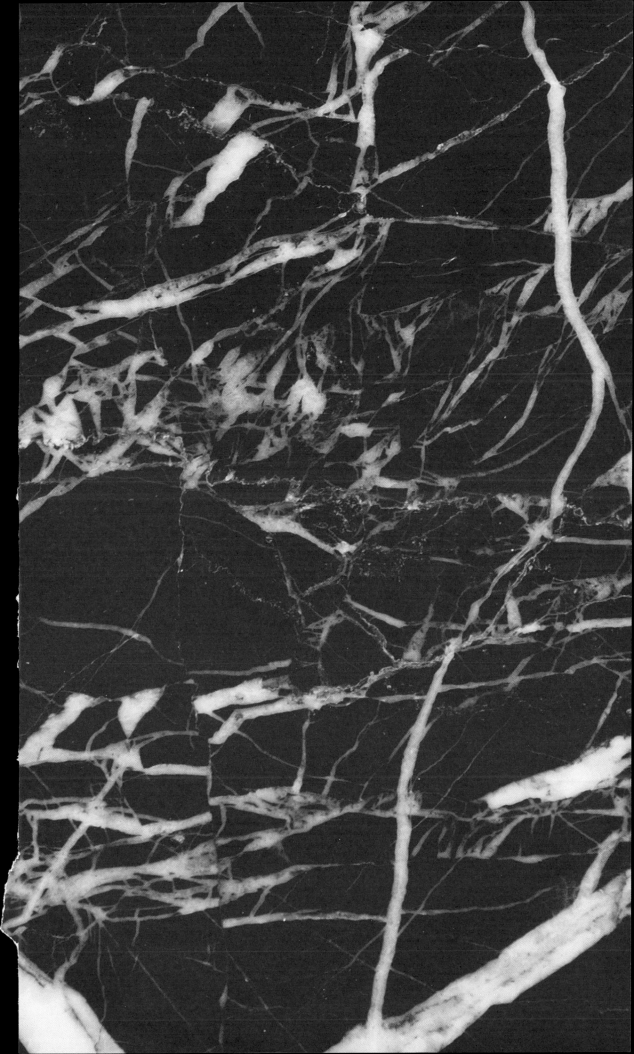

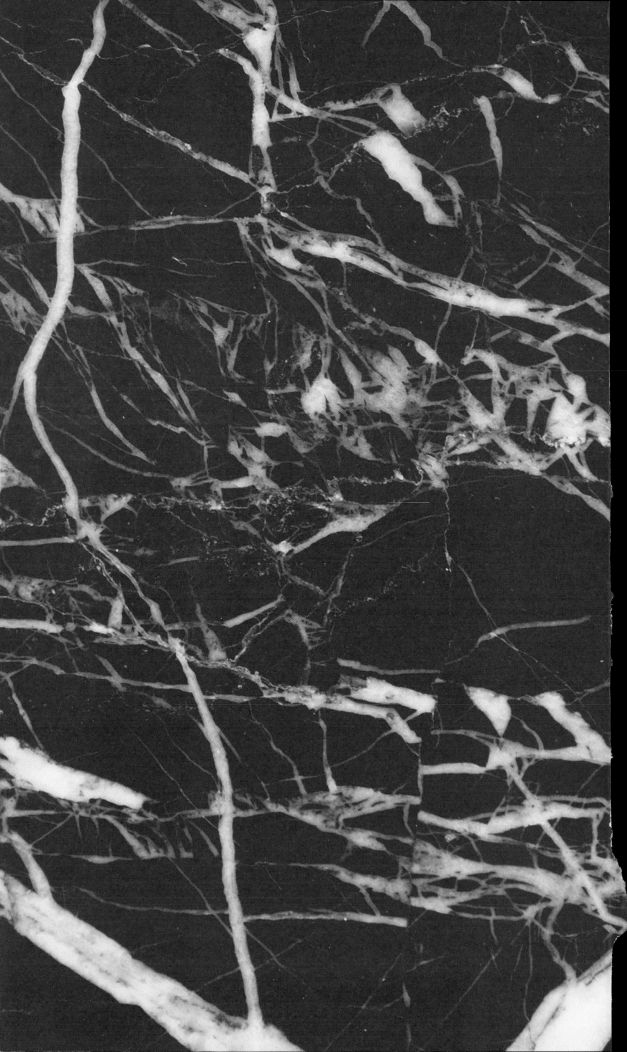